Computational Intelligence and Modern Heuristics

Computational Intelligence and Modern Heuristics

Editor

Brygida Cullen

Computational Intelligence and Modern Heuristics
Edited by **Brygida Cullen**

ISBN: 978-1-68117-199-9
Library of Congress Control Number: 2016934746

Notice

Reasonable efforts have been made to publish reliable data and views articulated in the chapters are those of the individual contributors, and not necessarily those of the editors or publishers. Editors or publishers are not responsible for the accuracy of the information in the published chapters or consequences of their use. The publisher believes no responsibility for any damage or grievance to the persons or property arising out of the use of any materials, instructions, methods or thoughts in the book. The editors and the publisher have attempted to trace the copyright holders of all material reproduced in this publication and apologize to copyright holders if permission has not been obtained. If any copyright holder has not been acknowledged, please write to us so we may rectify.

Preface

A heuristic technique, often called simply a heuristic, is any approach to problem solving, learning, or discovery that employs a practical method not guaranteed to be optimal or perfect, but sufficient for the immediate goals. Where finding an optimal solution is impossible or impractical, heuristic methods can be used to speed up the process of finding a satisfactory solution. Heuristics can be mental shortcuts that ease the cognitive load of making a decision. In computer science, artificial intelligence, and mathematical optimization, a heuristic is a technique designed for solving a problem more quickly when classic methods are too slow, or for finding an approximate solution when classic methods fail to find any exact solution. This is achieved by trading optimality, completeness, accuracy, or precision for speed. In a way, it can be considered a shortcut.A heuristic function, also called simply a heuristic, is a function that ranks alternatives in search algorithms at each branching step based on available information to decide which branch to follow. The objective of a heuristic is to produce a solution in a reasonable time frame that is good enough for solving the problem at hand. This solution may not be the best of all the actual solutions to this problem, or it may simply approximate the exact solution. But it is still valuable because finding it does not require a prohibitively long time. Heuristics may produce results by themselves, or they may be used in conjunction with optimization algorithms to improve their efficiency. Results about NP-hardness in theoretical computer science make heuristics the only viable option for a variety of complex optimization problems that need to be routinely solved in real-world applications. This book entitledComputational Intelligence and Modern

Heuristicshighlights on computational models using heuristic and meta-heuristic approaches.

Table of Contents

CHAPTER 1

Meta-Heuristic Optimization Techniques and Its Applications in Robotics

Alejandra Cruz-Bernal[1]

[1] *Polytechnic University of Guanajuato, Robotics Engineering Department, Community Juan Alonso, Cortázar, Guanajuato, Mexico*

INTRODUCTION

Robotics is the science of perceiving and manipulating the physical world. Perceive information on their envioronments through sensors, and manipulate through physical forces. To do diversity tasks, robots have tobe able to accomodate the ennormous uncertainty that exist in the physical world. The level of uncertainty depends on the application domain. In some robotics applications, such as assembly lines, humans can cleverly engineer the system so that uncertainty is only a marginal factor. In contrast, robots operating in residential homes, militar operates or on other planets will have to cope with substantial uncertainty. Managing uncertainty is possibly the most important step towards robust real-world robot system.

If considerate that, for reduce the uncertainty divide the problem in two problems, where is the first is to robot perception, and another, to planning and control. Likewise, path planning is an important issue in mobile robotics. It is to find a most reasonable collision-free path a mobile robot to move from a start location to a destination in an envioroment with obstacles. This path is commonly optimal in some aspect, such as distance or time. How to find a path meeting the need of such criterion and escaping from obstacles is the key problem in path planning.

Optimization techniques are search methods, where the goal is to find a solution to an optimization problem, such that a given quantity is optimized, possibly subject to a set of constraints. Modern optimization techniques start to demonstrate their power in dealing with hard optimization problems in robotics and automation such as manufacturing cells formation, robot motion planning, worker scheduling, cell assignment, vehicle routing problem, assembly line balancing, shortest sequence planning, sensor placement, unmanned-aerial vehicles (UAV) communication relaying and multirobot coordination to name just a few. By example, in particle, path planning it is a difficult task in robotics, as well as construct and control a robot. The main propose of path planning is find a specific route in order to reach the target destination.

Given an environment, where a mobile robot must determine a route in order to reach a target destination, we found the shortest path that this robot can follow. This goal is reach using bio-inspired techniques, as Ant Colony Optimization (ACO)and the Genetics Algorithms (GA).

A principal of these techniques, is by example, with a colony can solve problems unthinkable for individual ants, such as finding the shortest path to the best food source, allocating workers to different tasks, or defending a territory from neighbors. As individuals, ants might be tiny dummies, but as colonies they respond quickly and effectively to their environment. They do it with something called Swarm Intelligence.

These novel techniques are nature-inspired stochastic optimization methods that iteratively use random elements to transfer one candidate solution into a new, hopefully better, solution with regard to a given measure of quality.

We cannot expect them to find the best solution all the time, but expect them to find the good enough solutions or even the optimal solution most of the time, and more importantly, in a reasonably and practically short time. Modern meta-heuristic algorithms are almost guaranteed to work well for a wide rangeof tough optimization problems.

Previous Work

Path planning is an essential task navigation and motion control of autonomous robot. This problem in mobile robotic is not simple, and the same is attached by two distint approaches. In the *classical approaches* present by Raja et al.[49]cited according to [3]the C-space, where the representation of the robot is a simple point. The same approach is described by Latombe's book [4]. Under concept of C-space, are developedpath planning approaches with roadmap and visibility graph was introduced[5].Sparce envioroments considering to polygonal obstacles and their edges [6, 7]. The Voroni diagram was introduced [8]. Other approach for roadmap andrecient applications in [9,10]. Cell descomposition approach [11, 12, 13, 14, 15,16]. A efficente use of grids [17].

A related problem is when both, the map and the vehicle position are not know. This problem is usually nown as Simultaneous Localization and Map Building (SLAM), and was originally introduced [18]. Until recently,have been significative advances in the solution of the SLAM problem [19,20,21,22].

Kalman filter methods can also be extended to perform simultaneous localization and map building. There have been several applications in mobile robotic, such as indoors, underwater and outdoors. The potential problems with SLAM algorithm have been the computational requeriments. The complexy of original algortihm is of $O(N^3)$ but, can be reduced to $O(N^2)$ where, N will be the number of landmarks in the map [23].

In computational complexity theory, path planning is classified as an *NP* (nondeterministic polynomial time) complete problem [33]. *Evolutionary approaches* provide these solutions. Where, one of the high advantage of heuristic algorithms, is that it can produce an acceptable solution very quickly, which is especially used for solving *NP-complete problems*.

A first path planning approach of a mobile robot trated as non-deterministic polynomial time hard (*NP-hard*) problem is [31]. Moreover, even more complicated are the environment dynamic, the classic approaches to be incompetent [32]. Hence, evolutionary approaches such as Tabu Search (TS), Artificial Neural Network (ANN), Genetic Algorithm (GA), Particle Swarm Optimization (PSO), Ant Colony Optimization (ACO) and Simulated Annealing (SA), name a few, are employed to solve the path planning problem efficiently although not always optimal.

Genetic Algorithm (GA) based search and optimization techniques have recently found increasing use in machine learning, scheduling, pattern recognition, robot motion planning, image sensing and many other engineering applications. The first research of Robot Motion Planning (RMP), according to Masehian et al.[53] although GA, was first used in [42] and [43]. An approach for solution the problem of collision-free paths is presented in [44].GA was applied [45, 46, 48] in planning multi-path for 2D and 3D environment dimension and shortest path problem. A novel GA searching approach for dynamic constrained multicast routing is developed in [49]. Parallel GA [50], is used for search and constrained multi-objective optimization. Differentials optimization used hybrid GA, for path planning and shared-path protections has been extended in [51, 52]. In [62, 63, 64, 65], has been a compared of differential algorithms optimization GA (basically for dynamic environment), subjected to penalty function evaluation.

By other side, thetechnique PSO have some any advantages [35], such as simple implementationwith a few parameters to be adjusted. Binary PSO [37] withouta mutation operator[36]are used to optimize the shortest path. Planning in dynamic environment, that containing invalid paths (repair by a operator mutation), are subjected to penalty function evaluation [38]. Recently, [39] proposedwith multi-objective PSO and mutate operator path planning in dangerous dynamics environment.Finally, a newperspective global optimization is proposed [40].

Ant Colony Optimization (ACO) algorithms have been developed to mimic the behavior of real ants to provide heuristic solutions for optimization problems. It was first proposed by M. Dorigoin 1992 in his Ph. D. dissertation [54]. The first instance of the application of Ant Colony Optimization in Probabilistic Roadmap is the work [55, 56]. In [57] an optimal path planning for mobile robots based on intensified ACO algorithm is developed. Also in 2004, ACO was used to plan the best path [58]. ACRMP is presented in [60]. An articulated RMP using ACO is introduced in [59]. Also, a path planning based on ACO and distributed local navigation for multi robot systems is developed in [66]. Finally, an approach based on numerical Potential Fields (PF) and ACO is introduced in [61] to path planning in dynamic environment. In [66] a fast two-stage ACO algorithm for robotic path planning is used.

The notion of using Simulated Annealing (SA) for roadmap was initiated in [67]. PFapproach was integrated with SA to escape from local minimaand evaluation [68,70]. Estimates using SA for a multi-path arrival and path planning for mobile robotic based on PF, is introduced in [69, 72]. A path planning based on PF approach with SA is developed in [72].Finally, in [71] was presented a multi operator based SA approach for navigation in uncertain environments.

A case particle are militar applications, with an uninhabited combat air vehicle (UCAV). The techniques employed, have been proposed to solve this complicated multi-constrained optimization problem, solved contradiction between the global optimization and excessive information. Such techniques used to solution this problem are differential evolution [24], artificial bee colony [29], genetic algorithm [25], water drops optimization (chaotic and intelligent) [30] and ant colony optimization algorithm [26, 27, 28].

ROBOT NAVIGATION

Introduction

Mobile robots and manipulator robots are increasingly being employed in many automated envi;onments. Potential applications of mobile robots include a wide range such a service robots for elderly persons, automated guide vehicles for transferring goods in a factory, unmmaned bomb disposal robots and planet exploration robots. In all thes applications, the mobile robots perform their navegation task using the building blocks (see figure 1) [1], the same, is based on [2] known with the Deliberative Focus.

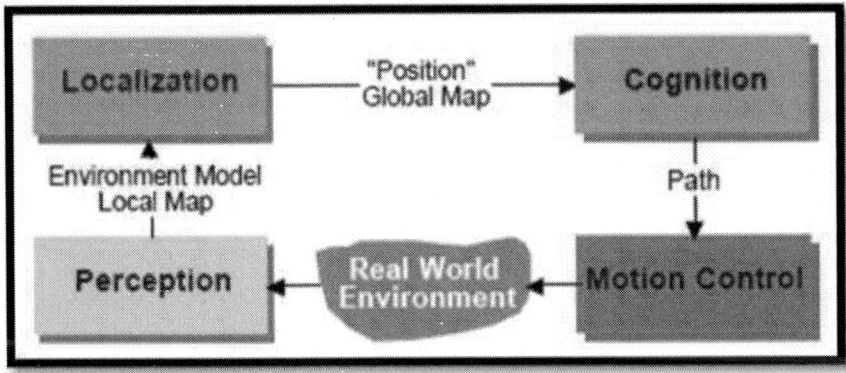

Figure 1. Deliberative Focus.

While perception of enviroment refers to understanding its sensory data, finding its pose or configuration in the surroundings is localization and map building. Planning the path in accordance with the task by using cognitive decision making is an essential phase before actually accomplishing the preferred trajectory by controlling the motion. As each of the building blocks is by itself a vast research field.

Map Representation Methods

When, Rencken in1993 [73] defined the map building problem as sensing capacity of robot, can be split in two, where robot know a pre-existing map or it has to build this, through information of the environment. According to above is assumed that the robot begins a exploration without having any knowledge of the environment, but with a exploration strategy, and it depends strongly on the kind of sensors used, the robot builds its own perception of environment[74].

A proposal of spatial representation is to sample discretely the two- or three-dimensional environment. This isa sample space in *cells* of a *uniform grid*for two-dimensional or considering the volume of elementsthat are used to represent objets named *voxel*.

Geometric maps are composed of the union of simple geometric primitives. Such maps are characterized by two key properties: the set of basic primitives used for describing objects, and the set of operators used to manipulate objects. The fundamentals problems with this technique are lack of stability, uniqueness, and potentiality.

Within the *geometric representations*, the *topological representation* can be used to solve abstract tasks that are not void reliance on error-prone metric, provided an explicit representation of connectivity between regions or objects. A topological representation is based on an abstraction of the environment in terms of discrete places (*landmarks*) with edges connecting them. In [76], present an example of this topological representation, where after of delimited region of interest, used a GA for the landmark search through the image.

Path Planning

The basic path planning problem refers to determining a path in configuration space between an initial configuration (start pose) of the robot and a final configuration (finish pose). Therefore, several approaches for path planning exist of course a particular problem of application, as well as the kinematic constraints of the robot [77]. Although is neccesary make a serie of simplied with respect to real environment, the techniques for path planning can be group in *discrete state space* and *continuum space*.

To efficiently compute of search the path through techniques as A* [14] and Dynamic programming. The difference between them usually, resides in the simplicity to define or compute the evaluation function, which hardly depends on the nature of the environment and the specific problem.Other techniques for mapping a robot's environment inside a discrete searchable space include visibility graphs and Voronoi diagrams [75].

Path planning in a *continuum space* is consideret as the determination of an appropriate trajectory within this continuum. Two of the most known techniques for continuous state space are the potentialfields [61] and the vector field histogram methods. Alternatively,these algorithms, can be based on the *bug algorithm*[41], guaranteed to find a path from the beginning until target, if such path exists. Unfortunately, these methods can be generated arbitrarytrajectories worse than the optimal path to the target.

POPULATION-BASED META-HEURISTIC OPTIMIZATION

Many results in the literature indicate that metaheuristics are the state-of-the-art techniques for problems for which there is non efficient algortihm. Thus, meta-heuristics approach approximationsallow solving complex optimization problems. Although these methods, cannot guarantee that the best solution found after termination criteria are satisfied or indeed its global optimal solution to the problem.

Optimization

The theory of optimization refers to the quantitative study of optima and the methods for finding them. Global Opitimization is the branch of applied mathematics and numerical analysis that focuses on well optimization. Of course T. Weise in [81], the goal of global optimization is to find to the best possible elements x* form a set according to a set criteria $F = \{f_1, f_2,..., f_n\}$ These criteria are expressed as mathematical function, that so-called objetive functions. In the Weise's book[81] the Objective Function is described as:

DefinitionObjective Function. An objective funtion f : [Math Processing Error] Y with Y [Math Processing Error] is a mathematical function which is subject to optimization.

Global optimization comprises all techniques that can be used to find the best elements x* in [Math Processing Error] with respect to such criteria $f \in F$.

Nature-Inspied Meta-Heuristic Optimization

The high increase in the size of the search space and the need of proccesing in real-time has motivated recent researchs to solving scheduling problem using nature inspired heuristic techniques. The principal components of any metaheuristic algorithms are: intensification and diversification, or explotation and exploration. The optimal combination of these will usually ensure that a global optimization is achievable.

---

1. Initialize the solution vectors and all parameters.
2. Evaluate the candidate solutions.
3. Repeat
   a. Generation a new candidate solutions via the nature o social behaviors.
   b. Evaluate the new candidate solutions.
4. Until meet optimal criteria.

Figure 2. General Description for Nature Inspired Algorithm

Genetic Algorithms

In essence, a Genetic Algorithm (GA) is a search method based on the abstraction Darwinian evolution and natural selection of biological systems and respresenting them in the mathematical operator: croosover or recombination, mutation, fitness, and selection of the fittest.

The application of GA to path planning problem requires development of a chromosome

representation of the path, appropriate constraint definition to minimize path distance, such that is providing smooth paths. It is assumed that the environment is static and known.

Cited according to [81], the genotypes are used in the reproduction operations whereas the values of the objective funtions f(or fitness function), where $f \in F$ are computed on basis of the phenotypes in the problme space [Math Processing Error] which are obtained via the genotype-phenotype mapping (gpm) [82, 83, 84, 85], where [Math Processing Error] is a space any binary chain.

$$\forall g \in G \ \exists x \in X : gpm\ (g) = x$$

(1)

$$\forall g \in G \ \exists x \in X : P(gpm\ (g) = x) > 0$$

(2)

1. Random Initial Population.
2. Repeat
    a. Calculated through evaluation to aptitude of individual (gpm) of course (2).
    b. Select two individuals, assign its aptitude probabilistic.
    c. Applied Genetic Operators. Only two gene with best aptitude probabilistic.
        i. Crossover to the couples.
        ii. Mutation to the equal rate probabilistic.
    d. Extinction (or null reproduction) by a poor aptitude probabilistic individual.
    e. Reproduction.
3. Until finish condition

Figure 3. GA Modified.

Swarm Intelligence

Swarm intelligence (SI) is based on the collective behavior. The collective behaviorthat emerges is a form of autocatalytic behaviour [34], self-organized systems. It's typically made up of a population of simple agents interacting locally with one another and with their environment. Natural examples of SI include ant colonies, bird flocking, animal herding, bacterial growth, and fish schooling. Nowadays swarm intelligence more generically, referes design complex adaptive systems.

Particle Swarm Optimization

Particle Swarm optimization (PSO), is a form of swarm intelligence, in which the behavior of a biological social systems. A particullaritied is when, looks for food, its individuals will spreadin the environment and move around independently. According to [81] Particle Swarm Optimization, a swarm of particles (individuals) in a n-dimensional search space [Math Processing Error] is simulated, where each particle q has a position genotype $p.g \in$ [Math Processing Error] $\subseteq R^n$ in this case n-dimensional is two, likewise a velocity $p.v \in R^n$. Therefore, each particle p has a memory holding its best position (can be reference to minim distance Euclidian of target (p) respect to q) best $(q) \in$ [Math Processing Error]. In order to realize the social component, the particle furthermore knows a set of topological neighbors $N(q)$ which could well be strong landmarks in the R^n.

This set could be defined to contain adjacent particles within a specific perimeter, using the Euclidian distance measure d_{euc} specified by

$$d_{euc} = \| p - q_i \|_2 = \sqrt{\sum_{i=1}^{n} (p - q_{(x,y)})^2}$$

(3)

$$d_{p.g} = \min \{d_{euc}(p, q_i) | q_i \in Q\}$$

(4)

$$p.v. = \omega \cdot v_i + \alpha \cdot rnd() \cdot (p_i - p.g) + \beta \cdot rnd() \cdot (g_{best} - p.g)$$

(5)

$$p.g = d_{p.g} + p.v.$$

(6)

Ant Colony Optimization

Ant Colony Optimization (ACO) has been formalized into a meta-heuristic for combinatorial optimization problems by Dorigo and co-workers [78], [79].This behavior, described by Deneubourg [80] enables ants to find shortest paths between food sources and their nest. While walking, they decide about a direction to go, they choose with higher probability paths that are marked by stronger pheromone concentrations.

ACO has risen sharply. In fact, several successful applications of ACO to a wide range of different discrete optimization problems are now available. The large majorities of these applications are to *NP-hard* problems; that is, to problems for which the best known algorithms that guarantee to identify an optimal solution have exponential time worst case complexity. The use of such algorithms is often infeasible in practice, and ACO algorithms can be useful for quickly finding high-quality solutions.

ACO algorithms are based on a parameterized probabilistic model the *pheromone model*, is used to model the chemical pheromone trails. Artificial ants incrementally construct solutions by adding opportunely defined solution components of the partial solutions in consideration. The first ACO algorithm proposed in the literature is called Ant System (AS) [56].

In the construction phase, an ant incrementally builds a solution by adding solution components to the partial solution constructed so far. The probabilistic choice of the next solution component to be added is done by means of transition probabilities, which in AS are determined by the following *state transition rule*:

$$P(c_r | s_a[c_i]) = \begin{cases} \dfrac{[\eta_r]^\alpha [\tau_r]^\beta}{\sum_{c_u \in J(s_a[c_i])} [\eta_u]^\alpha [\tau_u]^\beta} & \text{if } c_r \in J(s_a[c_i]) \end{cases} \tag{7}$$

$$\tau_j \leftarrow (1 - \rho) \cdot \tau_j + \sum_{a \in A} \Delta\tau^{s_a}{}_j$$

where

$$\Delta\tau^{s_a}{}_j = \begin{cases} F(s_q) & \text{if } c_j \text{ is a component of } s_a \\ 0 & \text{otherwise} \end{cases} \tag{8}$$

This pheromone update rule leads to an increase of pheromone on solution components that have been found in high quality solutions.

BEHAVIOR FUSION LEARNING

Representation and Initial Population

Applied the performance and success of an evolutionary optimization approach given by a set of objective functions F and a problem space [Math Processing Error] is defined by:

- Its basic parameter settings like the population size $\leq ms$ or the crossover and mutation rates.
- Whether it uses an archive Arc of the best individuals found.
- The fitness assignment process and the selection algorithm.
- The genotype-phenotype mapping connecting the search Space and the problem space.

Such that:

- All ants begin in the same node. Applied (6) to node initiate.
- Initial population with a frequency f.
- L^+ is evaluated with a cost of T^+. This implies, that j is visited only once.
- The transition rule is (9).

Genetic Operators and Parameters

The properties of their crossover and mutatrion operations are well known and an extensive body of search on them is avaible [87, 88].

The natural selections is improved the next form:

- Select four members to population (two best explorers and two best workers).
- Applied crossover.
- The probability the mutation is same to the percentage this.
- Performed null reproduction.

A string chromosome can either be a fixed-length tuple (9) or a variable-length list (10).

$$G = \forall \left(g[1], \ ..., \ g[n] : g[i] \in G_i \forall_i \in 1 \dots n \right) \tag{9}$$

$$G = \{ \forall \ lists(g : g[i] \in G_T \ \forall \ 0 \leq i \leq len(g) \} \tag{10}$$

Selection

Elitist selection, returns the $k<ms$(number of individuals to be placed into the mating)best elements from the list. In general evolutionary algorithms, it should combined with a fitness assignment processs that incorporates diversity information in order to prevent premature convergence.The elitist selection, force to the best individuals, according with to fitness function, to move to the next iteration.

Crossover

Amongst all evolutionary algortihms, genetic algorithms have the recombination operation which probably comes closest to the natural paragon.

Example of a fixed-length string, applied to the best explorer and the best workers obtain the two best gene.

1. Init random value α, β and γ.
2. Repeat
        a.   Applied mask.
        b.   The value is one, pass to gen father.
        c.   The value is cero, pass to gen mother.
3. Until ms.

Figure 4. Crossover Algorithm.

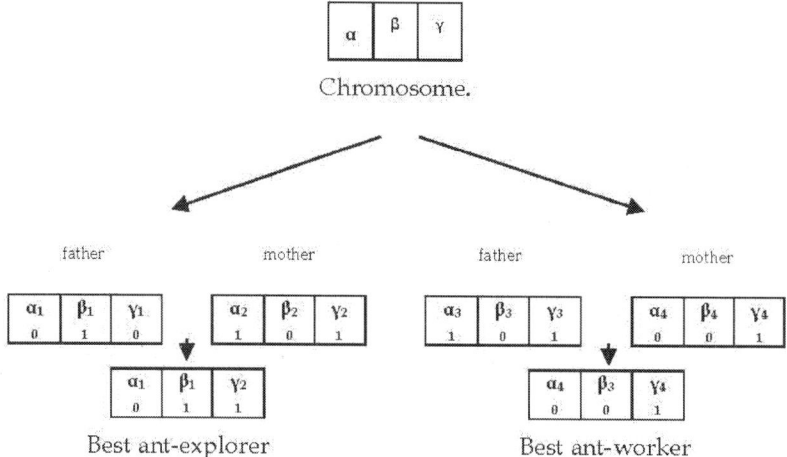

Chromosome.

Best ant-explorer   Best ant-worker

Figure 5. Example of Crossover, of course steps (b) and (c) crossover algorithm.

Mutation

The mutation operation applied, mutate [Math Processing Error] is used to create a new genotype g_n ∈ [Math Processing Error] by modifying an existing one. The way this modification is performed is application-dependent. It may happen in a randomized or in a deterministic fashion.

$$g_n = mutate(g) : g \in \mathbb{N}$$

(11)

Therefore, the mutation performed only one of the three gene in the chromosome offspring. The selected gen must a change in his allele (inherited), by a new random value in the same.

Null Reproduction or Extinction

After the crossover of the mutation (if is the case) is need performed a null reproduction or extinction of the ant workers or ant explorer. This allows the creation of fixed-length strings individuals means simple to create a new tuple of the structure defined by the genome and initialize it with random values.

Path Planning Through Evolutive ACO

The modifcation of the proposed ACO algorithm is applied. Due to in this modification have several parameters that determine the behavior of proposed algorithm, these parameters were optimized using a genetic algorithm.

A new transition rule. The rule of the equation(7) is modified by:

$$J = \begin{cases} argmax_{u \bar{\eta}_i^k} \{[\tau_{iu}(t)] \cdot [\eta_{iu}]\} & if\ q \le q_0 \\ J & if\ q > q_0 \end{cases}$$

(12)

This rule allows the exploration. Where, an ant k is move through i and j, with a distribution q in a range $[0, 1]$, q_0 is a parameter adjusts in the range $(0 \le q_0 \le 0)$ y $J \in$ [Math Processing Error], is a state select based ins

$$p_{ij}^k = \alpha \sin \beta \cdot \tau_{ij}^k(t) + \gamma$$

(13)

The transition rule, therefore is different (of equation 7) when $q \le q_0$, this meaning is a heuristic knowledge characterized in the pheromone. The amount of pheromone τ_{ij} [Math Processing Error], can be positive and negative, allowing a not-extinction, but rather to continue comparison between its these, is obtain the value more high of pheromone.

$$\Delta \tau_{ij}(t) = \frac{1}{L^+} where L^+ \in T^+$$

(14)

The update rule allows a global update, only for the path best.

1. For all edge (i, j) initialize $\tau_{ij} = c$ and $\Delta\tau_{ij}(t) = 0$.
2. Put the m-ant's created with a frequency f in the node top.
3. Repeat
    a. Natural selection (GA Modified)
    b. For k:=1 to ms
        i. Select to J with a transition rule $p_{ij}^k(t)$
        ii. Move to ant to J
        iii. Update the tour with (6)
        iv. If J was visited (Kill ant)
        v. Else (J is part of the list)
    c. Calculate the L^k of the k-th ant.
    d. Update the short best path.
    e. For any edge (i, j) applied (13) and (14).
4. Until Find a path.
5. If find a path finish else go to step 2.

Figure 6. ACO Modified.

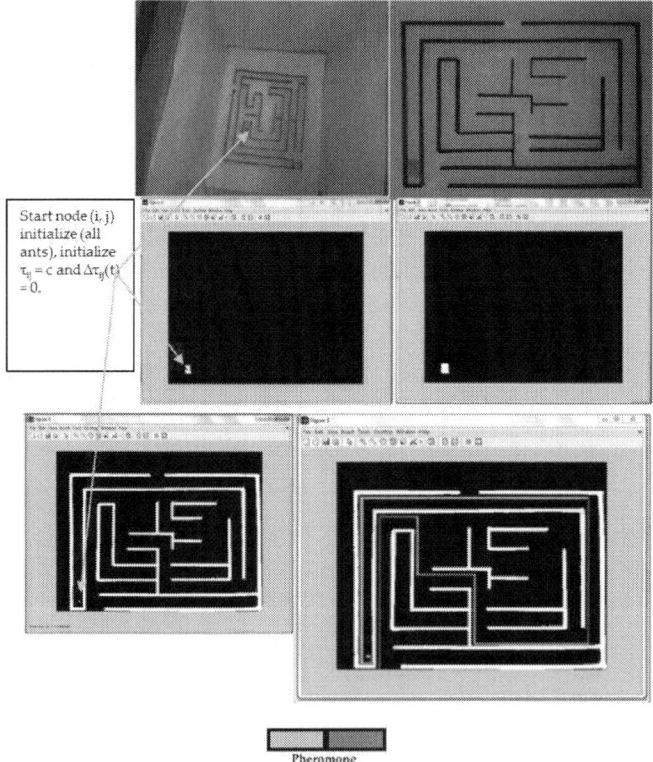

Start node (i, j) initialize (all ants), initialize $\tau_{ij} = c$ and $\Delta\tau_{ij}(t) = 0$.

Pheromone

Figure 7. Images of Solution Process applied the ACO-GA Algorithm and parameters of table 1.

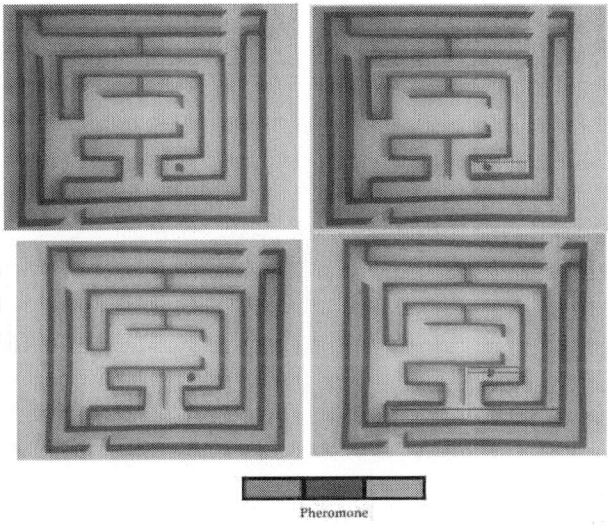

Figure 8. Images of Solution Process applied the ACO-GA Algorithm and parameters of table 1. Note that Different Start Node.

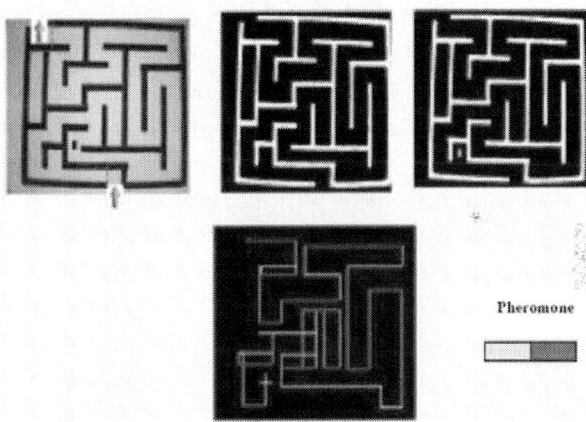

Figure 9. Solution to Labyrinth Applied ACO-algorithm.

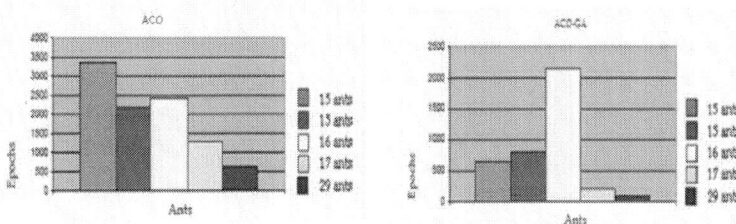

Figure 10. Example of a variation of epochs with $(0 < \alpha < 1)$; $\beta < \gamma$ or $\beta > \gamma$ applied ACO and ACO-GA.

Value	Configuration 1	Configuration 2	Configuration 3	Configuration 4	Configuration 5
α	-4.56	-3.69	0.1	4.4	-1.8
β	1.1	-2.69	-3.3	2.6	1.34
γ	-1.7	0.7	-2.02	1.7	3.49
Ants	29	15	15	17	16
Epoch	105	643	794	196	2166
t_{mseg}	13	273	277	40	860

Figure 11. Best value applied to the best ACO and the best ACO-GA

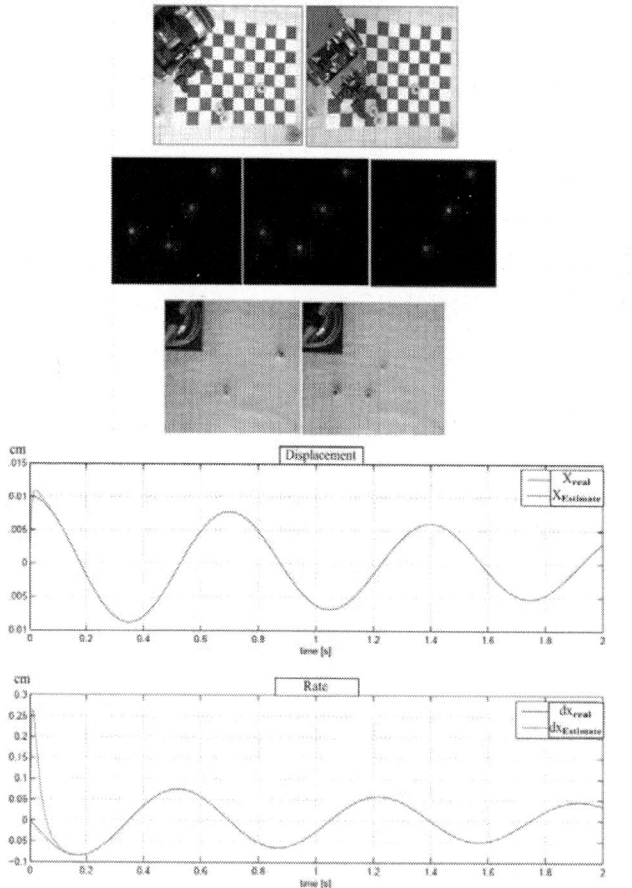

Figure 12. The robot manipulator select to piece of course to pheromone (green). Note, that the manipulator can't take the first piece. Graphics Displacement and Monitoring of the Manipulator respect to pheromone.

Figure 13. Besides using the ACO-Modified, is necessary calculate the distance of the first node through of de equation (6). Because the initial pose of manipulator can be "above" of the piece or well to distance not detectable by manipulator of the initial node, of manipulator makes it impossible of take.

EXPERIMENTAL RESULTS

In this section, the accuracy of the proposed algorithms described above. Present different start node, implying different grade of complexity.

Firtstly obtain the graph of each environment, the same applied of the landmarks of init and finish, also the centroid of the initial node. Finally, applied the algorithms over each graph, search the shortest path.

CONCLUSIONS AND PERSPECTIVES

This work was implemented with call selection natural and evolution natural, through basict two types of ants: jobs and explorer. The offspring are considered for the jobs, the best foragers and for the explorer, the offspring least have lost their path. Each ant have three gene, α, β and γ owned to transition function. The parameters was applied in the same algorithms, but, the best result is obtanained with the ACO-GA algorithm.

The implementation of the ACO-GA in a robot manipulator, is a job in prosess, but the first results prove the effectively of the same. See figures (9) and (10).

REFERENCES

1. R Siegwart, I. R Nourbakhsh, 2004Introduction to autonomous mobile robot. Massachusetts Institute of Technologypress, Cambridge, U.S.A.
2. J Borenstein, H. R Everett, L Feng, Where am I? Sensors and Methods for Mobile Robot Positioning, Ann Arbor, University of Michigan,1996184186available at http://www-personal.engin.umich.edu/~johannb/position.htm)
3. S Udupa, 1977Collision detection and avoidance in computer controlled manipulator. PHD thesis, California Institute Technology, California, USA.
4. J. C Latombe, 1991Robot motion planning. Kluwer Academic Publisher, Boston.
5. Lozano-Perez T Wesley MA (1979An algorithm for planning collision-free paths among polyhedral obstacles. Commun ACM., 2210560570
6. L Li, T Ye, M Tan, X Chen, 2002Present state and future development of mobile robot technology research. Robot., (24)5: 475-480.
7. R Siegwart, I. R Nourbakhsh, 2004Introduction to autonomus mobile robot. Massachusets Institute of Technology Press, Cambridge, U.S.A.
8. C. O Dunlaing, C. K Yap, 1985A retraction method for planning a motion of a disc. J. Algorithms, 6104111
9. E Masehian, , Katebi, (2007). Robot motion planning in dynamic environments with mobile obstacles and target. Int. J. Mech. Syst. Sci. Eng., 1(1): 20-25.
10. S Garrido, L Moreno, D Blanco, P Jurewicz, 2011Path planning for mobile robot navigation using Voronoi diagram and fast marching. Int. J. Robot. Autom., 214264
11. T Lozano-perez, 1983spatial planning: A configuration approach. IEEE T. Comput. C, 323108120
12. D. J Zhu, J. C Latombe, 1989New heuristic algorithms for efficient hierarchical path planning. Technical report STAN-CS-891279Computer Science Department, Stanford University, USA.

13. D. W Payton, J. K Rosenblatt, D. M Keirsey, 1993Grid-based for mobile robot. Robot Auton. Syst., 1111321

14. M Likhachev, D Ferguson, G Gordon, A Stentz, S Thrun, 2005Anytime dynamic A*: An anytime, replanning algorithm. Proceedings of the international Conference on Automated Planning and Scheduling, 262271

15. Hachour, 2008The processed genetic FPGA implementation for path planning of autonomous mobile robot. Int. J. Circ. Syst. Sig. Proc., 22151167

16. Hachour, 2008Path planning of autonomous mobile robot. Int. J. Syst. Appl. Eng. Dev., 42178190

17. T. G Zheng, H Huan, S Aaron, 2007Ant Colony System Algorithm for Real-Time Globally Optimal Path Planning of Mobile Robots. Acta. Automatica. Sinica, 333279285

18. P Cheeseman, R Smith, M Self, A stochastic map for uncertain spatial relationships. In 4th International Symposium on Robotic Research, MIT Press, 1987

19. Jhon Leonard and HFeder. Decoupled stochastic mapping for mobile robot and auto navigation. IEEE Journal Oceanic Engineering, 66, 442001561571

20. J Neira, and J. D Tardos, Data association in stchocastic mapping using joint compatibility test. IEEE Transaction Robotics and Automation, 8908972001

21. S Clark, G Dissanayake, P Newman, andH Durrant-whyte, A Solution Localization and Map Building (SLAM) Problem. IEEE Journal of Robotics and Automation, 173June 2001

22. Michael Montemerlo SebastianFastslam: A factored solution totje simultaneous localization and mapping problem,http://citeseer. nj.nec.com/503340.html.

23. E Nebot, 2002Simultaneous localization and mapping 2002 summer school. Australian centre field robotics.http://acfr.usyd.edu.au/ homepages/academic/enebot/

24. H. B Duan, X. Y Zhang, and C. F Xu, Bio-Inspired Computing", Science Press, Beijing, (2011

25. Y. V Pehlivanoglu, A new vibrational genetic algorithm enhanced with a Voronoi diagram for path planning of autonomous UAV", Aerosp. Sci. Technol., 1620124755

26. W Ye, D. W Ma, and H. D Fan, Algorithm for low altitude penetration aircraft path planning with improved ant colony algorithm", Chinese J. Aeronaut., 182005304309

27. H. B Duan, Y. X Yu, X. Y Zhang, and S Shao, Three-dimension path planning for UCAV using hybrid meta-heuristic ACO-DE algorithm", Simul. Model. Pract. Th., 18201011041115

28. H. B Duan, X. Y Zhang, J Wu, and G. J Ma, Max-min adaptive ant colony optimization approach to multi-UAVs coordinated trajectory replanning in dynamic and uncertain environments", J. Bionic. Eng., 62009161173

29. C. F Xu, H. B Duan, and F Liu, Chaotic artificial bee colony approach to uninhabited combat air vehicle (UCAV) path planning", Aerosp. Sci. Technol., 142010535541

30. H. B Duan, S. Q Liu, and J Wu, Novel intelligent water drops optimization approach to single UCAV smooth trajectory planning", Aerosp. Sci. Technol., 132009442449

31. J Canny, J Reif, 1987New lower bound techniques for robot motion planning problems. Proceedings of IEEE Symposium on the Foundations of Computer Science, Los Angeles, California, 4960

32. K Sugihara, J Smith, 1997Genetic algorithms for adaptive motion planning of an autonomous mobile robot. Proceedings of the IEEE International Symposium on Computational Intelligence in Robotics and Automation, Monterey, California, 138143

33. S Goss, S Aron, J. L Deneubourg, and J. M Pasteels, 1989Self-organized Shortcuts in the Argentine Ant. Naturwissenchaften 76pg. 579581Springer-Verlag.

34. M Dorigo, V Maniezzo, and A Colorni, 1996The Ant System: Optimization by a colony of cooperating agents", IEEE Transactions on Systems, Man and Cybernetics-Part-B, 261113

35. J Kennedy, R. C Eberhart, 1995Particle Swarm Optimization. Proceedings of the IEEE International Conference on Neural Networks, Perth, Australia, 19421948

36. Y. Q Qin, D. B Sun, N Li, Y. G Cen, 2004Path Planning for mobile robot using the particle swarm optimization with mutation operator. Proceedings of the IEEE International Conference on Machine Learning and Cybernetics, Shanghai, 24732478

37. Q. R Zhang, G. C Gu, 2008Path planning based on improved binary particle swarm optimization algorithm. Proceedings of the IEEE International Conference on Robotics, Automation and Mechatronics, Chendu, China, 462466

38. Z Nasrollahy, Javadi HHS (2009Using particle swarm optimization for robot path planning in dynamic environments with moving obstacles

and target. Third European Symposium on computer modeling and simulation, Athens, Greece, 6065

39. D. W Gong, J. H Zhang, Y Zhang, 2011Multi-objective particle swarm optimization for robot path planning in environment with danger sources. J. Comput., 6815541561

40. X. S Yang, 2011Optimization Algorithms. Comput., optimization, methods and algorithms. SCI 356, 1331

41. K Taylor, La Valle, M.S. "I-Bug: An Intensity-Based Bug Algorithm", Robotics and Automation, 2009ICRA'09. IEEE International Conference, 39813986

42. J. K Parker, A. R Khoogar, and D. E Goldberg, Inverse kinematics ofredundant robots using genetic algorithms". Proc. IEEE ICRA, 11989271276

43. Y Davidor, Robot programming with a genetic algorithm" Proc. IEEEInt. Conf. on Computer Sys. & Soft. Eng. (1990186191

44. J Solano, and D. I Jones, Generation of collision-free paths, a geneticapproach", IEEE Colloquium on Gen. Alg. for Control Sys. Eng. (19935

45. C Hocaoglu, and A. C Sanderson, Planning multi-paths usingspeciation in genetic algorithms", Proc IEEE Int. Conf. on Evolutionary Computation, (1996378383

46. M Gen, C Runwei, and W Dingwei, Genetic algorithms for solvingshortest path problems", Proc. IEEE Int. Conf. on Evolutionary Comput. (1997401406

47. Kumar PratiharD.; Deb, K.; Ghosh, A. "Fuzzy-genetic algorithms and mobile robot navigation among static obstacles" In Proc. CEC'99, 11999

48. S Zein-sabatto, and R Ramakrishnan, Multiple path planning for agroup of mobile robots in a 3D environment using genetic algorithms"Proc. IEEE Southeast, (2002359363

49. P Raja, And Pugazhenthi, S. "Optimal path planning of mobile robots: A review"International Journal of Physical Sciences 7131413202012

50. L. A Wilson, M. D Moore, J. P Picarazzi, and S. D. S Miquel, Parallel genetic algorithm for search and constrained multi-objectiveoptimization" Proc. Parallel and Distributed Processing Symp., (2004165

51. L Qing, T Xinhai, X Sijiang, and Z Yingchun, Optimum PathPlanning for Mobile Robots Based on a Hybrid Genetic Algorithm" InProc. HIS'06. (20065358

52. Z Qingfu, S Jianyong, X Gaoxi, and T Edward, EvolutionaryAlgorithms Refining a Heuristic: A Hybrid Method for Shared-PathProtections in

WDM Networks Under SRLG Constraints", IEEE Trans.on Systems, Man and Cybernetics, Part B, 37120075161

53. E Masehian, and D Sedighizadeh, Classic and Heuristic Approaches in Robot Motion Planning- A Chronogical Review", World Academy of Science, Engineering and Technology, (2007100106

54. M Dorigo, Optimization, learning and natural algorithms (in italian)," Ph.D. dissertation,Dipartimento di Elettronica, Politecnico di Milano, Italy, 1992

55. J Deneubourg, L Clip, P. -L Camazine, S. S Ants, buses androbots-self-organization of transportation systems", Proc. FromPerception to Action, (19941223

56. M Dorigo, and L. M Gambardella, Ant Colony System: A CooperativeLearning Approach to the Traveling Salesman Problem," IEEE Trans. InEvolutionary Comput., 1119975356

57. F Xiaoping, L Xiong, Y Sheng, Y Shengyue, and Z Heng, Optimalpath planning for mobile robots based on intensified ant colonyoptimization algorithm" Proc. IEEE on Rob. Intel. Sys. & Sig.Processing, 12003131136

58. H Ying-tung, C Cheng-long, and C Cheng-chih, Ant colonyoptimization for best path planning" Proc. IEEE/ISCIT'04, 2004109113

59. R Ramakrishnan, and S Zein-sabatto, Multiple path planning for agroup of mobile robot in a 2-D environment using genetic algorithms" InProc. Of the IEEE Int. Conf. on SoutheastCon'01, (20016571

60. M. M Mohamad, M. W Dunnigan, and N. K Taylor, Ant ColonyRobot Motion Planning" Proc. Int. Conf. on EUROCON'05. 12005213216

61. Na Lv and Zuren FNumerical Potential Field and Ant ColonyOptimization Based Path Planning in Dynamic Environment", IEEEWCICA'06, 2200689668970

62. P Raja, S Pugazhenthi, 2008Path planning for a mobile robot to avoid polyhedral and curved obstacles. Int. J. Assist. Robotics. Mechatronics, 923141

63. P Raja, S Pugazhenthi, 2009Path planning for a mobile robot using real coded genetic algorithm. Int. J. Assist. Robot. Syst., 1012739

64. P Raja, S Pugazhenthi, 2009Path planning for mobile robots in dynamic environments using particle swarm optimization. IEEE International Conference on Advances in Recent Technologies in Communication and Computing (ARTCom 2009), Kottayam, India, 401405

65. P Raja, S Pugazhenthi, 2011Path Planning for a mobile robot in dynamic environments. Int. J. Phys. Sci., 62047214731

66. XiongCh, Yingying K, Xian F, and Quidi W; "A fas two-stage ACO algorithm for robotic planning" Neural Comp., and App., Springer-Verlag (2011DOI:s00521-011-0682-7.

67. W. F Carriker, P. K Khosla, B. H Krogh, The use of simulated annealing to solve the mobile manipulator path planning problem", Proc. IEEE ICRA (1990204209

68. F Janabi-sharifi, and D Vinke, Integration of the artificial potential field approach with simulated annealing for robot path planning" Proc. IEEE Int. Conf. on Intel. Control (1993536541

69. D Blackowiak, S. D Rajan, Multi-path arrival estimates using simulated annealing: application to crosshole tomography experiment", IEEE J. Oceanic Eng. 2031995157165

70. M. G Park, and M. C Lee, Experimental evaluation of robot path planning by artificial potential field approach with simulated annealing", Proc. SICE (2002421902195

71. H Miao, A multi-operator based simulated annealing approach for robot navigation in uncertain environments. (2010Int. J. Comput. Sci. Secur., 415061

72. Z Qidan, Y Yongjie, and X Zhuoyi, Robot Path Planning Based on Artificial Potential Field Approach with Simulated Annealing" Proc. ISDA'06, (2006622627

73. W. D Rencken, 1993Concurrent Localization and Map Building for Mobile Robots Using Ultrasonic Sensors." Proceedings of the 1993 IEEE/RSJ International Conference on Intelligent Robotics and Systems, Yokohama, Japan, July 26-30, 21922197

74. T Edlinger, and E Puttkamer, 1994Exploration of an Indoor Environment by an Autonomous Mobile Robot." International Conference on Intelligent Robots and Systems (IROS'94). Munich, Germany, Sept. 12-16, 12781284

75. Y. V Pehlivanoglu, A new vibrational genetic algorithm enhanced with a Voronoi diagram for path planning of autonomous UAV", Aerosp. Sci. Technol., 1620124755

76. M Mata, J. M Armingol, F. J Rodríguez, 2004A deformable Model-Based Visual System for Mobile Robot Topologic Navigation" CICYT Project TAP 990214

77. G Dudek, M Jenkin, 2000Computational Principles of Mobile Robotics, Cambridge University Press, Cambridge, UK

78. M Dorigo, and G. Di Caro, "The Ant Colony Optimization meta-heuristic," in New Ideas in Optimization, D. Corne et al., Eds., McGraw Hill, London, UK, 11321999

79. M Dorigo, G Di, Caro, and L.M. Gambardella, "Ant algorithms for discrete optimization," Artificial Life, 521371721999

80. J. -L Deneubourg, S Aron, S Goss, and J. -M Pasteels, The self-organizing exploratory pattern of the argentine ant. Journal of Insect Behaviour, 31990159168

81. T Weise, Global Optimization Algorithms-Theory and Application"-Second edition e-book. (2007

82. E David, Goldberg. Genetic Algorithms in Search, Optimization, and Machine Learning.Addison-Wesley Longman Publishing Co., Inc. Boston, MA, USA, January 19890-20115-767-5

83. John Henry HollandAdaptation in Natural and Artificial Systems: An IntroductoryAnalysis with Applications to Biology, Control, and Artificial Intelligence. The University of Michigan Press, Ann Arbor, 1975. 047-2-08460-797-8Reprinted by MIT Press, April1992Net Library, Inc.

84. J°orgHeitk°otter and David Beasley, editors. Hitch-Hiker's Guide to EvolutionaryComputation: A List of Frequently Asked Questions (FAQ). ENCORE (The EvolutioNary Computation REpository Network), 1998. USENET: comp.ai.genetic. Onlineavailable at http://www.cse.dmu.ac.uk /~rij/gafaq/top.htm and http://alife.santafe.edu/~joke/encore/www/

85. Hui-Chin TangCombined random number generator via the generalized Chinese remainder theorem. Journal of Computational and Applied Mathematics, 1422377388May 15, 20020377-0427doi:10.1016/S0377 -0427(01)00424-1.Onlineavailable at http://dx.doi.org/10.1016/ S0377- 0427(01)00424-1

86. C Blum, and X Li, Swarm Intelligence: Swarm Intelligence in Optimization",Natural Computing Series, 2008Part I, 4385DOI:

87. E David, Goldberg. Genetic Algorithms in Search, Optimization, and Machine Learning.Addison-Wesley Longman Publishing Co., Inc. Boston, MA, USA, January 19890-20115-767-5

88. John Henry HollandGenetic Algorithms. Scientific American, 26714450July1992Online available athttp://www.econ.iastate.edu/ tesfatsi/holland.GAIntro.htmandhttp://www.cc.gatech.edu/~turk/bio_ sim/articles/genetic_algorithm.pdf.

Services Net Modeling For Dependability Analysis

Wojciech Zamojski[1] and Tomasz Walkowiak[1]

[1] NASA Langley Research Center, Poland

1. INTRODUCTION

Network technologies are being developed for many years. Most of large technical systems could be seen as a kind of network, for example: information, transport or electricity distribution systems. Networks are modelled as directed graphs with nodes, in which commodities and information media are being processed, and arcs as communication links (telecommunication channels, roads, pipelines, conveyors, etc.) for media transportation. Resources of networks could be divided into two classes: services (functionality resources) and technical infrastructures (hardware and software resources).

We propose to analyse the network system from the functional and user point of view, focusing on business service realized by a network system (Gold et al., 2004). Users of the network system realise some tasks in the system (for example: send a parcel in the transport system or buy a ticket in the internet ticket office). We assume that the main goal, taken into consideration during design and operation, of the network system is to fulfil the user requirements. Which could be seen as some quantitative and qualitative parameters of user tasks.

Network services and technical resources are engaged for task realization and each task needs a fixed list of services which are processed on the base of whole network technical infrastructure or on its part. Different services may be realized on the same technical resources and the same services may be realized on different sets of technical resources. Of course with different values of performance and reliability parameters. The last statement is essential when tasks are realized in the real network system surrounded by unfriendly environment that may be a source of threads and even intentional attacks. Moreover, the real networks are build of unreliable software and hardware components as well.

In (Avižienis et al., 2000) authors described basic set of dependability attributes (i.e. availability, reliability, safety, confidentiality, integrity and maintainability). This is a base of defining different dependability metrics used in dependability analysis of computer systems and networks. In this paper we would like to focus on more functional approach metrics which could be used by the operator of the network system. Therefore, we consider dependability of networks as a property of the networks to reliable process of user tasks, that is mean the tasks have to perform not only without faults but more with demanded performance parameters and according to the planned schedule.

We propose to concentrate the dependability analyse of the networks on fulfilling the user requirements. Therefore, it should take into consideration following aspects:

- specification of the user requirements described by task demands, for example certainty of results, confidentiality, desired time parameters etc.,
- functional and performance properties of the networks and theirs components,
- reliable properties of the network technical infrastructure that means reliable properties of the network structure and its components considered as a source of failures and faults which influence the task processing,
- process of faults management,
- threads in the network environment,
- measures and methods which are planned or build-in the network for elimination or limitation of faults, failures and attacks consequences; reconfiguration of the network is a good example of such methods,
- applied maintenance policies in the considered network.

As a consequence, a services network is considered as a dynamical structure with many streams of events generated by realized tasks, used services and resources, applied maintenance policies, manager decisions etc. Some network events are independent but other ones are direct consequences of previously history of the network life. Generally, event streams created by a real network are a mix of deterministic and stochastic streams which are strongly tied together by a network choreography. Modelling of this kind of systems is a hard problem for system designers, constructors and maintenance organizers, and for mathematicians, too. It is worth to point out some achievements in computer science area such as Service Oriented Architecture (Gold et al., 2004, Josuttis, 2007) or Business Oriented Architecture(Zhu & Zhang, 2006) and a lot of languages for network description on a system choreography level, for example WS-CDL (Yang et al., 2006), or a technical infrastructure level, for example SDL (Aime et al., 2007). These propositions are useful for analysis of a network from the designer point of view and they may been supported by simulation tools, for example modified SSF.Net simulator (Zyla & Caban, 2008), but it is difficult to find a computer tools which are combination of language models and Monte Carlo (Fishman, 1996) based simulators.

The chapter presents a step to a creation of a verbal and formal model of a net of services. It presents a generic approach to modelling performability (performance and reliability) properties of the services net. The Petri Nets will is used for the task realization process modelling. Moreover, an example of service net– the discrete transport system analysed by an event-driven simulator is presented.

2. SERVICE NETWORK – OVERVIEW

We can distinguish three main elements of any network system: users, services and technical resources. As it presented in the Figure 1 users are generating tasks which are being realized by the network system. The task to be realized requires some services presented in the system. A realization of the network service needs a defined set of technical resources. In a case when any resource component of this set is in a state "out of order" or "busy" then the network service may wait until a moment when the resource component returns to a state "available" or the service may try to create other configuration on the base of available technical resources.

Therefore, following problems should be taken into consideration:

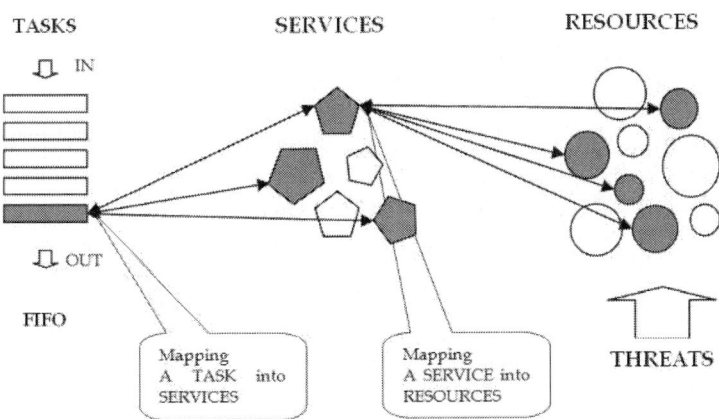

Figure 1. Task mapping on business services and technical resources.

- description and mapping a service net on existed net resources for each moment of its using;
- a prognoses process of the service net behaviour in a real life conditions – definition and selection of measures;
- finding relations between measures/criteria and functional, performance and reliability parameters of the service net;
- evaluation methods of choose measures of the service net;
- decision process of maintenance organization - decision steps as a reaction on appeared events, specially on threats;
- definition of measures and criteria of decision steps - risk of threats, and evaluation of decision risk and its cost.

An illustration of problems connected with functional – dependability modelling of services networks is shown in Figure 2.

3. FUNCTIONAL – DEPENDABILITY MODELS

The *ST model* (*State - Transition model*) is the most popular and useful methodology used in modelling of systems.

The system is considered as a union of its hardware, management system and involved personnel (administrators, users, support services etc.), so the system states depend on the states of all these elements. The system transitions are consequences of events connected with execution of system tasks and jobs, system faults and system reactions to them, incidents, attacks and system responses etc., i.e. system events are observable occurrences which change states of the system.

The functional – reliability model (Zamojski, 2005) of computer system S_C is a configuration of hardware H, software SP, men M, management system (operating system) MS, tasks (functions) J and system events E_S

$$S_C \subset H \times SP \times J \times M \times MS \times E_S$$

(1)

The system events includes those connected with tasks realization, occurrence of incidents (faults, viruses, and attacks) and system reactions to them (hardware and information renewals). The system events are very often described by their time parameters which are collected in so called *a chronicle* of the system.

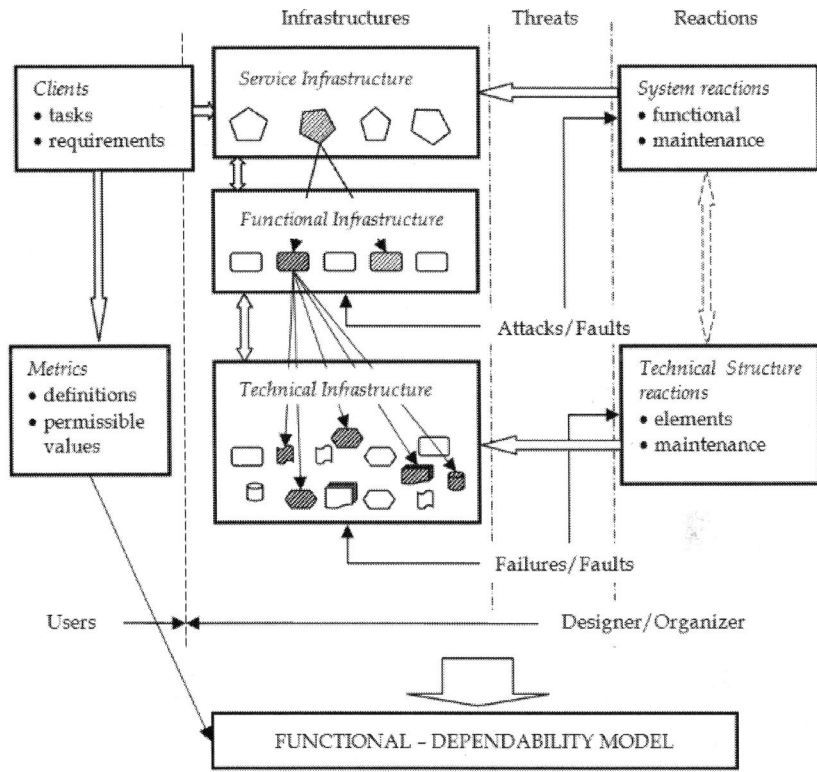

Figure 2. Basic terms and a functional - dependability model of a services network (Zamojski, 2009)

A functional configuration [Math Processing Error] of the computer system is a set of hardware and software resources that are allocated to realize *i*-th task [Math Processing Error]

$$\left(j^{(i)} \subset J \right) \Rightarrow \left(S_C^{(i)} \subset S_C \right)$$

(2)

$$S_C^{(i)} \subseteq H^{(i)} \times SP^{(i)} \times j^{(i)} \times M^{(i)} \times MS^{(i)} \times E_S^{(i)}$$

(3)

where superscript *(i)* fix subsets of system resources needed for execution *i*-th task.

A functional – reliability model in the system engineering is regarded as a structured representation of the functions, activities or processes, and events generated inside of the considered system and/or by its surroundings. The system events may be divided into two main classes: functional events and reliable (together with maintenance) events. In practice this classification is very often difficult to be made because a system reaction on an event may involve a lot of functional or/and maintenance reactions. Therefore, it is better to create one common class of functional–reliable events, so called *performability* events (Zamojski & Caban, 2006). Because of these reasons considered model of services network will be called *performability* model or *functional-dependability* model (Zamojski & Caban, 2007).

If the functional – reliability model is built as the ST model then the set of the system states is determined by the states of all resources involved in tasks realized at the moment. The system resource allocations are dynamic, modified due to the incoming tasks, occurring incidents and system reactions (especially reconfiguration).

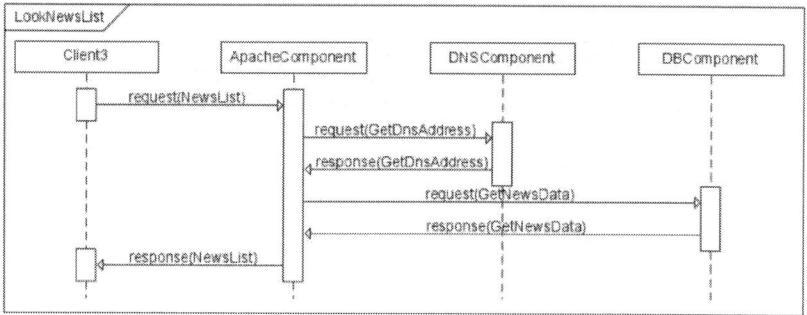

Figure 3. Exemplar choreography

4. FORMAL MODEL OF A SERVICE NET

4.1. A Service Net

A *services net* is a system of business services that are necessary for user (clients) tasks realization process. The services net are built on the bases of technical infrastructure (*technological resources*) and *technological services* which are involved into a task realization process according to decisions of a management system. The task realization process may include many sequences of services, functions and operations which are using assignment network resources - in the computer science this process of assignments and realization steps is called as *a choreography*. An example of choreography for web service is presented in Figure 3.

The functional – dependability model of a services network has to consider specificity of the network: nodes and communication channels, the ability of dynamic changes of network traffic (routing) and reconfiguration, and all other tasks realized by the network.

The service network could be defined as a tuple:

$$SNet = \langle J, BS, TR, MS, C \rangle,$$

(4)

where:

- $J = \left\{ J^{(i)}; i = 1, 2, \ldots \right\}$ – a set of tasks generated by users and realized by the service network,

- $BS = \left\{ BS^{(b)}; b = 1, 2, \ldots \right\}$ – a set of services which are available in the considered network,

- $TR = \left\{ TR^{(r)}; r = 1, 2, \ldots \right\}$ – technical infrastructure of the network which consists of technical resources as machines/servers, communication links etc,

- MS – management system (for example - operating system),

4.2. Tasks

The task [Math Processing Error] is understood as a sequence of actions and works performed by services network in a purpose to obtain desirable results in accordance with initially predefined time schedule and data results. In this way a single task $J^{(i)} = \langle J_{IN}^{(i)}, J_{OUT}^{(i)} \rangle$ may be defined as an ordered pair of so called input task $J_{IN}^{(i)}$, which is described by the input parameters (postulated results and prognosis time schedule) and the corresponding output task $J_{OUT}^{(i)}$ (real results and real time schedule).

$$J_{IN}^{(i)} = \langle R_P^{(i)}, A^{(i)}, C_P^{(i)} \rangle,$$

(5)

The input task is define as the triple:

where:

- $R_P^{(i)}$ - postulated results of the i-th task execution,

- $C_P^{(i)}$ - postulated chronicle of the task realization,

- $A^{(i)} = A^{(i)} \left(R_P^{(i)}, C_P^{(i)} \right)$ - a sequence of actions and works necessary to obtain postulated results in planned time.

The [Math Processing Error] may be described by a flowchart of actions and works, and its realization depends on an availability of network services and technical resources.

The output task is define as the pair:

$$J_{OUT}^{(i)} = \left\langle R_{real}^{(i)}, , C_{real}^{(i)} \right\rangle,$$

(6)

where:

- $R_{real}^{(i)}$ - real results of the i-th task execution,

- $C_{real}^{(i)}$ - real chronicle of the task realization.

The postulated results and chronicles are defined with assumed tolerance intervals ($\underline{R_P^{(i)}} \leq R_P^{(i)} \leq \overline{R_P^{(i)}}$ and $\underline{C_P^{(i)}} \leq C_P^{(i)} \leq \overline{C_P^{(i)}}$) and when the real results and chronicles are inside the intervals ($R_{real}^{(i)} \subset \left[\underline{R_P^{(i)}}, \overline{R_P^{(i)}} \right]$ and $C_{real}^{(i)} \subset \left[\underline{C_P^{(i)}}, \overline{C_P^{(i)}} \right]$) then the task is assumed to be correctly realised.

4.3. Services

The term service is understood as a discretely defined set of contiguously cooperating autonomous business or technical functionalities. Of course, a special mechanism to enable an access to one or more businesses and functionalities should be implemented in the system. The access is provided by a prescribed interface and is monitored and controlled according to constraints and policies as specified by the service description[1] - .

The service $BS^{(b)}$ is defined as a sequence of activities described by a set of capabilities (functionalities) $\left\{ F_k^{(b)}, k = 1,2,... \right\}$, a set of demanded input parameters of data and/or media $BS_{IN}^{(b)}$ and a set of output parameters $BS_{OUT}^{(b)}$

$$BS^{(b)} = \left\langle \left\{ F_k^{(b)}; k = 1, 2,... \right\}, BS_{IN}^{(b)}, BS_{OUT}^{(b)} \right\rangle.$$

(7)

Because the services have to cooperate with other services than protocols and interfaces between services and/or individual activities are crucial problems which have a big impact on the definitions of the services and on processes of their execution.

A service may be realized on the base of a few separated sets of functionalities $\left\{ F_{k1}^{(b)}, \, k1 = 1,2,... \right\}, \, \left\{ F_{k2}^{(b)}, \, k2 = 1,2,... \right\}$ with different costs which are the consequences of using different network resources.

4.4. Technical Infrastructures

Hardware is considered as a set of hardware resources (devices and communication channels) which are described by their technical, performance, reliability and maintenance parameters. The system software is described in the same way.

4.5. Management System

The management system of service network allocates the services and network resources to realized tasks, checks the efficient states of the services network, performs suitable actions to locate faults, attacks or viruses and minimize their negative effects. Generally the management system has two main functionalities:

- monitoring of network states and controlling of services and resources,
- creating and implementing maintenance policies which ought to be adequate network reactions on concrete events/accidents. In many critical situations a team of men and the management system have to cooperate in looking for adequate counter-measures, for instance in case of a heavy attack or a new virus.

The maintenance policy is based on two main concepts: detection of unfriendly events (attacks, faults, failures) and network responses to them. In general the network responses incorporate the following procedures:

- detection of incidents and identification of them,
- isolation of damaged network resources in order to limit proliferation of incident consequences,
- renewal of damaged services, processes and resources.

It is hard to predict all possible events (for example all new demands for a task realization) or incidents (for example failures, faults, attacks or an end of a renewal procedure) in the services network, especially it is not possible to predict all possible attacks or men faults, so system reactions are very often "improvised" by the management system, by its administrator staff or even by expert panels specially created to find a solution for the existing situation. The time, needed for the renewal, depends on the incident that has occurred, the system resources that are available and the renewal policy that is applied. The renewal policy is formulated on the basis of the required levels of system dependability and on the economical conditions (first of all, the cost of downtime and cost of lost achievements) (Zamojski & Caban, 2006, Zamojski & Caban, 2007).

Maintenance policy is based on maintenance rules that are understood as chains of decisions about allocation of services and network resources (hardware, software, information and service staff) that are undertaken to keep the system operational after an incident. These rules are very often connected with small fragments of the system, for example; replacement of a machine (a processor) or communication links. These local operations may have impact on the whole network, e.g. if a communication channel is down for a few minutes, then rates of medium (data) traffic of the network may violently change (Zamojski & Caban, 2007).

4.6. Chronicles

The set of system events is created by events connected with tasks realization, incidents occurrence (faults, viruses, and attacks) and system reactions (hardware and information renewals).

4.7 A Process of The Task Realization

The task realization process is supported by two-level decision procedures connected with selection and allocation of the network functionalities and technical resources. There are two levels of decision process: services management and resource management. The first level of decision procedure is connected with selection suitable services and creation a task configuration. Functional and performance task demands are the base for suitable services choosing from all possible network services. The goal of the second level of the decision process is to find needed components of the network infrastructure for each service execution and the next allocate them on the base their availability to the service configuration. If any component of technical infrastructure is not ready to support the service configuration then allocation process of network infrastructure is repeated. If the management system could not create the service configuration then the service management process is started again and other task configuration may be appointed. These two decision processes are working in a loop which is started up as a reaction on network events and accidences. On the beginning of a task realization procedure the task $J_{IN}^{(i)}$ is mapped on the network services and a subset of services $BS_s^{(i)}$ necessary for the task realization according to its postulated parameters is created; $J_{IN}^{(i)} \rightarrow BS_s^{(i)}$. Next, a demand of technical resources for each service realization is fixed: [Math Processing Error] . In a real services network the same task is very often realized on the base of various service subsets and the same service may involved different technical resources. Of course, this possible diversity of task realization is connected with the flowcharts $A^{(i)}$ and the availability of network resources is checking for each service. In this way a few task configurations service configurations, additionally described by appropriately defined cost parameters, may be fund for the i-th task realization.

5. THE PETRI NET MODEL

Petri Nets (Zhou & Kurapati, 1999) are a powerful and often used modelling tool. They allow to represent two aspects of a modelled system static and dynamic (thanks to the token evolution). A common definition of the Petri net is formulating as a triple:

where:

- P - set of places that represent deterministic states of processes, tasks, services, resources etc. of the considered system. The places are often complemented by tokens that are modeled abilities of these places.
- T – set of transitions that represent net events characterized by conditions necessary to come them into firing. The transitions are often described by firing time and other probabilistic characteristics etc.
- A – set of arches (directed and inhibited) that models routes on which events represented by tokens are passed by the net.

A state of the net, described by marking (tokens localization in the places) represents sufficient conditions for arising new events of a net's life. Net's events may be divided into many classes, for example functional, reliable or maintenance events, deterministic or probabilistic ones etc. The mention classification depends on assumed criteria.

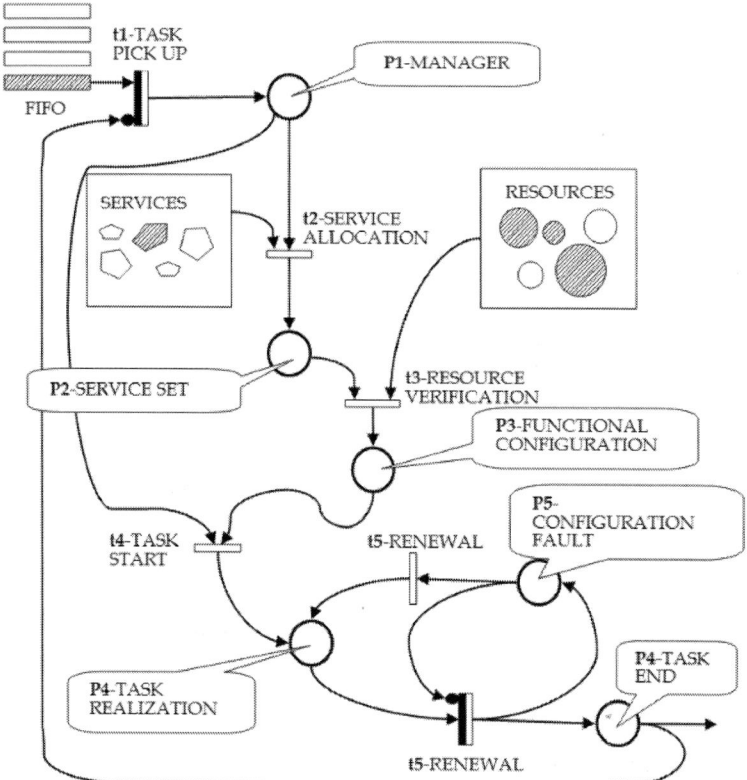

Figure 4. The Petri net model of a task realization in a services network.

The Petri net model of the ith task realization $(J_{IN}^{(i)})$ is shown in the Figure 4. It is assumed the input task $(J_{IN}^{(i)})$ is taken from the stack of waiting tasks (transition t1 and its firing time $\tau_{T1}^{(i)}$). The choice of the task may be based on the strategy FIFO (as it is illustrated on the Figure 2) and it is conditioned by ending of previously task (the transition t1 is guarded by inhibited arc from the place P6 (end of the task). The place P1 represents the management process of mapping the input task into a set of necessary services $(BS^{(b)})$ and when the services are ready then the transition t2 is fired (time $\tau_{T2}^{(i)}$). After checking if the chosen services may be activated on the base of needed efficient technical resources then a functional configuration of the task (place P3) is created (transition t3 with time $\tau_{T3}^{(i)}$) and at this moment the manager may take a decision about start of the task process realization (transition t4).

There is a build-in system of monitoring and detection of unfriendly accidences like faults and failures (place P5). When such unfriendly accidence is discovered then a renewal process of the functional configuration is started (transition t5 and renewal time $\tau_{T5}^{(i)}$) and the task realization process is broken (the inhibited input of the transition t6) till the end of renewal operations.

The firing process of each transition is described by conditions (tokens in input places for the transition) which may occur with probabilities, for example a probability of a machine failure, and time duration of transition firing may be a probabilistic function, too. Of course a transition may be many times fired during a task realization, because net events may need to repeat bigger or smaller loops of the net. The Petri net model shown in the Figure 4 is reduced and presented only to show the main idea of the proposed modelling method which may be useful for evaluation of dependability measures of services networks.

Real time of the ith task realization $T_{Jreal}^{(i)}$ that is modelled as a stochastic timed Petri net with k transitions and l loops and sub loops may be evaluated as:

$$T_{Jreal}^{(i)} \cong \sum_{l \in L} \Pr\left\{e_l^{(i)} = 1\right\} \left[\sum_k \Pr\left\{f_{Tk,l}^{(i)} = 1\right\} \tau_{Tk,l} \right],$$

$$(9)$$

where:

- $e_l^{(i)} = 1$ - an event (for example, a new task, an allocation a technical resource to the i-th task, an end of a renewal process etc.) which is started a loop or a sub loop in the Petri net model ascribed to the ith task realisation,
- $f_{Tk,l}^{(i)} = 1$ - an event; the k transition is fired during l loop connected with the i-th task realization.

Such dependability measures as a probability that the real time duration of the i-th task may be defined and evaluated on the base of the Petri net models as:

$$M_{Depend}^{(i)}(J_{IN}^{(i)}) = \Pr\left\{T_{Jreal}^{(i)} \leq T_{JOUT}^{(i)}\right\}.$$

$$(10)$$

6. DISCRETE TRANSPORT SYSTEM – SERVICE NET CASE STUDY

An example of service net could be a DTSCNTT - Discrete Transport System with Central Node and Time-Table (Walkowiak et al., 2007). This is a simplified case of the Polish Post transport system.

Following the definition (4) each elements of service net could be described as follows.

The business service (BS) provided the Polish Post and therefore DTSNTT service net is the delivery of mails. The technical infrastructure (TR) consists of a set of nodes placed in different geographical locations and set of vehicles and timetable. There are bidirectional routes between nodes marked by lines. There is distinguished one node called central mode. Mails are distributed among nodes by vehicles.

Each vehicle is described by following functional and reliability parameters: mean speed of a journey, capacity – number of containers which can be loaded, reliability function and time of vehicle maintenance.

Management system (MS) is defined by time table since vehicles distributing mails among system nodes operate according to the time-table exactly as city buses or intercity coaches. The time-table consists of a set of routes (sequence of nodes starting and ending in the central node, time of approaching each node in the route and the recommended size of a vehicle). The number of used vehicle, or the capacity of vehicles does not depend on temporary situation described by number of transportation tasks or by the task amount for example. It means that it is possible to realize the journey by completely empty vehicle or the vehicle cannot load the available amount of commodity (the vehicle is to small). Time-table is a fixed element of the system in observable time horizon, but it is possible to use different time-tables for different seasons or months of the year.

To reduce the complexity of the model we have decided to model the containers not separate mails (Walkowiak & Mazurkiewicz, 2009). Therefore, the tasks (J) of sending mails is modelled as a random process of containers generation. Each generated container has a destination address. The central node is the destination address for all containers generated in the ordinary nodes. Where containers addressed to in any ordinary nodes are generated in the central node. The generation of containers is described by Poisson process. In case of central node there are separate processes for each ordinary node. Whereas, for ordinary nodes there is one process, since commodities are transported from ordinary nodes to the central node or in opposite direction. Postulated result of any task is to transport a container to the destination node within a given time limit.

The process of any task realization could be described as follows. The container is generated in some node at a given time (according to Poisson process) and stored in the node waiting for the vehicle to be transported to the destination node. Each day a given time-table is realized, it means that at a time given by the time table a vehicle, selected randomly from vehicles available in the central node, starts from central node and is loaded with containers addressed to each ordinary nodes included in a given route. The loading is done in a service point. This is done in a proportional way. Since the number of service points is limited (parameter of the central node) and loading takes some time is there is no free service point vehicles has to wait in a queue. After loading the vehicle goes to a given ordinary node - it takes some time according to vehicle speed - random process and road length. After approaching the ordinary node the vehicle is waiting in an input queue if there is any other vehicle being loaded/unloaded at the same time. The containers addressed to given node are unloaded and empty space in the vehicle is filled by containers addressed to a central node. The operation is repeated in each node on the route and finally the vehicle is approaching the central node when is fully unloaded and after it is available for the next route. The process of vehicle operation could be stopped at any moment due to a failure (described by a random process). After the failure, the vehicle waits for a maintenance crew (if it is not available due to repairing other vehicles), is being repaired (random time) and after it continues its journey (Walkowiak & Mazurkiewicz, 2009).

As suggested in the introduction the simulator tool for analysing DTSCNTT service net was developed. The tool was adopting the event simulation approach, which is based on a idea of event, which could be described by time of event occurring, type of event (in case of DTSCNTT it could be a vehicle failure) and element or set of elements of the system on which event has its influence. The simulation is done by analyzing a queue of event (sorted by time of event occurring) while updating the states of system elements according to rules related to a proper type of an event. (Walkowiak et al., 2007)

We proposed for the case study analysis an exemplar DTSCNTT based on Polish Post regional centre in Wroclaw. We have modelled a system consisting of one central node (Wroclaw regional centre) and twenty two other nodes - cities where there are local post distribution points in Dolny Slask Province. The length of roads were set according to real road distances between cities used in the analyzed case study. The intensity of generation of containers for all destinations were set to 4,16 per hour in each direction giving in average 4400 containers to be transported each day. The vehicles speed was modelled by Gaussian distribution with 50 km/h of mean value and 5 km/h of standard deviation. The average loading time was equal to 5 minutes. There were two types of vehicles: with capacity of 10 and 15 containers. The MTTF of each vehicle was set to 2000. The average repair time was set to 5h (Gaussian distribution). (Walkowiak & Mazurkiewicz, 2009)

The simulation time was set to 100 days and each simulation was repeated 10.000 times. We have calculated the dependability measure defined by (10), the probability that the duration time of a task (delivery of some container) will be longer then a given time limit using Monte-Carlo approach (Fishman, 1996). The achieved results are presented in Figure 5.

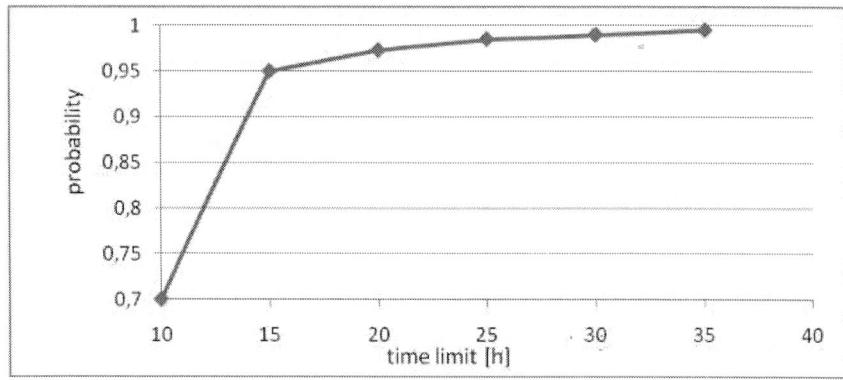

Figure 5. The probability of containers to be transported within a given limit time.

7. CONCLUSION

We have given a verbal and formal model of a service net. The formal model consists of a tuple mathematical model and the Petri Nets one. We hope that the proposed Petri net model will be very useful in the synthesis process of the service net. Of course there are a lot problems with building the Petri net model of the real services net in which exist a large number of services and technical resources that are mapped to many concurrent realized tasks. We have also presented an exemplar case study of service net a discrete transport system service net – a simplified case of Polish Post transport system. It was analysed by a usage of a discrete transport system simulator.

We plan to develop a simulation tool for a generic service nets with a functionality similar to presented discrete transport system simulator or BS.SSF simulator (Walkowiak, 2009) together with graphical tool for modelling and simulation. We also plan to use high level languages like for examples Business Process Modeling Notation (White & Miers 2008) for a graphical representation for specifying business processes in a workflow. We hope that it could be possible to map BPMN into a Petri net model or a general purpose service net simulator allowing to perform a service net dependability analysis.

REFERENCES

1. M. Aime, A. Atzeni, P. Pomi, 2007 Ambra- Automated Model-Based Risk Analysis, Proceedings of the 3rd International Workshop on Quality of Protection, 43 48 , Alexandria, ACM, New York

2. A. . Avižienis, J. . Laprie, B. Randell, 2000 Fundamental Concepts of Dependability. Proceedinggs of 3rd Information Survivability Workshop, Boston
3. G. Fishman, 1996 Monte Carlo: Concepts, Algorithms, and Applications, Springer-Verlag, New York
4. N. Gold, C. . Knight, A. Mohan, M. Munro, 2004 Understanding service-oriented software. IEEE Software, 21 71- 77
5. N. Josuttis, 2007 SOA in Practice: The Art of Distributed System Design, O'Reilly
6. T. Walkowiak, 2009 Information systems performance analysis using task-level simulator, Proceedings of International Conference on Dependability of Computer Systems, 218 225 , Brunow, IEEE Computer Society Press, Los Alamitos
7. T. . Walkowiak, J. Mazurkieiwicz, 2009 Analysis of critical situations in discrete transport systems, Proceedings of International Conference on Dependability of Computer Systems, 364 371 , Brunow, IEEE Computer Society Press, Los Alamitos
8. T. . Walkowiak, J. Mazurkiewicz, K. Kaplon, 2007 Functional analysis of discrete transport system realized by SSF simulation tool. Advances simulation of systems. Proceedings of the XXIXth International Autumn Colloquium, 103 108 , Sv. Hostyn, MARQ, Ostrava
9. S. A. White, D. Miers, 2008 BPMN Modeling and Reference Guide, Future Strategies Inc., Lighthouse Pt
10. H. Yang, X. Zhao, Z. Qiu, G. Pu, S. Wang, 2006 A Formal Model for Web Service Choreography Description Language (WS-CDL). Proceedings of the IEEE international Conference on Web Services, IEEE Computer Society, Washington
11. W. Zamojski, 2005 Functional-reliability model of computer-human system. Computer engineering, 278 297 , Eds. Wojciech Zamojski, WKL, Warszawa (in Polish)
12. W. Zamojski, 2009 Dependability of services networks. Proceedings of the Third Summer Safety and Reliability Seminars, 387 396 , Gdnask-Sopot, Polish Safety and Reliability Association, Gdansk
13. W. Zamojski, D. Caban, 2006 Introduction to the dependability modelling of computer systems. Proceedings of International Conference on Dependability of Computer Systems, 100 109 , Szklarska Poreba, IEEE Computer Society Press, Los Alamitos
14. W. Zamojski, D. Caban, 2007 Maintenance policy of a network with traffic reconfiguration. Proceedings of International Conference on

Dependability of Computer Systems, 213 220 , Szklarska Poreba, IEEE Computer Society Press, Los Alamitos

15. J. Zhu, L. Z. Zhang, 2006 A Sandwich Model for Business Integration in BOA (Business Oriented Architecture). Proceedings of the 2006 IEEE Asia-Pacific Conference on Services Computing, 305 310 , IEEE Computer Society, Washington

16. M. Zhou, V. Kurapati, 1999 Modeling, Simulation, & Control of Flexible Manufacturing Systems: A Petri Net Approach. World Scientific Publishing

17. M. Zyla, D. Caban, 2008 Dependability Analysis of SOA systems. Proceedings of International Conference on Dependability of Computer Systems, 301 306 , Szklarska Poreba, IEEE Computer Society Press, Los Alamitos

Service Based Information Systems Analysis Using Task-Level Simulator

Tomasz Walkowiak[1]

[1] *Wroclaw University of Technology, Poland*

1. INTRODUCTION

Complex information systems (CIS) are nowadays the core of a large number of companies. And therefore, there is a large need to analyze various system configuration and chose the optimal solution during design and even operation of the information system.

In this paper we propose a common approach (Birta & Arbez, 2007) based on modelling and simulation. The aim of simulation is to calculate some performance metrics which should allow to compare different configuration taking into consideration technical (like performance) and economical (like price) aspects.

There is a large number of event driven computer network simulators, like OPNET, NS-2, QualNet, OMNeT++ or SSFNet/PRIME SSF(Liu, 2006, Nicol et al., 2003). However, they are mainly focused on a low level simulation (TCP/IP packets).

It is obvious that increasing the system details causes the simulation becoming useless due to the computational complexity and a large number of required parameter values to be given. On the other hand a high level of modelling could not allow to record required data for system measure calculation. Therefore, the level of system model details should be defined by requirements of the system measure calculation (Walkowiak, 2009).

Modelling and simulation based on TCP/IP packets level results in a large number of events during simulation and therefore in a long simulation time. It is a very good approach if one plans to analyze the influence of the traffic on the network performance. However in modern information systems high speed local networks are used. In a result for a large number of information systems (except media streaming ones) the local network traffic influence on the whole system performance is negligible.

Therefore, we want to propose a novel approach based on a higher level then TCP/IP packets. We will focus on a business service realized by an information system (Gold et al., 2007) and functional aspects of the system, i.e. performance aspects of business service realized by an information system (like buying a book in the internet bookstore). We assume that the main goal, taken into consideration during design and operation of the CIS, is to fulfil the user requirements, which could be seen as some requirements to perform a user tasks within a given time limit. Therefore, the presented in the chapter modelling and simulation will be focused on a process of execution of a user request, understand as a sequence of task realised on technical services provided by the system.

The structure of the chapter is as follows. In Section 2, a model of information system is given. In Section 3, information on simulator implementation is given, next exemplars information system is analysed and simulation results are presented. It is followed by information on graphical user interface. Finally, there are conclusions and plans for further work.

2. COMPUTER INFORMATION SYSTEM MODELLING

As it was mentioned in the introduction we decided to analyze the CIS from the business service point of view. Generally speaking users of the system are generating tasks which are being realized by the CIS. The task to be realized requires some services presented in the system. A realization of the system service needs a defined set of technical resources. Moreover, the services has to be allocated on a given host. Therefore, we can model CIS as a 4-tuple (Walkowiak, 2009):

$$CIS = \langle Client, BS, TI, Conf \rangle$$

(1)

$Client$ – finite set of clients,

BS – business service, a finite set of service components,

TI – technical infrastructure,

Conf – information system configuration.

During modelling of the technical infrastructure we have to take into consideration functional aspects of CIS. Therefore, the technical infrastructure of the computer system could be modelled as a pair:

$$TI = \langle H, N \rangle$$

(2)

where: *H* - set of hosts (computers); *N* – computer network.

We have assumed that the aspects of TCP/IP traffic are negligible therefore we will model the network communication as a random delay. Therefore, the *N* is a function which gives a value of time of sending a packet form one host (v_i) to another (v_i). The time delay is modelled by a Gaussian distribution with a standard deviation equal to 10% of mean value.

The main technical infrastructure of the CIS are hosts. Each host is described by its functional parameters:

- server name (unique in the system),
- host performance parameter – the real value which is a base for calculating the task processing time (described later),
- set of technical services (i.e. apache web server, tomcat, MySQL database), each technical service is described by a name and a limit of tasks concurrently being executed.

We have distinguished a special kind of technical service witch models a load balancer (Aweya et al., 2002). A load balancer is described by its name and a limit of tasks (like all technical services) and additionally by a list of technical services, it sends requests to.

The *BS* is a set of services based on business logic, that can be loaded and repeatedly used for concrete business handling process (i.e. ticketing service, banking, VoIP, etc). Business service can be seen as a set of service components and tasks, that are used to provide service in accordance with business logic for this process (Michalska & Walkowiak, 2008).

Therefore, *BS* is modelled as a set of business service components (*BSC*), (i.e. authentication, data base service, web service, etc.), where each business service component is described a name, reference to a technical service and host describing allocation of business service component on the technical infrastructure and a set of tasks. Tasks are the lowest level observable entities in the modelled system. It can be seen as a request and response form one service component to another. We have distinguished two kinds of task: local and external. If request is send to service component and this component is able to respond without asking other service component than this tasks is assumed to be local. If request is send to service component and this component must ask another service component for response then than this tasks is assumed to be external. Each task is described by its name, task processing time parameter and in case of external task by a sequence of task calls. Each task call is defined by a name of business service component and task name within this business service component and time-out parameter.

System configuration (*Conf*) is a function that gives the assignments of each service components to a technical service and therefore to hosts since a technical set is placed on a given host. In case of service component assigned in a configuration to a load balancing technical service the tasks included in a given service component are being realised on one of technical services (and therefore hosts) defined in the load balancer configuration.

The client model [Math Processing Error] consist of set of users where each user is defined by its allocation (host name), replicate parameter (number of concurrently ruing users of given type), set of activities (name and a sequence of task calls) and inter-activity delay time (modelled by a Gaussian distribution).

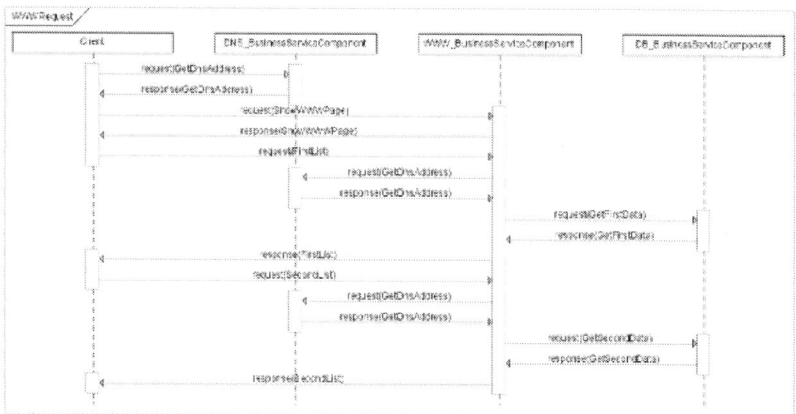

Figure 1. Task and business services interaction.

Summarising, a user initiate the communication requesting some tasks on a host, it could require a request to another host or hosts, after the task execution hosts responds to requesting server, and finally the user receives the respond. Requests and responds of each task gives a sequence of a user task execution as presented on exemplar Fig. 1.

The user request execution time in the system is calculated as a sum of times required for TCP/IP communication and times of tasks processing on a given host.

The request is understood as correctly answered if answers for each requests in a sequence of a user task execution were given within defined time limit (time-out parameter of each request in *BS* model) and if a number of tasks executed on a given technical service is not exceeding the limit parameter (parameter of *TI* model).

The user request execution time in the system is calculated as a sum of times required for TCP/IP communication (modelled by a random value) and times of tasks processing on a given host. The task processing time is equal to the task processing time parameter multiplied by a number of other task processed on the same host in the same time and divided by a the host performance parameter. Since the number of tasks is changing in simulation time, the processing time is updated each time a task finish the execution or a new task is starting to be processed.

Let [Math Processing Error] be a time moments when a task [Math Processing Error] with some execution time ([Math Processing Error]) is starting or finishing processing on a host [Math Processing Error] . Let [Math Processing Error] denotes a number of task being processed at time [Math Processing Error] on host *h*. It is not taking into account tasks which requests tasks on other hosts and waits for responses. Therefore, the time when task [Math Processing Error] finishes its execution [Math Processing Error] has to fulfill a following rule:

$$\sum_{k=2}^{e} (\tau_k - \tau_{k-1}) \frac{performance(h)}{number(h, \tau_{k-1})} = executiontime(t_j^i)$$

(3)

Having above notation the task processing time is equal to:

$$pt(t_j^i) = \tau_e - \tau_1 .$$

(4)

3. TASK-LEVEL SIMULATOR

Once a model has been developed, it is executed on a computer. It is done by a computer program which steps through time. One way of doing it is so called event-simulation. Which is based on a idea of event, which could is described by time of event occurring, type of event (in case of CIS it could be host failure) and element or set of elements of the system on which event has its influence. The simulation is done by analyzing a queue of event (sorted by time of event occurring) while updating the states of system elements according to rules related to a proper type of event.

As it was described in section 2, the network connections are modelled as a random delays. Therefore, we were not able to use mentioned in the introduction computer network simulators but we have to develop a new one (Walkowiak, 2009). The event-simulation program could be written in general purpose programming language (like C++), in fast prototyping environment (like Matlab) or special purpose discrete-event simulation kernels.

One of such kernels, is the Scalable Simulation Framework (SSF) (Nicol et al., 2003) which is a used for SSFNet (Nicol et al., 2003) computer network simulator. SSF is an object-oriented API - a collection of class interfaces with prototype implementations. It is available in C++ and Java. SSF API defines just five base classes: Entity, inChannel, outChannel, Process, and Event. The communication between entities and delivery of events is done by channels (channel mappings connects entities).

For the purpose of simulating CIS we have used Parallel Real-time Immersive Modeling Environment (PRIME) (Liu, 2006) implementation of SSF due to much better documentation then available for original SSF. We have developed a generic class (named BSObject) derived from SSF Entity which is a base of classes modeling CIS objects: host and client which models the behavior of CIS presented in section 2. Each object of BSObject class is connected with all other objects of that type by SFF channels what allows communication between them. In the first approach we have realized each client as a separated object. However, in case of increasing of the number of replicated clients the number of channels increases in power of two resulting in a large memory consumption and a long time for initialization simulation objects. Therefore, we have changed the implementation, and each replicated client is represented by one object.

The developed simulator is called SSF.BS (from SSF – the simulation framework and BS – business service).

4. COMPUTER INFORMATION SYSTEM SIMULATION ANALYSIS

4.1. First Case Study

For testing purposes of presented CIS system model (section 2) and developed extension of SSF (SSF.BS, section 3) we have analysed a case study information system. It consists of one type of client placed somewhere in internet, firewall, three hosts (Figure 2), three technical services and three business service components. An interaction between a client and tasks of each business service component is presented on UML diagram in Figure 1. The CIS structure as well as other functional parameters were described in a DML file (see example in Figure 3). The Domain Modeling Language (DML) (Nicol et al., 2003) is a SSF specific text-based language which includes a hierarchical list of attributes used to describe the topology of the model and model attributes values.

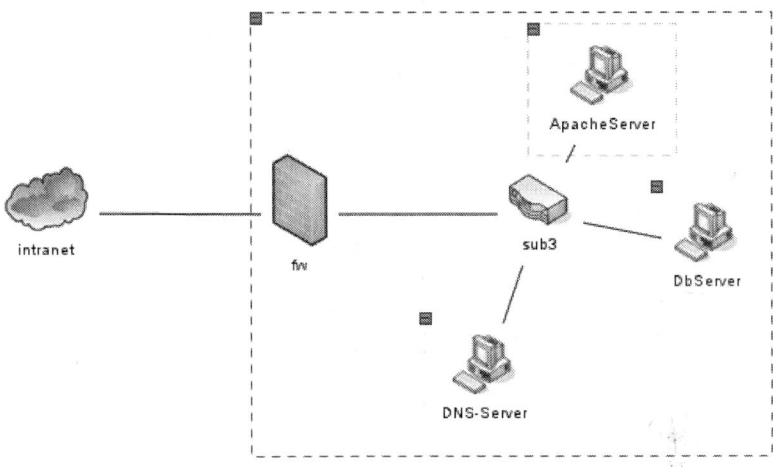

Figure 2. Case study system overview.

```
Net [
Host [
   Name DNS-Server
      Service [
         Name DNSService
         Limit 110
         LocalTask [
            Name GetDnsAddress
            Time 0.01]]]
     ...
Client [
   Name Client
   Replicate 1000
   Sleep 10.0
   Activity [
      Name WWWRequest
      TaskCall [
         Host DNS-Server
         Service DNSService
         Task GetDnsAddress
         ]
   ...
```

Figure 3. Exemplar CIS description in DML file.

In the presented information system we have observed the response time to a client request in a function of number of clients. The achieved results are presented in Figure 4.

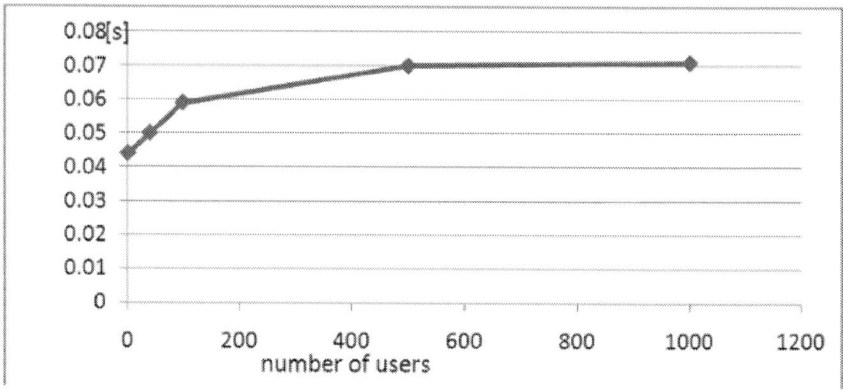

Figure 4. Response time to users requests in a function of number of concurrent users.

4.2. Simulator Performance Analysis

Next, we have tested the SSF.BS simulator performance and scalability. We calculated the time of running one batch of simulation of the exemplar IS described in previous chapter on a 2.80 GHz Intel Core Duo machine. We have compared the performance results with PWR.SSF.Net simulator (Zyla & Caban 2008) developed in Java. The CIS model used in PWR.SSF.Net differs from SSF.BS mainly in a method of calculation a task performance time and therefore the results of simulating cannot be compared. As it could be noticed on Figure 5& 6 the presented in the paper simulator (SSF.BS) simulates the CIS in shorter time, and a difference with PWR.SSF.Net is increasing with an increase of number of users.

For a number of concurrent users less than 300 (Figure 5) the SSF.BS is 10 times faster than PWR.SSF.Net. The main reason of this difference is the level of modelling details. In both cases simulators perform similar number of events per second. However, PWR.SSF.Net simulates the transmission of TCP/IP packets whereas SSF.BS works on higher level the tasks and therefore in case of presented here approach the number of events is smaller.

Not, only computational complexity of SSF.BS is lower than PWR.SSF.Net but also the usage of memory for SSF.BS is much smaller. For a case study example the SSF.BS requires 1.8 Mbytes for 0.1 client requests per second upto 4.8 Mbytes for a 1000 concurrent users. In case of PWR.SSF.Net it is hard to state the memory usage due to the memory management techniques in Java. This is the problem of enlarging the difference of speed between analysed simulators. For number of clients more then 300 (Figure 6) Java based PWR.SSF.Net starts to have problems with memory management and large number of processing time is used by JVM garbage collector (even Java based simulator was started 1 Gbyte memory limit). It results in 1000 faster simulation of SSF.BS in case of 1000 concurrent users.

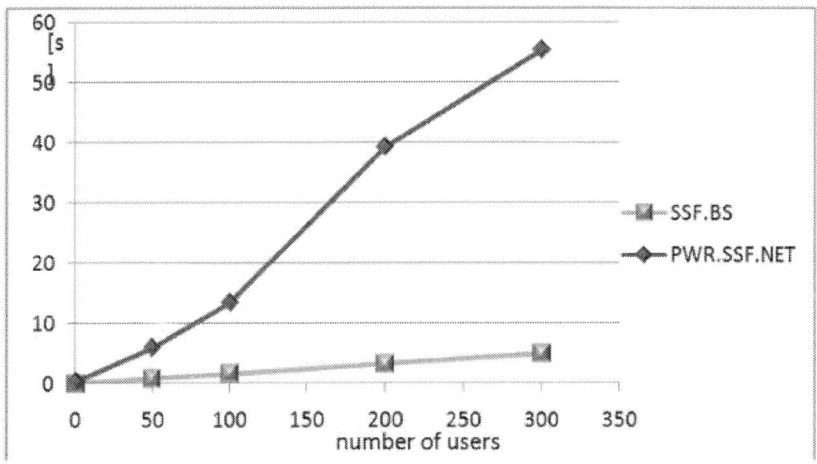

Figure 5. Simulation time (time of running the simulator) for case study system in a function of number of users (till 300 concurrent users).

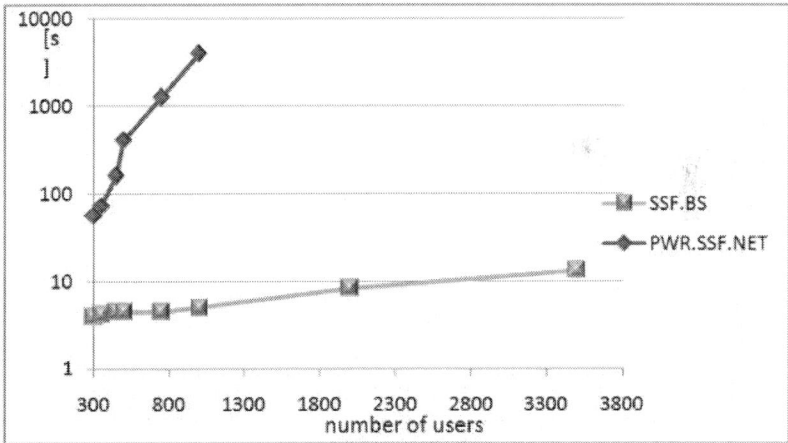

Figure 6. Simulation time (time of running the simulator) for case study system in a function of number of users (for more than 300 users).

4.3. Second Case Study – Load Balancer

A very common technique of achieving height availability of their services in CIS is using a load balancer. Load balancer allows a traffic distribution among replicated services on a server farm. Therefore, the most common load balancing algorithm – *round robin* (Aweya, et al. 2002) - has been implemented in the SSF.BS.

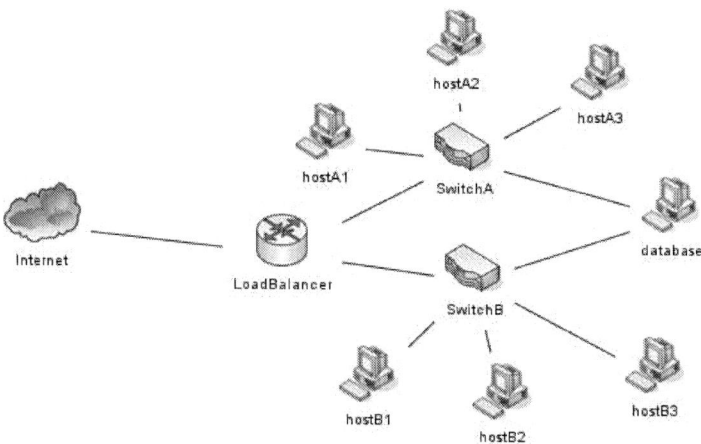

Figure 7. Load balancer case study system overview.

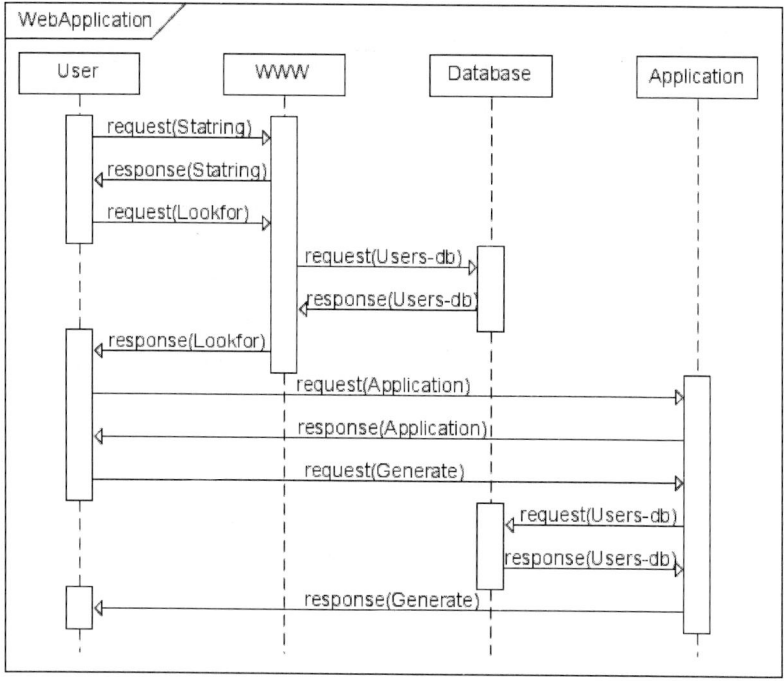

Figure 8. Task and business services interaction for case study.

For the case study analysis of CIS with load balancing we propose an exemplar service system illustrated in Fig.7. Essentially the test-bed system consists of two server farms A (included host„hostA1"-„hostA3") and B (included host„hostB1"-„hostB3") and a database server. Both farms are connected with LoadBalancer as a gate to internet users. For the case study, let us imagine, that this system is responsible for some Web Application that allows searching the database and executes a Tomcat based application. Fig. 8 shows choreography of this service, based on three service components. WWW service component has been replicated on hosts: A1-A3, Application of on hosts: B1- B3 and Database is not replicated is placed on one host. For this scenario two configuration has been proposed: first (I) standard and second (II) with all hosts with doubled performance parameter.

The achieved simulation results, the response time to user requests in a function of number of concurrent users is presented in Figure 9. The simulation time was set to 1000 seconds. The limit of concurrent tasks for all technical services was equal to 1000, whereas the inter-activity delay time equal to 1 s. As it could be expected the response time for configurations II is almost twice shorter than for configuration I. However, if we slightly change configuration II, setting the performance of database host equal to the value used in configurations I the resulting response time will be very similar to results of configuration I. These small experiment shows the ability of simulator to compare performance of different system configurations.

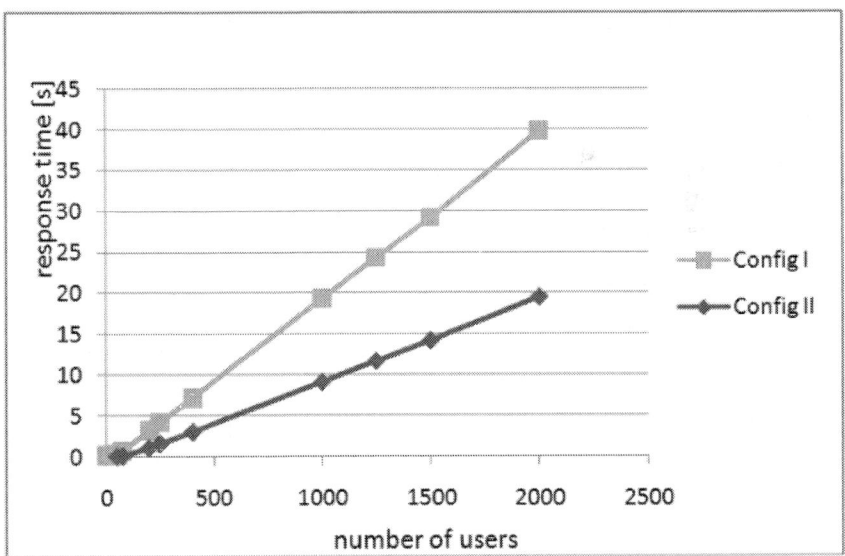

Figure 9. Response time to users requests in a function of number of concurrent users for two configurations of load balancer case study.

5. GRAPHICAL INTERFACE

The previous section showed the possibilities of using SSF.BS simulator and its good computational performance capabilities. However, nowadays the practical usage of any computer tool requires a good graphical interface. As it was mentioned in the section 3, all input information of modelled CIS is described in DML text file. Even the DML file format is simple (Figure 3), it is difficult for a human being to describe a CIS with large number of host and sophisticated service interaction without any error in text file.

Within the framework of DESEREC EU grant (http://www.deserec.eu) a Java based graphical tool called "Integrated Analysis Environment" (IAE) was developed (Michalska & Walkowiak, 2008b) for a usage of PWR.SSF.Net simulator. After a few changes in IAE it was adopted to SSF.BS simulator.

In IAE we took into consideration an inconvenient format of Domain Modelling Language and we proposed its XML representation with all supplements attributes of proposed extended simulation framework - called XDML. Creation of XDML language gave many processing possibilities. IAE framework using JAXB techniques and implemented translation methods creates one model (XDML) from other modelling languages: system infrastructure from SDL (System Description Language, http://www.positif.org/) and task interaction from WS-CDL (WebServices Choreography Description Language, http://www.w3.org/). This XDML model is visualized showing the structure of the network and it's element (Figure 10). Each network element has several functional parameters and user can graphically edit this information. In proposed framework user is able to put its own variables and attributes based on XDML specification or use extend models (i.e. consumption model, operational configuration model) to simplified its work.

After setting up all parameters of network elements and service components the user is able to perform simulation. It is done by transforming XDML into DML. The resulting DML file is then simulated. Simulation is integrated into IAE since both tools are developed in Java therefore user can see on the screen text output from the simulator on-line. The results from simulation (output file from simulator) are caught by IAE and response time to user requests is calculated and displayed.

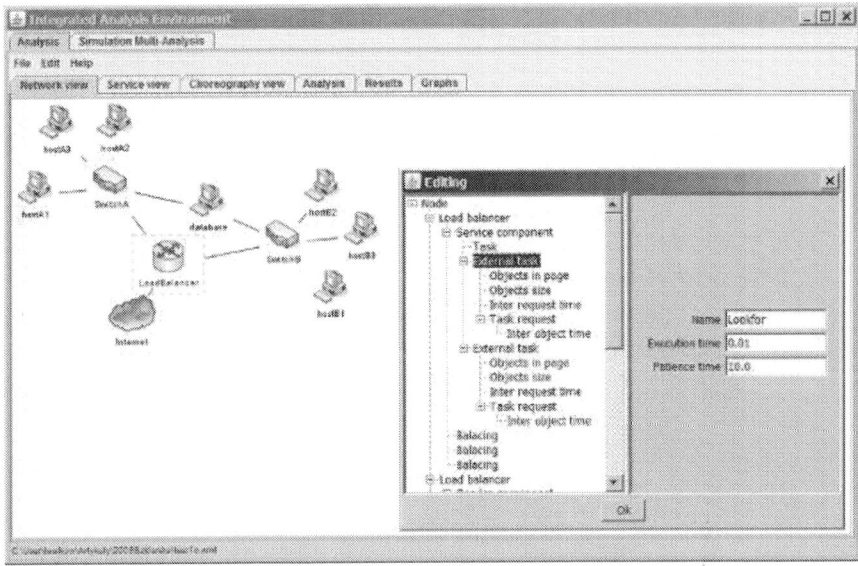

Figure 10. Integrated Analysis Environment - screenshot.

6. CONCLUSION

We have presented a simulation approach to functional analysis of complex information systems. Developed simulation software allows to analyze the effectiveness (understood in given exemplar as a the response time to a client request) of a given configuration of computer system. Changes in a host performance or in a number of clients can be easily verified. Also, some economic analysis could be done following the idea presented in (Walkowiak & Mazurkiewicz, 2005). The implementation of CIS simulator done based on SSF allows to apply in a simple and fast way changes in the CIS model. Also the time performance of SSF kernel results in a very effective simulator of CIS.

We are now working on implementing other load balancing algorithms what should allow to analyze a wider range of enterprise information systems and compare different load balancing algorithms.

We also plan to extend the model and simulator to include the reliability model of technical infrastructure components. It should allow to measure the availability of a business service in a function of functional and reliability parameters of information systems components.

REFERENCES

1. A. . Avižienis, J. . Laprie, B. Randell, 2000 Fundamental Concepts of Dependability. Proceedinggs of 3rd Information Survivability Workshop (ISW-2000), Boston, Massachusetts

2. J. Aweya, M. Ouellette, D. Montuno, B. Doray, K. Felske, 2002 An adaptive load balancing scheme for web servers. International Journal of Network Management, 12

3. L. . Birta, G. Arbez, 2007 Modelling and Simulation: Exploring Dynamic System Behaviour. Springer, London

4. N. Gold, C. . Knight, A. Mohan, M. Munro, 2004 Understanding service-oriented software. IEEE Software, 21 71 77 - 77

5. J. Liu, 2006 Parallel Real-time Immersive Modeling Environment (PRIME), Scalable Simulation Framework (SSF), User's maual. Colorado School of Mines Department of Mathematical and Computer Sciences, 2006, [Online]. Available: http://prime.mines.edu/

6. D. . Nicol, J. Liu, M. . Liljenstam, Y. Guanhua, 2003 Simulation of large scale networks using SSF. Proceedings of the 2003 Winter Simulation Conference, 1 650 657 , New Orleans,

7. K. . Michalska, T. Walkowiak, 2008 Hierarchical Approach to Dependability Analysis of Information Systems by Modeling and Simulation. Proceedings of the 2008 Second international Conference on Emerging Security information, Systems and Technologies,, 356 361 Cap Esterel, IEEE Computer Society, Washington

8. T. . Walkowiak, J. Mazurkiewicz, 2005 Reliability and Functional Analysis of Discrete Transport System with Dispatcher. Advances in Safety and Reliability, European Safety and Reliability Conference-ESREL 2005, Gdynia, 2017 2023 , Taylor & Francis Group, London

9. T. Walkowiak, 2009 Information systems performance analysis using task-level simulator, Proceedings of International Conference on Dependability of Computer Systems, 218 225 , Brunow, IEEE Computer Society Press, Los Alamitos

10. M. Zyla, D. Caban, 2008 Dependability Analysis of SOA systems. Proceedings of International Conference on Dependability of Computer Systems, 301 306 , Szklarska Poreba, IEEE Computer Society Press, Los Alamitos

A Comparative Study of Modern Heuristics on the School Timetabling Problem

Iosif V. Katsaragakis [1,2], Ioannis X. Tassopoulos [1] and Grigorios N. Beligiannis [1,*]

[1] Department of Business Administration of Food and Agricultural Enterprises, University of Patras, G. Seferi 2, 30100 Agrinio, Greece

[2] School of Science and Technology, Hellenic Open University, Parodos Aristotelous 18, 26335 Patra, Greece

ABSTRACT

In this contribution a comparative study of modern heuristics on the school timetabling problem is presented. More precisely, we investigate the application of two population-based algorithms, namely a Particle Swarm Optimization (PSO) and an Artificial Fish Swarm (AFS), on the high school timetabling problem. In order to demonstrate their efficiency and performance, experiments with real-world input data have been performed. Both algorithms proposed manage to create feasible and efficient high school timetables, thus fulfilling adequately the timetabling needs of the respective high schools. Computational results demonstrate that both algorithms manage to reach efficient solutions, most of the times better than existing approaches applied to the same school timetabling input instances using the same evaluation criteria.

Keywords:school timetabling; particle swarm optimization; artificial fish swarm

1. INTRODUCTION

The problem faced in this contribution belongs to the wide family of educational timetabling problems which are NP-complete in their general form [1,2]. The main categories of educational timetabling problems are examination timetabling, university course timetabling, and high school timetabling [3]. In the current work we focus on the high school timetabling problem, which involves the weekly scheduling for all lecturers of a high school.

The school timetabling problem requires the assignment of lectures (events) to timeslots in such a way that no teacher/class (resources) is involved in more than one lecture simultaneously, while many other significant constraints are satisfied. These constraints include both hard and soft constraints. Hard constraints must be satisfied by all means, while soft constraints represent preferences and are used to evaluate the solution's quality [4]. The goal is to find a feasible solution, satisfying all hard constraints, which is as qualitative as possible; that is, satisfying the maximum number of soft constraints. In the respective literature many variants of the high school timetabling problem have been presented, which mainly differ due to the educational system of each country [5,6,7]. In recent years many papers have been published describing specific techniques applied to the high school timetabling problem [8,9,10,11,12,13,14,15].

In current work, we investigate the application of two heuristic algorithms on the Greek school timetabling problem, namely Particle Swarm Optimization (PSO) [16] and Artificial Fish Swarm (AFS) [17]. The algorithms are applied to real-world input data coming from different Greek high schools. Simulation results demonstrate that both algorithms proposed manage to create feasible and efficient high school timetables, thus adequately fulfilling the timetabling needs of the respective high schools. The specific input instances used for their performance evaluation and comparison is the well-established Beligiannis benchmark [7]. This school timetabling data set has been already used as a benchmark by many researchers in the respective literature [6,7,18,19].

The PSO algorithm presented in current contribution is based on the PSO algorithm introduced in [20]. However, the proposed algorithm has significant differences compared to that algorithm, which are the following (see Section 3.1 for more details):

- The population size of the PSO based algorithm used in current work equals 15, while in [20] equals to 50.
- In the proposed PSO algorithm procedure SwapWithProbability() is applied to two randomly selected timeslots, while in [20] one of the two timeslots selected should have a hard clash (if such a timeslot exists).

- In [20] procedure SwapWithProbability() accepts swaps causing a hard clash with probability 2.2%, while in current algorithm this probability is set to 50%.
- In the proposed PSO algorithm procedure SwapWithProbability() accepts swaps which cause a raise in the individual's fitness with probability equal to 0.5% while in [20] this probability is set to 2.2%.
- In [20] the probability of exiting the While Loop Structure which purpose is to produce a new particle with at least equal fitness value to the fitness value of the global best of the current generation, for each given particle, is set to 1.1%, while in current algorithm this probability is set to 1.08%.

These differences enable the proposed PSO algorithm to perform significantly better compared to the PSO algorithm presented in [20] as demonstrated by experimental results in Section 4.

The AFS algorithm presented in current contribution is a novel approach since, although there are plenty of population based algorithms applied to timetabling problems in the literature, there is no specific AFS based approach, to the best of our knowledge, applied to the high school timetabling problem. The structure of the proposed AFS approach is given in detail in Section 3.2.

Both algorithms presented in the current contribution use the same formalism for modeling the timetabling problem, try to minimize the same fitness function, and use the same performance criteria in order to evaluate the quality of each resulted timetable. Thus, a straightforward comparison of their experimental results is fair. Additionally, since there are other approaches in the respective literature using the same fitness function and evaluation criteria [21], a comparison of the proposed heuristic approaches with other approaches can be also performed on a fair basis.

All timetables created by the proposed algorithms are compared on the basis of three criteria, which are well-established school timetabling performance criteria in the respective literature. The first criterion, which investigates how evenly each teacher's hours are distributed among the days she/he is available at school, is the teachers' teaching hours' distribution. The second criterion, which presents how uniformly distributed are the hours of the same lesson for each class among its teaching days, is the lessons' hours' distribution. Finally, the third criterion, which checks whether there are idle hours between teaching hours of each teacher, is the teachers' gaps.

Numerical results demonstrate that both proposed algorithms achieve very satisfactory results and justify that modern heuristics constitute a very useful family of algorithms to cope effectively with this kind of problems. Moreover, one major advantage of the proposed algorithms lies in their inherent adaptive behavior. More specifically, both algorithms, by assigning weights that can be defined by the user to each specific constraint that should be satisfied, are able to fulfill adequately different timetabling needs of each respective school.

This paper is organized as follows. In Section 2 we present the mathematical model of the school timetabling problem faced. Section 3 describes the structure and operation of the proposed algorithms. Section 4 assesses and compares the performance of the proposed algorithms to each other and to that of existing approaches. Finally, Section 5 summarizes the conclusions and presents future work.

2. PROBLEM DEFINITION

The problem faced in this contribution is the weekly high school timetabling. This problem is affected by many parameters and has to satisfy a large number of hard and soft constraints [21]. Hard constraints are the ones that have to be fulfilled in order a timetable to be feasible, while soft constraints are the ones that affect the quality of a timetable. In Section 2.1, we list all hard and soft constraints considered in current contribution, while in Section 2.2 we present the mathematical model of the problem at hand.

2.1. Constraints

The hard constraints considered in current contribution are the following:

- Teachers' clash: each teacher can teach to only one class at a given time period.
- Classes' clash: each class can be taught only one lesson at a given time period.
- Teachers-classes-lessons assignment: each teacher can teach a limited number of hours and lessons to each class, which is predefined by input data.
- Teachers' availability: each teacher can teach only in periods he/she is available, which is predefined by input data.
- Classes' idle timeslots: must be only at the last hour of a day.
- Co-teaching restrictions: two or more teachers who teach the same lesson to the same class must be assigned to it at the same time period. For example, one class can be firstly joined with another class and then divided into two sub-classes, one for "English language for beginners" and one for "English language for intermediates" [22].
- Sub-classes restrictions: two or more teachers who teach different lessons to the same class at the same time period must be simultaneously assigned to it. For example, one class can be divided into two sub-classes, one for "Gymnastics" and one for "Economics" [22].

For a more detailed description of how both presented algorithms deal with co-teaching and sub-classes cases the interested reader can refer to [20]. The soft constraints considered in current contribution are the following:

- Teachers' teaching hours' distribution: checks how evenly each teacher's hours are distributed among the days she/he is available at school.
- Lessons' hours' distribution: checks how uniformly distributed are the hours of the same lesson for each class among its teaching days.
- Teachers' gaps: checks whether there are idle hours between teaching hours of each teacher.

2.2. Mathematical Model

The necessary data sets needed for the problem's model definition are the following:

- $T=\{1,...,TeachersNo\}$; the set of teachers

- $C=\{1,...,ClassesNo\}$; the set of classes

- $D=\{1,...,DaysNo\}$; the set of teaching days in a week

- $L=\{1,...,LessonsNo\}$; the set of lessons

- $H=\{1,...,HoursNo\}$; the set of teaching hours in a day

- $Hnot\ available t$; the set of time periods teacher t is not available at school

- $Hlast$; the set of the last hours of all days

- U; a set of tuples (m,n) for $m,n \in P{:}n \geq m+1$

- E; a set of meetings (events) such that to each meeting $e \in E$ a class-teacher pair and a given number of lessons that must be scheduled is preassigned [23]

- Et; a set of meetings assigned to teacher t

The necessary variables and functions needed for the problem's model definition are the following:

- $$x_{tcdh} = \begin{cases} 1, & \text{if teacher } t \text{ teaches class } c \text{ at day } d \text{ at hour } h \\ 0, & \text{if teacher } t \text{ does not teach class } c \text{ at day } d \text{ at hour } h \end{cases}$$

- $$y_{lcdh} = \begin{cases} 1, & \text{if lesson } l \text{ is taught at class } c \text{ at day } d \text{ at hour } h \\ 0, & \text{if lesson } l \text{ is not taught at class } c \text{ at day } d \text{ at hour } h \end{cases}$$

- $$\chi(\alpha) = \begin{cases} 1, & \text{if } \alpha \text{ is true} \\ 0, & \text{if } \alpha \text{ is false} \end{cases}$$

- *UpperBoundtd* is the maximum number of teaching hours that can be assigned to teacher t at day d so as his/her teaching hours are uniformly distributed

- *LowerBoundtd* is the minimum number of teaching hours that can be assigned to teacher t at day d so as his/her teaching hours are uniformly distributed

- *Subjectstc* is the number of different subjects that teacher t teaches to class c

- *Teacher_Total_Hourst* is the total number of teaching hours that teacher t can teach in a week

- *Class_Total_Hoursc* is the total number of teaching hours that class c can be taught in a week

- *ztdnm* is the idle times of teacher t between time slots m and n on day d [23]

- $v_{edp} = \begin{cases} 1, \text{if the event } e \text{ is scheduled to timeslot } (d, p) \\ 0, \text{otherwise} \end{cases}$

Except for that, the following soft constraint costs are defined:

- *teachers'teaching hours' distribution cost*= $scw_1 \times \sum_{t=1}^{T} \sum_{d=1}^{D} D_{td},$ where

$$D_{td} = \begin{cases} 1, if \left(\sum_{c=1}^{C} x_{tcdh} \geq UpperBound_{td} + 1 \right) \vee \left(\sum_{c=1}^{C} x_{tcdh} \leq Loweround_{td} - 1 \right) \\ 0, \quad otherwise \end{cases}$$

and

$\forall t \in 1, `q \in D`$ $\forall \in ll$ is the respective soft constraint weight as described in [20].

- *lessons'hours' distribution cost*= $scw_2 \times \sum_{t=1}^{T} G_{tcd},$ where

$G_{tcd} = \begin{cases} 1, if \sum_{h=1}^{H} x_{tcdh} \geq Subjects_{tc}, \forall t \in T, c \in C, d \in D \\ 0, otherwise \end{cases}$ otherwise

and scw2 is the respective soft constraint weight as described in [20]. If $\sum_{h=1}^{H} x_{tcdh} \geq$ Subjectstc, this means that teacher t teaches, at least one subject, at class c at day d more than one hour (twice or more).

- *teachers'gaps cost* $= scw_3 \times \sum_{d=1}^{D} \sum_{(m,n) \in U} z_{tdmn},$

where $z_{tdmn} \geq 0, \forall t \in T, d \in D, (m, n) \in U$ and

$z_{tdmn} \geq (n - m - 1) \times \left(-1 + \sum_{e \in E_t} (v_{edm} + v_{edn} - \sum_{m < p < n} v_{edp}) \right), \forall t \in T, d \in D, (m, n) \in U.$

The latter inequality defines that, if variables are activated and there are no teaching periods between them, the value of *ztdmn* equals (*n*−*m*−1) which is the number of idle times between m and n. This formulation of teachers' gaps cost was firstly presented in [23]. Accordingly, $scw3$ is the respective soft constraint weight as described in [20].

Thus, the mathematical model of the problem can be expressed as follows:

$Min(teachers'teaching\ hours'\ distribution\ cost+lessons'hours'distribution\ cost$
$+teachers'gaps\ cost)\ ,\forall\ t\in T,\ c\in C,\ l\in L,\ d\in D,\ h\in H,\ e\in E$ under the following constraints:

- $\sum_{t=1}^{T} \chi\left(\left(x_{tc_idh} = 1\right)\wedge\left(x_{tc_jdh} = 1\right)\right) \leq 0, \forall\ c_i, c_j \in C\ (i \neq j), d \in D, h \in H;\ (Teachers'\ clash)$
- $\sum_{c=1}^{C} \chi\left(\left(x_{l_icdh} = 1\right)\wedge\left(x_{l_jcdh} = 1\right)\right) \leq 0, \forall\ l_i, l_j \in L\ (i \neq j), d \in D, h \in H;\ (Classes'\ clash)$
- $\sum_{c=1}^{C}\sum_{d=1}^{D}\sum_{h=1}^{H} x_{tcdh} = Teacher\_Total\_Hours_t, \forall\ t \in T\ ;\ (Teachers'\ teaching\ hours)$
- $\sum_{t=1}^{T}\sum_{d=1}^{D}\sum_{h=1}^{H} x_{tcdh} = Class\_Total\_Hours_c, \forall\ c \in C\ ;\ (Classes'\ teaching\ hours)$
- $\sum_{t=1}^{T} \chi\left((x_{tcdh} = 1)\wedge h \in H_t^{not\ available}\right) \leq 0, \forall\ c \in C\ , d \in D, h \in H;\ (Teachers'\ availability)$
- $\sum_{c=1}^{C} \chi\left((y_{lcdh} = 0)\wedge h \notin H_{last}\right) \leq 0, \forall\ l \in L\ , d \in D,\ h \in H;\ (Classes'\ idle\ time\ slots)$
- $\sum_{c=1}^{C} \chi\left((x_{t_icdh} = 1)\wedge\left(x_{t_jcdh} = 1\right)\wedge teachers\ t_i\ and\ t_jare\ not\ involved\ in\ co-teaching\right) \leq 0.$
  $\forall\ t_i, t_j \in T\ (i \neq j), d \in D,\ h \in H;\ (co\text{-}teaching/sub\text{-}classes)$

3. THE PROPOSED ALGORITHMS

3.1. Description of the Proposed PSO Algorithm

The population of the proposed PSO algorithm consists of 15 particles, each one comprising a two-dimensional array. Although the population size is different from the one used in [20], the particle encoding is the same. The number of rows of each particle equals the number of different classes of each school, while the number of columns is 35, since the timeslots of a weekly Greek school timetable are 35 at the most [20]. Each particle's cell contains a number ranging from 1 to the number of different teachers of each school or a "–1" value. For example, if cell [i,j] equals "4", that means that the 4th teacher teaches one of his/her lessons at the i-th class at the j-th timeslot. If cell [i,j] equals "–1", that means that the i-th class has the j-th timeslot empty. The interested reader can find more details about the particle encoding used in [20].

In Algorithm 1, the pseudo code of the proposed PSO based algorithm is presented. The algorithm is a hybrid one consisting of two basic components:

- the main algorithm, which has four basic differences compared to the algorithm presented in [20].
- a local search procedure, which is the same as the one used in [20] and aims improving the quality of the resulted timetable after the execution of the main algorithm.

In this contribution we limit our description on the main algorithm, since all differences between the proposed PSO algorithm and the one presented in [20] lie there. The interested reader can find more details about the local search procedure in [20]. The differences of the proposed PSO algorithm compared to the algorithm presented in [20] are the following:

- Line 3: The number of particles P (i.e., the population size) is set to 15, while in [20] equals 50.
- Lines 22–23: Procedure SwapWithProbability() is applied to two randomly selected timeslots, while in [20] one of the two timeslots selected should have a hard clash (if such a timeslot exists).
- Line 22: Procedure SwapWithProbability() accepts swaps causing a hard clash with probability equal to 50%, while in [20] this probability equals 2.2%.
- Line 22: Procedure SwapWithProbability() accepts swaps which cause a raise in the individual's fitness with probability equal to 0.5%, while in [20] this probability equals 2.2%.
- Lines 29–34: The probability of exiting the While Loop Structure is set to 1.08%, while in [20] equals 1.1%.

Algorithm 1: The pseudo code of the main PSO algorithm.

In what follows, P is the number of particles (i.e., the population size), particle(p) is the p-th particle of the population, Personal_best(p) is the personal best achieved by particle p till current generation and Global_best is the globally best particle among all particles till current generation, i.e. the particle with smallest fitness. In addition, F() stands for the fitness function, F stands for the fitness function value, while auxiliary_particle is a structure identical to any particle's structure that serves for temporal storage of a particle.

1. Start of hybrid PSO algorithm

2. **Start of main algorithm**

3. Read input Data; // i.e. teachers, classes, hours, co teachings etc.

4. Initialize P particles with random structure;

5. **For** each particle(p) {

6.   Personal_best(p) ← particle(p);

7.   F(Personal_best(p)) ← F(particle(p));

8. } // end **For**

9. Global_best ← the particle with smallest fitness;

10. F(Global_best) ← the smallest fitness among all particles;

11. **While** generation < numOfGenerations { // numOfGenerations is set to 10,000

12.   **For** each particle(p) {

13.   F(particle(p)) ← compute fitness of particle(p);

14.   **If** (F(particle(p)) <= F(Personal_best(p)) **then** {

15.   F(Personal_best(p)) ← F(particle(p));

16.   Personal_best(p) ← particle(p);

17.   **If** F(particle(p) <= F(Global_best) **then** {

18.   F(Global_best) ← F(particle(p));

19.   Global_best ← particle(p);

20.   } // end **If**

21.   } // end **If**

22.   Select two different timeslots t_1, t_2 at random;

23.   Execute procedure SwapWithProbability(particle(p), t_1, t_2);

24.   Select a timeslot t at random;

25.   Execute procedure InsertColumn(Personal_best(p), particle(p), t);

26.   Select a timeslot t at random;

27.   Execute procedure InsertColumn(Global_best, particle(p), t);

28.   F_before_entering_While_Loop_Structure ← F(particle(p));

29.   auxiliary_particle ← particle(p);

30.   **While** F(particle(p)) > F(Global_best) { // this is the While Loop Structure

31.   Every 10 loop-cycles break the while-loop with a fixed probability;

32.   Select a timeslot t at random;

33.   Execute procedure InsertColumn(Global_best, particle(p), t);

34. $F(particle(p)) \leftarrow$ compute fitness of particle(p);

35.   } // end **While** (While Loop Structure)

36.   **If** $F(particle(p))$ > $F\_before\_entering\_While\_Loop\_Structure$ **then** {

37.   particle(p) $\leftarrow$ auxiliary_particle;

38. $F(particle(p)) \leftarrow F\_before\_entering\_While\_Loop\_Structure$;

39.   } // end **If**

40.   } // end **For** each particle(p)

41.   ++generation;

42.   } // end **While** termination criterion is not met

43.   **End of main algorithm** //Here is where the main algorithm ends and the refinement phase starts

44.   Execute **Local Search Procedure**

45.   output $\leftarrow$ Global_best;

46. End of hybrid PSO algorithm

As seen in Algorithm 1, the main algorithm involves three major components, namely, procedure SwapWithProbability(), procedure InsertColumn() and a While Loop Structure. These three parts are described in short in the next paragraphs. The interested reader can find more details about these parts in [20].

The parameters of procedure SwapWithProbability() are the current particle (particle(p)) and two different timeslots t_1 and t_2. Procedure SwapWithProbability() investigates all swaps between the cells of timeslots (columns) t_1 and t_2 for all classes of particle(p). Swaps that cause no hard clash and result in a smaller or equal fitness value are always accepted and executed. Swaps that cause hard constraint violations are accepted with probability equal to 50%. Swaps that do not cause hard constraint violations but lead to larger (worse) fitness function values are accepted with probability equal to 0.5%. Note that the acceptance of invalid swaps permits the algorithm to escape from local optima in most cases.

Procedure InsertColumn() is used in order to substitute timeslots of current particle (particle(p)) with timeslots either from the personal best of current particle (Personal_best(p)) or the global best (Global best) found until that point. Procedure InsertColumn() takes as first parameter either Personal_best(p) or Global_best, as second parameter particle(p) and as third parameter a random selected timeslot t, which is the timeslot to be replaced in particle(p) either from Personal_best(p) or Global_best.

The While Loop Structure tries to discover, for each particle(p), a particle with the best or equal fitness value to the fitness value of Global_best, by applying procedure InsertColumn() between Global_best and particle(p). In order to avoid being trapped into an infinite loop, the algorithm can exit the While Loop Structure with a probability set to 1.08%, no matter what the achieved fitness value is, every time 10 more loops are executed.

3.2. Description of the Proposed AFS Algorithm

The population of the proposed AFS algorithm consists of 24 fish, each of which is a two-dimensional array. The fish encoding is the same as the one used by the proposed PSO algorithm and the algorithm presented in [20]. In Algorithm 2, the pseudo code of the proposed AFS based algorithm is presented. The algorithm is a hybrid one consisting of two basic components:

- the main algorithm, which will be analytically described in the following paragraphs.
- a local search procedure, which is the same as the one used by the proposed PSO algorithm and the algorithm presented in [20].

As seen in Algorithm 2, the main algorithm includes seven basic procedures, namely, procedure Prey(), procedure InnerPrey(), procedure SwarmNChace(), procedure CreateNeighborhood(), procedure CalculateLocalCentre(), procedure Leap(), and procedure Turbulence(). A detailed description of these procedures is given in the following paragraphs.

We define Distance d between two fish f1 and f2 as the number of cells of the timetable, in which the two fish differ. The distance of two fish takes values in the interval [0, number of classes × 35]. As the algorithm progresses, the fish tend to approach each other and the population loses the desired diversity that allows fish to seek in wider solutions area. To prevent this phenomenon, a shaker process of the space of solutions is used, called procedure Turbulence() (Algorithm 2, Line 10). This procedure is activated when the current maximum distance between all fish becomes less than MIN_DISTANCE_COEF × number of classes × 35, wherein MIN_DISTANCE_COEF is a user defined "proximity factor", constant throughout the execution of the algorithm. During the execution of procedure Turbulence(), a random fish, a random row (class) in this fish and two random periods are selected. Then, utilizing procedure Swap() teachers assigned to classes during these periods are swapped. Procedure

Swap() is repeated a specified number of times, which in all experiments conducted was equal to TURBULENCE_ITERATIONS × NUMBER_OF_FISH, wherein TURBULENE_ITERATIONS is user defined, constant throughout the execution of the algorithm. The parameters of procedure Swap(), as used in procedure Turbulence(), are four: a randomly selected fish, a randomly selected row in this fish and two randomly selected columns in this fish. It swaps the values of two cells in the same row of the timetable of fish f, i.e., teachers who are assigned to the two periods of the timetable of the class.

Procedure Leap() (Algorithm 2, Line 12) is activated when the improvement of the obtained best fitness (i.e., the fitness of minGlobalFish) during the last LEAP_NUMBER generations is less than MIN_IMR, where LEAP_NUMBER and MIN_IMR are user defined and constant. Parameter MIN_IMR is the threshold of desired fitness' rate improvement in order to start the Leap() procedure (see also Table 1). Procedure Leap() is triggered repeatedly until the desired rate of improvement has been achieved or 10 executions without desired rate of improvement have been completed. In order to achieve the desired rate of improvement, procedure Leap() applies procedure Approach(), which forces current fish f to approach minGlobalFish. Procedure Approach() is repeated, inside the body of procedure Leap(), a specified number of times, which in all experiments conducted was equal to NUMBER_OF_FISH.

The Approach() procedure is one of the two procedures (the second one is RandomApproach() presented below) aimed to make fish f_1 approach fish f_2. To achieve this, it initially identifies the cells in which the teachers in the timetables of the two fish are different. So, if the values of fish f_1 and f_2 in cell (i, j) are different, say, "Professor A" and "Professor B", then it is certain that there will be another cell (i, m), having "Professor B" as a value (this is assured by the initialization procedure) (Figure 1). Procedure Approach() is executed as follows: It randomly selects two cells of f_1 among cells in which timetables of f_1 and f_2 are different. Then, utilizing procedure Swap(), the algorithm switches the values of cells (i, j) and (i, m) of f_1 (Figure 2). In this way, the distance of the two fish is decreased by at least one unit (perhaps two if A = C). By choosing to make alternations per class (horizontal) we ensure that the number of hours that each teacher is assigned to each class is not violated (this has been ensured by the initialization procedure). The process ends when the percentage of the initial distance between the two fish becomes less than (1.0-STEP_RATIO), where STEP_RATIO is a user defined variable, constant throughout the execution of the algorithm.

Algorithm 2: The pseudo code of the main AFS algorithm.

In what follows, minGlobalFish is the fish with best fitness in population, personalBest is the current personal best structure of each fish, localBest is the fish with the best fitness in a neighborhood, localCentre is the centre of a fish neighborhood and leapNumber is the number of generations every which the algorithm checks whether an improvement over threshold

MIN_IMPR in the fitness of minGlobalFish has occurred.

1. Start of hybrid AFS algorithm

2. **Start of main algorithm**

3. Read input Data; //i.e., teachers, classes, hours, co teachings etc.

4. Create initial population of fish, each fish having random structure; //The size of initial population is set to NUMBER_OF_FISH=24

5. minGlobalFish ← The fish with best fitness in population;

6. Make every fish best position equal to current (i.e., initial) position;

7. generation ← 0;

8. **While** generation < NUM_OF_GENERATIONS { // NUM_OF_GENERATIONS is set to 10,000

9. **If** maximum distance between all fish is less than a minimum threshold **then**

// Threshold is equal to MIN_DISTANCE_COEF × number of classes × 35

10. Execute procedure Turbulence(); // perturb the population of fish

11. **If** the % improvement of minGlobalFish fitness for the last LEAP_NUMBER generations is not bigger than MIN_IMPR **then**

// MIN_IMPR is a threshold in the improvement of minGlobalFish

12. Execute procedure Leap(); // make fish approach minGlobalFish

13. **For** each fish f {

14. **If** fitness of fish f has been improved **then**

15. update fish f personalBest;

16. Execute procedure CreateNeighborhood(f); // recreate the neighborhood of fish f

17. Execute procedure CalculateLocalCentre(f); // calculate the localCentre of the neighborhood of fish f and the localBest of fish f

18. **If** the neighborhood of fish f is sparse **then** // if it contains less than SPARSE_COEF × NUMBER_OF_FISH members

19. Execute procedure Prey(f); // try to find for fish f, in the whole

population, a structure having better fitness

20.　　　　**Else**

21.　　　　**If** the neighborhood of fish f is dense **then** // if it contains more than DENSE_COEF × NUMBER_OF_FISH members

22. Execute procedure InnerPrey(f); // try to find for fish f, in its neighborhood, a structure having better fitness

23.　　　　**Else**

24. Execute procedure SwarmNChase(f); // try to find for fish f a structure having better fitness in case its neighborhood is neither dense nor sparse

25.　　　　**If** fitness of fish f < = fitness of minGlobalFish **then**

26.　　　　　　minGlobalFish ← f;

27.　　　} // end **For** each fish f

28. **++**generation;

29. } // end **While** generation < numOfGenerations

30. **End of main algorithm** //Here is where the main algorithm ends and the refinement phase starts

31. Execute **Local Search Procedure**

32. output ← minGlobalFish;

33. End of hybrid AFS algorithm

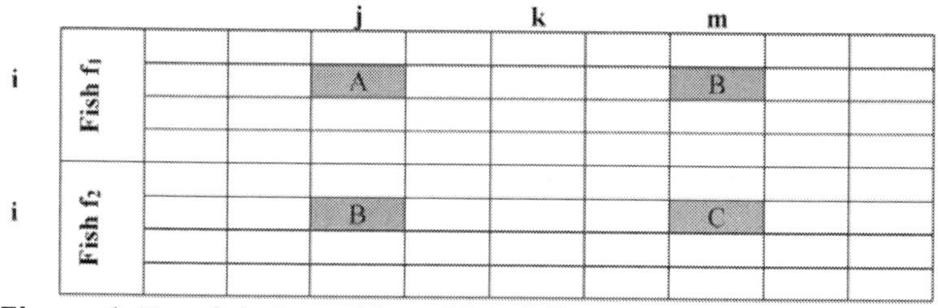

Figure 1. Two fish timetables which differ in cell (i, j).

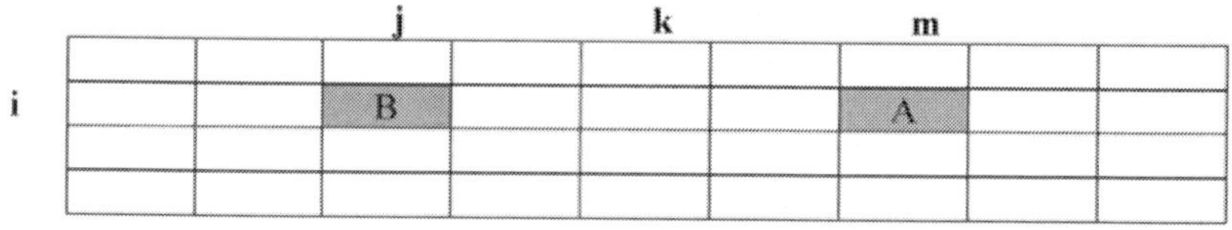

Figure 2. The fish f_1 after switching cells (i, j) and (i, m).

Procedure CreateNeighborhood() (Algorithm 2, Line 16) plays a major role in the operation of the algorithm. Its aim is to create in every generation, the current neighborhood for each fish. As the fish move, their mutual distances change and they are getting far or near to each other. Thus, neighborhoods of fish evolve dynamically during execution of the algorithm. This means that in every generation, the neighborhood of each fish is recalculated. We define that a fish f_1 is assumed to lie in the neighborhood of a fish f, if its distance from f is less than minDist + (maxDist – minDist) × VISUAL_SCOPE_COEF, where minDist and maxDist are respectively the minimum and maximum distance among all fish of current population and VISUAL_SCOPE_COEF is a user defined parameter, constant throughout the execution of the algorithm. As stated before, the distance d between two fish f1 and f2 is the number of cells of the timetable, in which the two fish differ. Procedure CreateNeighborhood() sets the fish in the neighborhood of f in increasing fitness order. So, the first fish in the neighborhood of f, is the one with the best fitness among all its neighbors. A neighborhood is considered "sparse" (Algorithm 2, Line 18), when containing less than SPARSE_COEF × NUMBER_OF_FISH members, where NUMBER_OF_FISH is the population size and SPARSE_COEF is a user defined parameter, constant throughout the execution of the algorithm. A neighborhood is considered "dense" (Algorithm 2, Line 21), when containing more than DENSE_COEF × NUMBER_OF_FISH members, where DENSE_COEF is a user defined parameter, constant throughout the execution of the algorithm.

One of the key behaviors simulated in AFS algorithms is the tendency of fish to gather in flocks to maximize food-finding and survival chances [17]. The concentration of fish in flocks (swarming) is simulated by moving the fish to a "notional" fish located in the "center" of their neighborhood (localCentre fish). This "notional" fish, is created by procedure CalculateLocalCentre() (Algorithm 2, Line 17). At first, the localCentre fish is set equal to the first fish in the neighborhood of f, which is the fish with the best fitness among all fish in the neighborhood, since procedure CreateNeighborhood() sets the fish in the neighborhood of f in increasing fitness order (see above). Then, for an arbitrary number of times (without, of course, exceeding the size of the neighborhood), the localCentre approaches the i-th fish of the neighborhood, utilizing procedure Approach(), with a diminishing step given by the expression $\frac{1}{i+1}$. For example, the localCentre approaches the second fish of the neighborhood with step equal to $\frac{1}{3}$, the third fish of the neighborhood with step equal to $\frac{1}{4}$ and so on. As a result, the participation of the most robust fish of the neighborhood to the creation of the localCentre is greater than that of the less robust.

A fish f performs Prey() (Algorithm 2, Line 19) procedure when its neighborhood is not rich enough in fish, towards which it could make a move. A fish f1, different from f, is selected randomly from fish population and if it has better fitness than f then fish f moves towards f1 using procedure RandomApproach() and procedure Prey() is completed. If the fitness of f1 is not better than the fitness of f then f can also move towards f1 using procedure RandomApproach() with probability equal to $e-dcgen$ (dc is the difference in fitness between fish f_1 and fish f and gen is the current generation) and procedure Prey() is completed, too. This is repeated PREY_TRY_NUMBER times at most, where PREY_TRY_NUMBER is a user defined parameter, constant throughout the execution of the algorithm. In case none of the above situations take place, fish f moves towards its personalBest using procedure RandomApproach(). The pseudo code of procedure Prey() is presented in Algorithm 3.

Algorithm 3: The pseudo code of procedure Prey().

0. **For** try ← 1 to PREY_TRY_NUMBER {

1. accept ← random number between 0 and 1;

2. pick a random fish f_1 among the population;

3. dc ← fitness of f_1 – fitness of f;

4. **If** fitness of f_1<= fitness of f then {

5. Execute procedure RandomApproach(); // move randomly fish f towards fish f_1;

6. **return;**

7. } // end **If**

8. **If** (fitness of f_1> fitness of f) **and** $(e-dcgen$> = accept) **then** {

9. Execute procedure RandomApproach(); // move randomly fish f towards fish f_1;

10. **return;**

11. } // end **If**

11. } // end **For**

13. Execute procedure RandomApproach(); // move randomly fish f towards its personalBest

14. **return;**

Procedure RandomApproach() works the same way as procedure Approach(), with the difference that it completes its operation if f_1 achieves better fitness than its initial one during the execution of the procedure. It chooses, just like in procedure Approach(), a random location at which both fish differ and then tests all the swaps (in same way as described in procedure Approach()). From all these swaps, procedure RandomApproach() finally chooses to carry out the one that leads to better fitness among all available swaps. The process ends either when a better fitness to f_1 is achieved or when—as in procedure Approach()—the percentage of the initial distance between the two fish becomes less than (1.0-STEP_RATIO).

A fish f performs InnerPrey() (Algorithm 2, Line 22) procedure when its neighborhood is quite rich in fish, towards which it could make a move. A fish f_1, different from f, is selected randomly from the neighbors of f and if it has better fitness than f then fish f moves towards f_1 using procedure RandomApproach() and procedure InnerPrey() is completed. If the fitness of f_1 is not better than the fitness of f then f can also move towards f_1 using procedure RandomApproach() with probability equal to $e-dcgen$ (dc is the difference in fitness between fish f_1 and fish f and gen is the current generation) and procedure InnerPrey() is completed, too. This is repeated PREY_TRY_NUMBER times at most. In case none of the above situations take place, procedure Prey() is executed. The pseudo code of procedure InnerPrey() is presented in Algorithm 4.

Algorithm 4: The pseudo code of procedure InnerPrey().

0. **For** try ← 1 to PREY_TRY_NUMBER {

1. accept ← random number between 0 and 1;

2. pick a random fish f_1 among its neighbors;

3. dc ← fitness of f_1 – fitness of f;

4. **If** fitness of f_1<= fitness of f then {

5. Execute procedure RandomApproach(); // move randomly fish f towards fish f_1;

6. **return;**

7. } // end **If**

8. **If** (fitness of f1 > fitness of f) **and** ($e-dcgen$> = accept) **then** {

9. Execute procedure RandomApproach(); // move randomly fish f towards fish f_1;

10. **return;**

11. } end **If**

12. } end **For**

13. Execute procedure Prey(); // See Algorithm 3

14. **return;**

A fish f performs procedure SwarmNChase() (Algorithm 2, Line 24) when its neighborhood is neither "dense" nor "sparse" in fish. Fish f, using procedure RandomApproach(), moves separately and independently to the localCentre fish of its neighborhood (swarm behavior) and to the localBest fish of its neighborhood (chase behavior) [17,24]. The fish f finally takes the move which gives better fitness. If neither of the two moves lead to better fitness, the fish executes procedure InnerPrey(). The pseudo code of procedure SwarmNChase() is presented in Algorithm 5.

Algorithm 5: The pseudo code of procedure SwarmNChase().

0. tempFish$_1 \leftarrow$ f;

1. tempFish$_2 \leftarrow$ f;

2. move randomly fish tempFish$_1$ towards the localBest of neighborhood of f;

3. move randomly fish tempFish$_2$ towards the localCentre of neighborhood of f;

4. tempFish $\leftarrow$ fish with the minimum fitness among tempFish$_1$ and tempFish$_2$;

5. **If** fitness of tempFish <= fitness of f **then**

6. f $\leftarrow$ tempFish;

7. **Else**

8. InnerPrey(f); // See Algorithm 4

9. **Return;**

As seen in the previous paragraphs, the proposed AFS algorithm uses many user-defined parameters that affect the algorithm's convergence and efficiency. In Table 1 all user defined parameters are summarized and explained. Moreover, their values used in all experiments are listed. Although the adjustment of user-defined parameter values remains an open issue, as there is no obvious way to tune them, after having conducted exhaustive experiments we decided to use the values presented in Table 1, since this combination of values resulted in the best performance of the proposed AFS algorithm.

Table 1. The user defined parameters used by the proposed AFS algorithm.

Parameter	Value	Comments
NUMBER_OF_GENERATIONS	10,000	The number of generations the algorithm is executed
NUMBER_OF_FISH	24	The size of the population of fish. We decided to use 24 fish since, after exhaustive experiments we came to the conclusion, that this is the minimum number of fish which guarantees that the AFS algorithm will always reach feasible solutions.
VISUAL_SCOPE_COEF	0.7	A fish f_1 is located in the neighborhood of a fish f, if its distance from f is less than minDist + (maxDist – minDist) × VISUAL_SCOPE_COEF, where minDist and maxDist are respectively the minimum and maximum distance among all fish of current population
SPARSE_COEF	0.1	A neighborhood is considered "sparse", when containing less than SPARSE_COEF × NUMBER_OF_FISH individuals
DENSE_COEF	0.8	A neighborhood is considered "dense", when containing more than DENSE_COEF × NUMBER_OF_FISH individuals
STEP_RATIO	0.047	Fish approaching factor. Procedures *Approach()* and *RandomApproach()* are completed when the percentage of the initial distance between two fish becomes less than (1.0 - STEP_RATIO)
PREY_TRY_NUMBER	3	Number of iterations for procedures *Prey()* and *InnerPrey()*
MIN_DIST_COEF	0.01	The turbulence activating factor. Procedure *Turbulence()* is activated when the current maximum distance between all fish becomes less than MIN_DISTANCE_COEF × number of classes × 35
LEAP_NUMBER	100	Procedure *Leap()* is activated when the improvement of the obtained best fitness (i.e. the fitness of minGlobalFish) during the last LEAP_NUMBER generations is less than MIN_IMR
TURBULENCE_ITERATIONS	5	Number of repetitions of procedure *Swap()* inside procedure *Turbulence()* equals to TURBULENCE_ITERATIONS × NUMBER_OF_FISH
MIN_IMPR	0.01	Threshold of desired fitness rate improvement in order to start the *Leap()* procedure

4. COMPUTATIONAL RESULTS

The proposed algorithms were coded in C++ and run on i7–4770, 3.40 GHz and 16 GB of RAM, under the Windows 7 (64 bit) OS. All results presented in this section were accomplished using the same set of PSO parameters' values for the PSO algorithm and the same set of AFS parameters' values for the AFS algorithm. This was adopted in order to have a fair comparison of both algorithm's efficiency and performance. Both algorithms use the same encoding described in Section 3.1 and Section 3.2. The population size was set to 15 for the PSO algorithm and to 24 for the AFS algorithm.

The fitness function used for both algorithms is the one presented in [20], incorporates all hard and soft constraints listed in Section 2 and has the following form:

- f = cases_of_teachers'_unavailability × HCW × BASE^3

- + cases_of_classes'_empty_periods × HCW × (2 × BASE)^{BASE}

- $+$ cases_of_parallel_teaching $\times$ HCW $\times$ BASE[k]

- $+$ cases_of_wrong_co-teaching $\times$ HCW $\times$ (2 $\times$ BASE)[BASE]

- $+$ cases_of_class_lessons'_dispersion $\times$ ICDW $\times$ HOURS $\times$ BASE[DAYS]

- $+$ cases_of_teachers'_empty_spaces $\times$ TEPW $\times$ HOURS $\times$ BASE[DAYS]

- $+$ cases_of_teacher_lessons'_dispersion $\times$ ITDW $\times$ absolute_error $\times$ BASE[DAYS]

Moreover, all hard and soft constraints weights used in it have the exact same values as the ones used in [20] and are described as follows:

- Hard Constraints' Weight (HCW). This weight is utilized by the algorithm in order to distinguish feasible and infeasible timetables. Its value is set to 10.
- Ideal Classes' Dispersion Weight (ICDW). This weight is relevant to the classes' lessons dispersion and its value is set to 0.95.
- Teachers' Empty Periods Weight (TEPW). This weight relates to teachers' idle hours during any working day and its value is set to 0.06.
- Ideal Teachers' Dispersion Weight (ITDW). This weight is relevant to the teachers' teaching hours' dispersion and its value is set to 0.6.

Apart from the above weights, another major parameter that affects the behavior of the evaluation function is the exponential rise base (BASE). This is a real number (typically between 1 and 2) that is used as a base for the exponential rise of the sub-costs corresponding to violations of a certain constraints in a timetable. For all experiments conducted, its value was set to 1.3. In order to demonstrate the performance and efficiency of both computational intelligence algorithms, their experimental results are compared with the respective results of four different heuristics that have been applied to the school timetabling problem in the literature [19,20,25,26]. The approach presented in [19] is a simulated annealing (SA)-based algorithm with a newly-designed neighborhood structure. Its main innovation is that, in search for the best neighbor, the heuristic performs a sequence of swaps between pairs of timeslots, instead of swapping two assignments, as in the standard simulated annealing. The algorithm presented in [25] is an evolutionary one (EA). The algorithm uses linear ranking selection, two mutation operators, and an elitism schema, while, on the other hand, it does not use any crossover operator. The third algorithm presented in [20] comprises a hybrid particle swarm optimization (PSO)-based algorithm. It consists of a main PSO algorithm and a refining local search procedure which is applied in order to improve the best solution found by the main PSO algorithm. Finally, the fourth algorithm is a genetic algorithm selection perturbative hyper-heuristic (GASPHH) [26]. It comprises a two-phase approach, with the first phase focusing on hard constraints and the second phase on soft constraints. Both phases employ the same genetic algorithm selection perturbative hyper-heuristic, with the low-level heuristics differing for each phase.

Both algorithms presented in the current contribution, as well as the four algorithms mentioned in the previous paragraph, use the same formalism for modeling the timetabling problem and the same three performance criteria (teachers' teaching hours' distribution, lessons' hours' distribution and teachers' gaps) in order to evaluate their performance. Thus, we can perform a straightforward and fair comparison of their experimental results. The input instances selected in order to compare the performance of these algorithms is the well-established Beligiannis data set [6,7], which has been used by all these algorithms in the respective contributions. In Table 2 the major characteristics of the input instances used in the experimental results are presented. The interested reader can find a thorough description of these instances in [19,20].

Table 2. Major characteristics of the Beligiannis school timetabling data set.

Instance	Number of Classes	Number of Teachers	Number of Teaching Hours	Number of Teachers Involved in Co-Teaching	Number of Co-Teachings
1	11	34	385	9	36
2	11	35	385	17	67
3	6	19	210	0	0
4	7	19	245	6	31
5	6	18	184	0	0
7	13	35	455	17	70

The proposed PSO algorithm as well as the proposed AFS algorithm is by nature stochastic. As a result, different computational results may be obtained in different runs. So, in order to demonstrate their efficiency, in Table 3 we present not only the best but also the worst and the average fitness achieved together with the respective standard deviation. The average execution time as well as the respective standard deviation is presented, too. In Table 4 the best performance of both proposed algorithms is compared with the best timetables created by the evolutionary algorithm (EA) presented in [25], the simulated annealing (SA) algorithm presented in [19], the hybrid PSO algorithm presented in [20] and the genetic algorithm selection perturbative hyper-heuristic (GASPHH) presented in [26].

The reason why we decided to present and compare the best timetables constructed by the proposed algorithms is that in [19,20,25,26] also the best timetables constructed are presented. Thus, in this way a fair comparison between all algorithms can be performed. Note that in order to compare two different timetables, we define one unit cost for any soft constraint violation among the three performance criteria (teachers' teaching hours' distribution, lessons' hours' distribution, and teachers' gaps) [26]. All values presented in Table 3 and Table 4 are computed after executing 100 Monte Carlo runs. For both algorithms the average fitness achieved is very close to the best fitness. Additionally, the respective STD value is rather small. This demonstrates their efficiency and stability. Furthermore, the STD concerning the execution time of both algorithms is really small. This also demonstrates that both algorithms are very stable.

Table 3. Experimental results of the proposed stochastic algorithms.

Instance	Proposed PSO							Proposed AFS					
	Fitness				Execution Time			Fitness				Execution Time	
	Best	Worst	Average	STD	Average (min)	STD	Best	Worst	Average	STD	Average (min)	STD
1	11	21	16	2.5	36.6	3.4	16	31	22	3.8	120.8	2.4
2	18	28	22.1	2.5	47.8	0.8	21	34	28.6	3.7	147.2	1.1
3	5	9	7.4	0.8	3.9	0.4	7	13	10.4	1.7	28.5	0.9
4	5	13	8.8	2.2	13.2	0.3	9	23	14.5	3.2	59.4	2.6
5	0	0	0	0	3	0.1	0	5	2.4	1.5	24	0.1
7	23	49	35.8	6.8	57.9	0.4	32	88	56.1	14.1	168	3

Table 4. Comparing the performance of both proposed algorithms with the algorithms presented in [19,20,25,26].

Instance	Proposed PSO	Proposed AFS	EA [25]	SA [19]	Hybrid PSO [20]	GASPHH [26]
1	11	16	70	29	15	60
2	18	21	76	37	17	66
3	5	7	20	8	7	17
4	5	9	41	28	8	43
5	0	0	10	3	0	15
7	23	32	109	40	32	83

In Table 4 the best results, which are the best results found until today for the respective instances, are written in bold. The proposed PSO algorithm appears to have the best performance among all other algorithms achieving the best results ever found for 5/6 instances of the Beligiannis data set. The proposed AFS algorithm has also very satisfactory performance, since it surmounts the heuristic algorithms presented in [19,25,26]. At this point we have to call to mind that the proposed AFS algorithm uses many user-defined parameters that affect the algorithm's convergence and efficiency. As a result, it is still an open issue to investigate thoroughly the best set of parameter values for the proposed AFS approach which will assist it to achieve even better results.

5. CONCLUSIONS

In this contribution a comparative study of modern heuristics on the school timetabling problem is presented. Two new heuristic population based algorithms, a particle swarm optimization (PSO) and an artificial fish swarm (AFS) one, are introduced and applied to the school timetabling problem. In order to demonstrate their efficiency and good performance both algorithms have been applied to the Beligiannis school timetabling data set and their results are compared with the results of four other heuristic algorithms published in the respective literature. Experimental results showed that the proposed PSO algorithm achieves the best results ever found for 5/6 instances of the Beligiannis data set. On the other hand, the proposed AFS algorithm, which comprises the first attempt to apply an AFS algorithm to the school timetabling problem in the literature, exhibits also very satisfactory results, since it overcomes three other heuristic approaches. This demonstrates that the AFS algorithm can be effectively applied to solve the school timetabling problem. The application of the proposed PSO algorithm to problems belonging to other timetabling or scheduling domains, as well as the investigation of the best set of parameter values for the proposed AFS approach, will be the main issues of our future work.

REFERENCES

1. Cooper, T.B.; Kingston, J.H. The complexity of timetable construction problems. In Practice and Theory of Automated Timetabling; Springer Berlin Heidelberg: Berlin, Germany, 1996.

2. De Werra, D. The combinatorics of timetabling. Eur. J. Op. Res. **1997**, 96, 504–513.

3. Johnes, J. Operational research in education. Eur. J. Ope. Res. **2015**, 243, 683–696.

4. Tassopoulos, I.X.; Beligiannis, G.N. Using particle swarm optimization to solve effectively the school timetabling problem. Soft Comput. **2012**, 16, 1229–1252.

5. Pillay, N. Hyper-heuristics for educational timetabling. In Proceedings of the Ninth International Conference on the Practice and Theory of Automated Timetabling (PATAT 2012), Son, Norway, 28–31 August 2012; pp. 316–340.

6. Pillay, N. A survey of school timetabling research. Ann. Op. Res. **2014**, 218, 261–293. [Google Scholar] [CrossRef]

7. Kristiansen, S.; Stidsen, T.R. A Comprehensive Study of Educational Timetabling—A Survey; Technical University of Denmark: Copenhagen, Denmark, 2013.

8. Raghavjee, R.; Pillay, N. A comparison of genetic algorithms and genetic programming in solving the school timetabling problem. In Proceedings of the Fourth World Congress on Nature and Biologically Inspired Computing (NaBIC 2012), Mexico City, Mexico, 5–9 November 2012; pp. 98–103.

9. Sørensen, M.; Kristiansen, S.; Stidsen, T.R. International timetabling competition 2011: An adaptive large neighborhood search algorithm. In Proceedings of the Ninth International Conference on the Practice and Theory of Automated Timetabling (PATAT 2012), Son, Norway, 28–31 August 2012; pp. 489–492.

10. Domrös, J.; Homberger, J. An evolutionary algorithm for high school timetabling. In Proceedings of the Ninth International Conference on the Practice and Theory of Automated Timetabling (PATAT 2012), Son, Norway, 28–31 August 2012; pp. 485–488.

11. Kalender, M.; Kheiri, A.; Özcan, E.; Burke, E.K. A greedy gradient-simulated annealing selection hyper-heuristic. Soft Comput. **2013**, 17, 2279–2292. [

12. Kheiri, A.; Özcan, E.; Parkes, A.J. A stochastic local search algorithm with adaptive acceptance for high-school timetabling. Ann. Op. Res. **2014**.

13. Da Fonseca, G.H.G.; Santos, H.G.; Toffolo, T.A.M.; Brito, S.S.; Souza, M.J.F. GOAL solver: A hybrid local search based solver for high school timetabling. Ann. Op. Res. **2014**, 1–21.

14. Ahmed, L.N.; Özcan, E.; Kheiri, A. Solving high school timetabling problems worldwide using selection hyper-heuristics. Expert Syst. Appl. **2015**, 42, 5463–5471.

15. Al-Yakooba, S.M.; Sheralib, H.D. Mathematical models and algorithms for a high school timetabling problem. Comput. Op. Res. **2015**, 61, 56–68.

16. Kennedy, J.; Eberhart, R.C.; Shi, Y. Swarm Intelligence; Morgan Kaufmann: Burlington, MA, USA, 2001.

17. Neshat, M.; Sepidnam, G.; Mehdi, S.; Toosi, A.N. Artificial fish swarm algorithm: a survey of the state of the art, hybridization, combinatorial and indicative applications. Artif. Intell. Rev. **2014**, 42, 965–997.

18. Raghavjee, R.; Pillay, N. A study of genetic algorithms to solve the school timetabling problem. In Advances in Soft Computing and Its Applications; Castro, F., Gelbukh, A., Mendoza, M.G., Eds.; Springer Berlin Heidelberg: Berlin, Germany, 2013; pp. 64–80.

19. Zhang, D.; Liu, Y.; M'Hallah, R.; Leung, C.H.S. A simulated annealing with a new neighborhood structure based algorithm for high school timetabling problems. Eur. J. Op. Res. **2010**, 203, 550–558.

20. Tassopoulos, I.X.; Beligiannis, G.N. A hybrid particle swarm optimization based algorithm for high school timetabling problems. Appl. Soft Comput. **2012**, 12, 3472–3489.

21. Tassopoulos, I.X.; Beligiannis, G.N. Solving effectively the school timetabling problem using particle swarm optimization. Expert Syst. Appl. **2012**, 39, 6029–6040.

22. Beligiannis, G.N.; Moschopoulos, C.N.; Likothanassis, S.D. A Genetic Algorithm Approach to School Timetabling. J. Op. Res. Soc. **2009**, 60, 23–42.

23. Dorneles, Á.P.; de Araújo, O.C.B.; Buriol, L.S. The Impact of compactness requirements on the resolution of high school timetabling problem. In Proceedings of the XLIV Simpósio Brasileiro de Pesquisa Operacional (SBPO 2012), Rio de Janeiro, Brazil, 24–28 September 2012; 2012; pp. 3336–3347.

24. Rocha, A.M.A.C.; Fernandes, E.M.G.P.; Martins, T.F.M.C. Novel fish swarm heuristics for bound constrained global optimization problems. In Computational Science and Its Applications (ICCSA 2011); Springer Berlin Heidelberg: Berlin, Germany, 2011.

25. Beligiannis, G.N.; Moschopoulos, C.N.; Kaperonis, G.P.; Likothanassis, S.D. Applying evolutionary computation to the school timetabling problem: The Greek case. Comput. Op. Res. **2008**, 35, 1265–1280.

26. Raghavjee, R.; Pillay, N. A genetic algorithm selection perturbative hyper-heuristic for solving the school timetabling problem. ORiON **2015**, 31, 39–60

An Automatic Multilevel Image Thresholding Using Relative Entropy and Meta-Heuristic Algorithms

Yun-Chia Liang * and Josue R. Cuevas

Department of Industrial Engineering and Management, Yuan Ze University, 320, Taiwan

ABSTRACT

Multilevel thresholding has been long considered as one of the most popular techniques for image segmentation. Multilevel thresholding outputs a gray scale image in which more details from the original picture can be kept, while binary thresholding can only analyze the image in two colors, usually black and white. However, two major existing problems with the multilevel thresholding technique are: it is a time consuming approach, i.e., finding appropriate threshold values could take an exceptionally long computation time; and defining a proper number of thresholds or levels that will keep most of the relevant details from the original image is a difficult task. In this study a new evaluation function based on the Kullback-Leibler information distance, also known as relative entropy, is proposed. The property of this new function can help determine the number of thresholds automatically. To offset the expensive computational effort by traditional exhaustive search methods, this study establishes a procedure that combines the relative entropy and meta-heuristics. From the experiments performed in this study, the proposed procedure not only provides good segmentation results when compared with a well known technique such as Otsu's method, but also constitutes a very efficient approach.

Keywords: thresholding; relative entropy; Gaussian mixture models; meta-heuristics; virus optimization algorithm

1. INTRODUCTION

Segmenting an image into its constituents is a process known as thresholding [1,2]. Those constituents are usually divided into two classes: foreground (significant part of the image), and background (less significant part of the image). Several methods for thresholding an image have been developed over the last decades, some of them based on entropy, within and between group variance, difference between original and output images, clustering, etc. [3,4,5].

The process of thresholding is considered as the simplest image segmentation method, which is true when the objective is to convert a gray scale image into a binary (black and white) one (that is to say only one threshold is considered). However, in the process of segmentation, information from the original image will be lost if the threshold value is not adequate. This problem may even deteriorate as more than one threshold is considered (i.e., multilevel thresholding) since not only a proper number of thresholds is desirable, but also a fast estimation of their values is essential [1].

It has been proven over the years that as more thresholds are considered in a given image, the computational complexity of determining proper values for each threshold increases exponentially (when all possible combinations are considered). Consequently, this is a perfect scenario for implementing meta-heuristic tools [6] in order to speed up the computation and determine proper values for each threshold.

When dealing with multilevel thresholding, in addition to the fast estimation, defining an adequate number of levels (thresholds that will successfully segment the image into several regions of interest from the background), has been another problem without a satisfactory solution [7]. Therefore, the purpose and main contribution of this study is to further test the approach proposed in [8] which determines the number of thresholds necessary for segmenting a gray scaled image automatically. The aforementioned is achieved by optimizing a mathematical model based on the Relative Entropy Criterion (REC).

The major differences between this work and the one presented in [8], are that a more extensive and comprehensive testing of the proposed method was carried out. To verify the feasibility of the proposed model and reduce the computational burden, when carrying out the optimization process, this study implements three meta-heuristic tools and compares the output images with that delivered by a widely known segmentation technique named Otsu's method. In addition to the aforementioned, further tests with images having low contrast and random noise are conducted; this intends to probe the robustness, effectiveness and efficiency of the proposed approach.

The following paper is organized as follows: Section 2 introduces the proposed procedure, which consists of the mathematical model based on the relative entropy and the meta-heuristics—virus optimization algorithm, genetic algorithm, and particle swarm optimization as the solution searching techniques. Section 3 illustrates the performance of the proposed method when the optimization is carried out by three different algorithms VOA, GA and PSO, where six different types of images were tested. Lastly, some concluding remarks and future directions for research are provided in Section 4.

2. THE PROPOSED MULTILEVEL THRESHOLDING METHOD

A detailed explanation of the proposed method is given in this section, where the mathematical formulation of the model is presented and the optimization techniques are introduced.

2.1. Kullback-Leibler Information Distance (Relative Entropy)

The Kullback-Leibler information distance (known as Relative Entropy Criterion) [9] between the true and the fitted probabilities is implemented in [8] for estimating an appropriate model that will best represent the histogram coming from the gray level intensity of an image. According to McLachlan and Peel [10] this information distance is defined as in Equation (1). However, the gray levels (intensity) are discrete values between [0, 255]; therefore, Equation (1) can be rewritten as in Equation (2):

$$J(d) = I\{p(i); p(i; \theta_d)\} = \int \left(p(i)\log_e \left[\frac{p(i)}{p(i; \theta_d)} \right] \right) di$$

(1)

$$J(d) = I\{p(i); p(i; \theta_d)\} = \sum_{i=0}^{255} \left(p(i)\log_e \left[\frac{p(i)}{p(i; \theta_d)} \right] \right)$$

(2)

where p(*i*) and p(*i;θ$_d$*) are the probabilities values from the image histogram and fitted model respectively. These probabilities are estimated using Equations (3) and (4), where the value of *i*∈[0, 255] and represents the gray intensity of a pixel at location (x, y) on the image of size *M*×*N* pixels; while $\sum_{j=1}^{d} g(i, \theta_j)$ in Equation (4) is a mixture of "d" distributions which are used to estimate the value of p(i;θ$_d$). Lastly, θ$_j$ is a vector containing the parameters of each distribution in the mixture:

$$p(i) = \frac{Total\ number\ of\ pixels\ with\ gray\ intesity\ of\ i}{Size\ of\ the\ image}$$

(3)

$$p\left(i;\theta_d\right) = g\left(i,\theta_1\right) + \cdots + g\left(i,\theta_d\right) = \sum_{j=1}^{d} g\left(i,\theta_j\right)$$

(4)

In this study Gaussian distributions are used to estimate Equation (4), where the central limit theorem (C.L.T.) is the motivation of using this type of distributions [11]. Therefore, Equation (4) can be expressed as in Equation (5):

$$p\left(i;\theta_d\right) = \sum_{j=1}^{d} \left(\frac{w_j}{\sqrt{2\pi}\sigma_j} e^{-\frac{1}{2}\left[\left(\frac{i-\mu_j}{\sigma_j}\right)^2\right]} \right)$$

(5)

where θ$_j$ contains the prior probability (or weight) w$_j$, mean μ$_j$, and variance σ^2$_j$, of the jth Gaussian distribution. The minimization of the Relative Entropy Criterion function can be interpreted as the distance reduction between the observed and estimated probabilities. This should provide a good description of the observed probabilities p(i) given by the gray level histogram of the image under study. However, finding an appropriate number of distributions, i.e., "d", is a very difficult task [12,13,14,15]. Consequently, the addition of a new term in Equation (2), which is detailed in the following subsection helps to automatically determine a suitable amount of distributions.

2.2. Assessing the Number of Distributions in a Mixture Model

The purpose of assuming a mixture of "d" distributions in order to estimate p(*i;θ$_d$*) in the REC function, is that the number of thresholds for segmenting a given image, can be easily estimated using Equation (6). The value of each threshold is the gray level intensity "i" that minimizes Equation (7), where "i" is a discrete unit (i.e., an integer $\in [0,255]$):

$$number\ of\ thresholds = d - 1$$

(6)

$$Threshold_k = \arg\min_i \left\{ \left\| p\left(i;\theta_k\right) - p\left(i;\theta_{k+1}\right) \right\| \right\} \quad k = 1, 2, ..., d-1$$

(7)

Given that estimating a suitable number of distributions "d" is a very difficult task, the method proposed in [8] attempts to automatically assess an appropriate value for "d" combining Equations (2) and (8) as a single objective function. The vector **w** contains all the prior probabilities (weights) of the mixture model. The values max(**w**) and min(**w**) are the maximum and minimum weights in **w** respectively:

$$P(d) = \frac{1}{d-1} \sum_{j=1}^{d} \left(\frac{\max(\mathbf{w}) - w_j}{\max(\mathbf{w}) - \min(\mathbf{w})} \right)$$

(8)

Equation (8) compares each prior probability w_j with respect to the largest one in the vector **w**. The result is normalized using the range given by [max(**w**)-min(**w**)] in order to determine how significant w_j is with respect to the probability that contributes the most in the model, i.e., max(**w**). Therefore, Equation (8) will determine if the addition of more Gaussian distributions is required for a better estimation of $p(i;\theta_d)$. The term $1d-1/$ will avoid Equation (8) overpowering Equation (2) when the mathematical model of Equation (9) is minimized:

$$\Theta(d) = \sum_{i=0}^{255} \left(p(i) \log_e \left[\frac{p(i)}{p(i;\theta_s)} \right] \right) + \frac{1}{d-1} \sum_{j=1}^{d} \left(\frac{\max(\mathbf{w}) - w_j}{\max(\mathbf{w}) - \min(\mathbf{w})} \right)$$

(9)

Therefore, by minimizing Equation (9), the introduced approach will not only determine an appropriate number of distributions in the mixture model, but will also find a good estimation (fitting) of the probabilities $p(i)$ given by the image histogram. An appropriate value for "d" is determined by increasing its value by one, i.e., $d^l = d^{l-1} + 1$ where "l" is the iteration number, and minimizing Equation (9) until a stopping criterion is met and the addition of more distributions to the mixture model is not necessary. Therefore, not all the possible values for "d" are used.

2.3. Mathematical Model Proposed for Segmenting (Thresholding) a Gray Level Image

The mathematical model which is used for segmenting a gray level image is presented as in Equation (10). By minimizing Equation (10) with "d" Gaussian distributions, J(d) is in charge of finding a good estimation (fitting) of the image histogram, while P(d) determines whenever the addition of more distributions to the model is necessary:

$$Min \; \Theta(d) = Min\left[J(d) + P(d)\right] = \underset{\theta_d, d}{Min}\left[\sum_{i=0}^{255}\left(p(i)\log_e\left[\frac{p(i)}{p(i;\theta_d)}\right]\right) + \frac{1}{d-1}\sum_{j=1}^{d}\left(\frac{\max(\mathbf{w})-w_j}{\max(\mathbf{w})-\min(\mathbf{w})}\right)\right]$$

(10)

subject to:

$$\sum_{j=1}^{d} w_j = 1$$

(11)

$$w_j > 0 \;\; \forall j \in [1, d]$$

(12)

$$\sigma_j^2 > 0 \;\; \forall j \in [1, d]$$

(13)

Equation (11) guarantees a summation of the prior probabilities equal to 1, while Equations (12) and (13) ensure positive values to all prior probabilities and variances in the mixture model, respectively. To minimize Equation (10), a newly developed meta-heuristic named Virus Optimization Algorithm [16], the widely known Genetic Algorithm [17] and Particle Swarm Optimization algorithm [18] are implemented in this study.

The flowchart at Figure 1 details the procedure of the proposed method using VOA (the similar idea is also applied to GA and PSO). As can be observed, the optimization tool (VOA, GA, or PSO) will optimize Equation (10) with d^l Gaussian distributions until the stopping criterion of the meta-heuristic is reached. Once the algorithm finishes optimizing Equation (10) with d^l distributions, the proposed method decides if the addition of more components is necessary if and only if $\Theta(d^l) \geq \Theta(d^{l-1})$ is true; otherwise a suitable number of distributions (thresholds) just been found and the results coming from $\Theta(d^{l-1})$ are output.

The purpose of using three algorithmic tools, is not only to reduce the computation time when implementing the proposed approach, but also, to verify if the adequate number of thresholds suggested by optimizing Equation (10) with different optimization algorithm remains the same. The reason of the aforementioned is because different algorithms may reach different objective function values. However, if the proposed method (Figure 1) is robust enough, all algorithms are expected to stop iterating when reaching a suitable number of thresholds, and this number has to be the same for all the optimization algorithms.

2.4. Algorithmic Optimization Tools

2.4.1. Virus Optimization Algorithm (VOA)

Inspired from the behavior of a virus attacking a host cell, VOA [8,16] is a population-based method that begins the search with a small number of viruses (solutions). For continuous optimization problems, a host cell represents the entire multidimensional solution space, where the cell's nucleolus denotes the global optimum. Virus replication indicates the generation of new solutions while new viruses represent those created from the strong and common viruses.

The strong and common viruses are determined by the objective function value of each member in the population of viruses, i.e., the better the objective function value of a member the higher the chance to be considered as strong virus. The number of strong viruses is determined by the user of the algorithm, which we recommend to be a small portion of the whole population (strong and common).

To simulate the replication process when new viruses are created, the population size will grow after one complete iteration. This phenomenon is controlled by the antivirus mechanism that is responsible for protecting the host cell against the virus attack. The whole process will be terminated based on the stopping criterion: the maximum number of iterations (i.e., virus replication), or the discovery of the global optimum (i.e., cell death is achieved).

The VOA consists of three main processes: Initialization, Replication, and Updating/Maintenance. The Initialization process uses the values of each parameter (defined by the user) to create the first population of viruses. These viruses are ranked (sorted) based on the objective function evaluation $\Theta(d)$ to select strong and common members. Here the number of strong members in the population of viruses is a parameter to be defined by the user, without considering the strong members; the population of viruses is the number of common viruses.

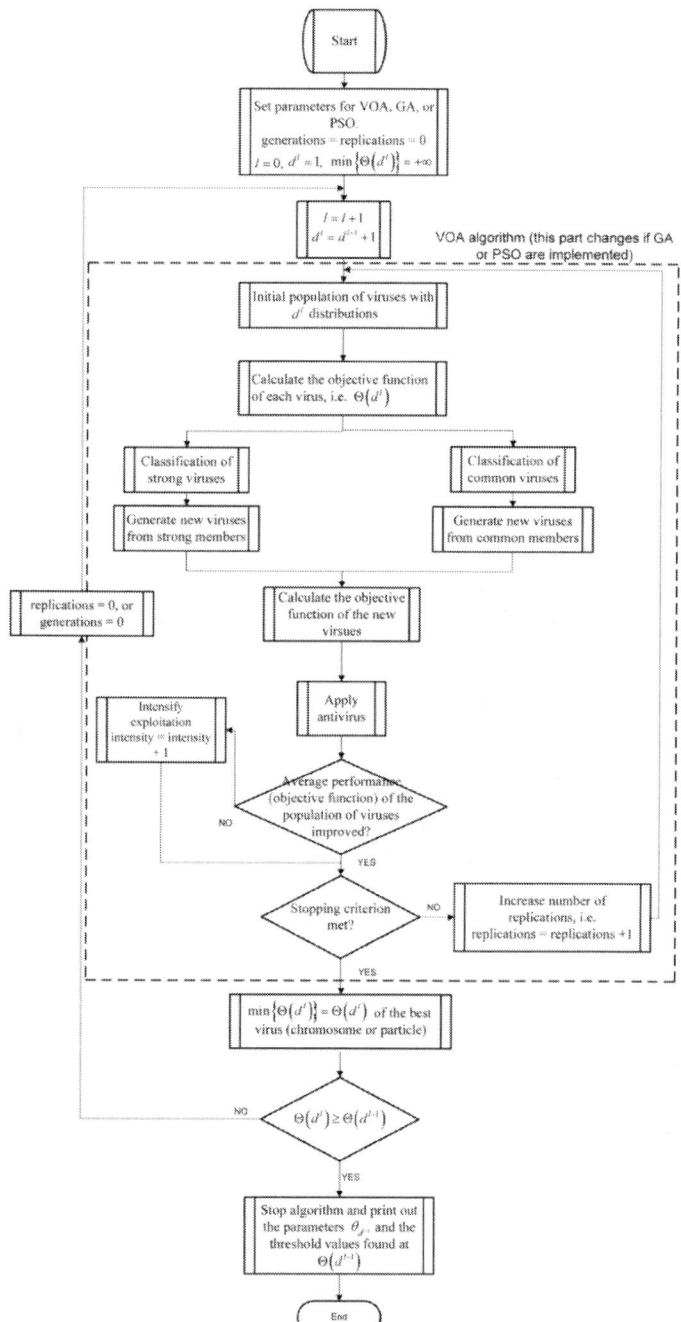

Figure 1. Flowchart of the proposed optimization procedure.

The replication process is performed using the parameters defined by the user in the Initialization stage described above, where a temporary matrix (larger than the matrix containing the original viruses) will hold the newly-generated members. Here, Equations (14) and (15) are used to generate new members, where "vn" stands for the value of the variable in the nth dimensional space, for viruses in the previous replication. "svn" stands for the value of the variable in the nth dimensional space generated from the strong viruses in the current replication, and "cvn" is the value of the variable in the nth dimensional space generated from the common viruses in the current replication:

$$svn = vn \pm \left(\frac{rand()}{intensity} \right) \times vn \tag{14}$$

$$cvn = vn \pm rand() \times vn \tag{15}$$

The intensity in Equation (14) above reduces the random perturbation that creates new viruses from the strong members. This will allow VOA to intensify exploitation in regions more likely to have a global optimum (i.e., areas where the strong viruses are located). The initial value for the intensity is set as one, which means that the random perturbation for strong and common viruses is the same in the early stages. Therefore, the exploration power of VOA is expected to be enhanced during the program's early stages. The intensity value increases by one when the average performance of the population of viruses in VOA; that is to say, the average objective function value of the whole population of viruses, did not improve after a replication. The flowchart of the proposed procedure is illustrated in Figure 1. Note that the VOA part can be easily switched to other optimization algorithms such as GA and PSO.

2.4.2. Genetic Algorithm (GA)

The basic concept of Genetic Algorithms or GAs [17] is to simulate processes in natural systems, necessary for evolution, especially those that follows the principles first laid down by Charles Darwin of survival of the fittest. In GA a portion of existing population (solutions) is selected to breed a new population (new solutions), individuals are selected to reproduce (crossover) through a fitness-based process (objective function). Mutation takes place when new individuals are created after crossover to maintain a diverse population through generations. The standard GA is summarized in Figure 2, for the selection of the parents the roulette wheel is used in this study, as for the population maintenance mechanism the best members in the pooled population (parents and offspring) survive. The crossover operators implemented in this paper are the geometric and arithmetic means [Equations (16) and (17)] for the creation of the first and second child respectively. Note that only the integer part is taken by the program since the chromosome contains only integers. The mutation operator in GA uses Equation (18), which is the floor function of a random number generated between $[T_{i-1}, T_i]$ where $T_0 = 0$ and $T_d = 255$:

$$T_i^{child1} = \left(T_i^{parent1} + T_i^{parent2} \right) \Big/ 2 \; , \; i \in [1, 2, \cdots, d-1] \tag{16}$$

$$T_i^{child2} = \left(T_i^{parent1} \times T_i^{parent2} \right)^{1/2}, \; i \in [1, 2, \cdots, d-1] \tag{17}$$

$$T_i = \left\lfloor rand \left[(T_{i-1}+1), (T_i-1) \right] \right\rfloor, \; i \in [1, 2, \cdots, d] \tag{18}$$

Procedure GA_meta-Heuristic

Set parameters

Generate initial population of individuals

Evaluate the fitness of individuals

While (not termination)

    Select the best-fit individuals for crossover

    Breed a new population through crossover

    Mutate the population of offspring

    Evaluate the fitness of offspring

    Replace least-fit individuals in the previous

    generation with best-fit individuals in the new

    population of offspring

End while

End procedure

Figure 2. Genetic Algorithm (GA) overview.

2.4.3. Particle Swarm Optimization (PSO)

Particle Swarm Optimization was inspired by social behavior of bird flocking or fish schooling [18]. Each candidate solution (known as particle) keeps track of its coordinates in the problem space which are associated with the best solution (fitness) it has achieved so far, also known as the particle's best (pbest). The swarm on the other hand, also keeps track of the best value, obtained until now, by any particle in the neighbor of particles. This is known as the global best (gbest). The basic concept of PSO consists of changing the velocity of each particle towards the gbest and pbest location. This velocity is weighted by a random term, with separated random numbers being generated for acceleration toward the pbest and gbest locations. The standard PSO is summarized as in Figure 3. For this study, the velocity of the particle is bounded to a V_{max} value which is only 2% of the gray intensity range; that is to say, $V_i \in [-V_{max}, V_{max}]$ where $V_{max} = 0.02 \times (255 - 0)$.

Procedure PSO_metaHeuristic

    Set parameters

    Generate initial swarm of particles (locations and velocities)

    Initialize the particle's best known position (pbest)

    Initialize the swarm's best known position (gbest)

    While (not termination)

        Update particle's velocities for each dimension

        If (velocities $\notin [-V_{max}, V_{max}]$)

            Generate velocities within $[-V_{max}, V_{max}]$

        End if

        Update particle's position for each dimension

        Update particle's best known position

        Update swarm's best known position

    End while

Figure 3. Overview of the Particle Swarm Optimization (PSO) algorithm.

The stopping criterion for the VOA (GA or PSO) is when two consecutive replications (generations) did not improve the objective function value of the best virus (chromosome or particle). Once the VOA (GA or PSO) stops searching and the best value for the $\Theta(d^l)$ is determined, the proposed method will decide if the addition of more distributions is necessary when the condition $\Theta(d^l) < \Theta(d^{l-1})$ is satisfied; otherwise the proposed method will automatically stop iterating. After the proposed method stops, the parameters contained inside the vector $\theta_d{}^{l-1}$ that represents the set of parameters of the best result in the previous iteration are output.

In order to avoid computational effort of calculating the threshold values that minimize Equation (7), the VOA (GA and PSO) will code each threshold value inside each solution, i.e., each virus (chromosome or particle) will have a dimensionality equal to the number of thresholds given by Equation (6). During the optimization, when the search of the best value for $\Theta(d^l)$ is in process, each threshold will be treated as a real (not an integer) value, which can be considered as the coded solution of the VOA and PSO.

In the case of GA a chromosome containing integer values is used for encoding the solution. In order to evaluate the objective function of each virus or particle, the real values coded inside each member will be rounded to the nearest integer, whereas in GA this is not necessary since each chromosome is an array of integers. The parameters for each Gaussian distribution are computed as in Equations (19)–(21), which is considered as the decoding procedure of the three meta-heuristics implemented in this study:

$$w_j = \sum_{i=T_{j-1}}^{T_j} p(i), \quad \forall j \in [1, d] \tag{19}$$

$$\mu_j = \sum_{i=T_{j-1}}^{T_j} \left(i \times \frac{p(i)}{w_j} \right), \quad \forall j \in [1, d] \tag{20}$$

$$\sigma_j^2 = \sum_{i=T_{j-1}}^{T_j} \left((i - \mu_j)^2 \times \frac{p(i)}{w_j} \right), \quad \forall j \in [1, d] \tag{21}$$

In Equations (19)–(21), T_j represents the jth threshold in the solution. The values for $T_0 = 0$ and $T_d = 255$, which are the lower and upper limits for thresholds values. During the optimization process, special care should be taken when generating new solutions (viruses, offspring, or particles). The details are as follows:

Condition 1: The threshold should be in increasing order when coded inside each solution, and two thresholds cannot have the same value, i.e., $T_0 < T_1 < T_2 < ... < T_d$.

Condition 2: The thresholds are bounded by the maximum ($T_d = 255$) and minimum ($T_0 = 0$) intensity in a gray level image, i.e., $0 \leq T_j \leq 255$, $\forall j \in [1, d-1]$.

Equation (22) is checked to ensure that the first condition is satisfied, where $\forall i, j \in [1, d-1]$ and $i < j$. If Equation (22) is not satisfied then the solution (virus, chromosome, or particle) is regenerated using Equation (23), where $\lfloor \cdot \rfloor$ is the floor function. The second condition is also checked and whenever any threshold value is outside the boundaries, i.e., $T_j \leq 0$ or $T_j \geq 255$, the virus (chromosome or particle) is regenerated using Equation (23).

$$T_i - T_j \leq 0 \qquad (22)$$

$$VOA, PSO: \ T_i = \left(rand() \times \frac{255}{d} \right) + \left((i-1) \times \frac{255}{d} \right)$$

$$GA: \ T_i = \left\lfloor \left(rand() \times \frac{255}{d} \right) + \left((i-1) \times \frac{255}{d} \right) \right\rfloor \qquad (23)$$

3. EXPERIMENTAL RESULTS

In order to further test the method proposed in [8], five different types of images were tested. The first image, which has a known number of three thresholds as in this case is tested (Figure 4a); secondly, an image containing text on a wrinkled paper which will cause lighting variation is tested (Figure 5a). Thirdly, the Lena image (Figure 6a) [1,12,13,14,15] is tested, which is considered as a benchmark image when a new thresholding technique is proposed.

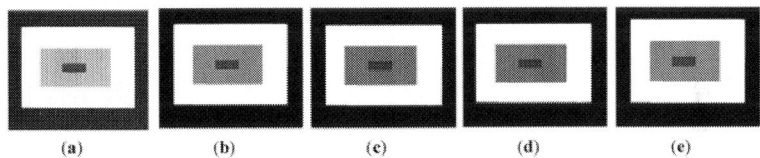

(a)   (b)   (c)   (d)   (e)

Figure 4. Test image 1: **(a)** Original image; Thresholded image implementing **(b)** VOA, **(c)** GA, **(d)** PSO, and **(e)** Otsu's method using three thresholds.

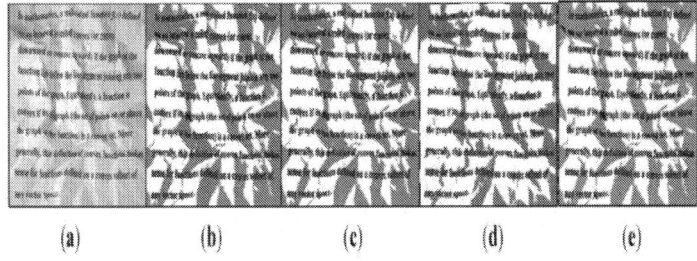

(a)   (b)   (c)   (d)   (e)

Figure 5. Test image 2: **(a)** Original image; Thresholded image implementing **(b)** VOA, **(c)** GA, **(d)** PSO, and **(e)** Otsu's method with two thresholds.

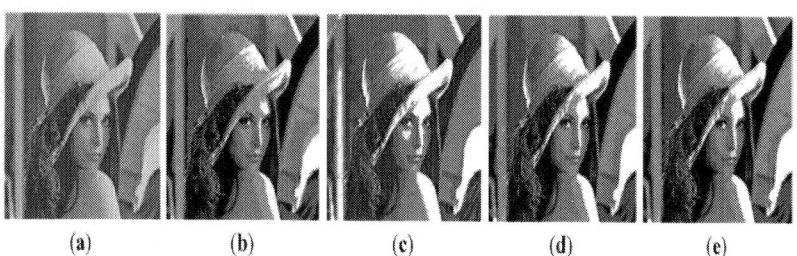

(a) (b) (c) (d) (e)

Figure 6. Test image 3: **(a)** Original image; Thresholded image implementing **(b)** VOA, **(c)** GA, **(d)** PSO, and **(e)** Otsu's method with four thresholds. (taken from [1])

3.1. Algorithmic Setting (VOA, GA, and PSO)

The setting of VOA, GA, and PSO was determined by using Design of Experiments (DoE) [19,20] to know which values for the parameters are suitable when optimizing Equation (10). The full factorial design, i.e., 3-levels factorial design was performed for the four parameters of the VOA; in other words, 3^4 combinations of the four parameters were tested. Table 1 shows the results after performing DoE, where the final setting of the VOA is presented in bold.

Table 1. Parameter values (factor levels) used during the for the VOA.

Parameter	Low Level	Medium Level	High Level
Initial population of viruses	5	**10**	30
Viruses considered as strong	1	**3**	10
Growing rate of strong viruses	2	**5**	10
Growing rate of common viruses	1	**3**	6

Similarly, a 3^3 full factorial design was implemented in order to set the population size (ps), crossover and mutation probabilities (pc and pm) respectively for GA. Table 2 summarizes the experimental settings and the final setting (in bold) of the GA. As for PSO a 3^4 full factorial design determined the values for the swarm size, inertia weight (w), cognitive and social parameters (c1 and c2) respectively. Table 3 summarizes the experimental settings of the PSO algorithm and the final setting is also highlighted in bold.

Table 2. Parameter values (factor levels) used during the DoE for the GA.

Parameter	Low Level	Medium Level	High Level
Population size (ps)	5	10	**30**
Probability of crossover (pc)	0.8	**0.9**	0.99
Probability of mutation (pm)	0.05	**0.1**	0.15

Table 3. Parameter values (factor levels) used during the DoE for the PSO.

Parameter	Low Level	Medium Level	High Level
Swarm size	5	10	**30**
Inertia weight (w)	0.5	**0.8**	0.99
Cognitive parameter (c1)	2	2.1	2.2
Social parameter (c2)	2	2.1	2.2

The basic idea of the DoE is to run the 3^4, 3^3, and 3^4, parameters combinations for VOA, GA, and PSO respectively, to later select which level (value) yielded the best performance (lower objective function value). Once the values for each parameter which delivered the best objective function value are identified, each test image is segmented by optimizing Equation (10) with each of the algorithmic tools used in this study.

The advantage of using DoE is that it is a systematic as well as well-known approach when deciding the setting yielding the best possible performance among all the combinations used during the full factorial design. In addition to the aforementioned, it is also a testing method that has been proven to be quite useful in many different areas such as tuning algorithm parameters [20].

3.2. Segmentation Results for the Proposed Model Using Meta-Heuristics as Optimization Tools

A comprehensive study of the proposed model implementing the three meta-heuristics introduced above is detailed in this part of Section 3. Additionally, a well-known segmentation method (Otsu's) is implemented, where only the output image is observed in order to verify if the segmentation result given by the optimization algorithms used to minimize Equation (10) is as good as the one provided by Otsu's method. The reason of the above mentioned, is because in terms of CPU time Otsu's is a kind of exhaustive search approach; therefore, it is unfair to compare both ideas (the proposed approach and the Otsu's method) in terms of computational effort.

Table 4, Table 5, Table 6, Table 7 and Table 8 detail the performance of the methods used for optimizing Equation (10) over different images. The results (objective function, threshold values, means, variances, and weights) are averaged over 50 independent runs, where the standard deviations of those 50 runs are not shown because they are in the order of 10^{-17}.

By testing the image in Figure 4a it is observed that implementing the three meta-heuristics previously introduced for the optimization of Equation (10), the correct number of thresholds needed for segmenting the image is achieved (which is three). The computational effort and parameters of each Gaussian distribution $(\theta_j = \{w_j, \mu_j, \sigma_j^2\})$ are summarized in Table 4. Here, the number of iterations for the algorithms were four, i.e., the proposed method optimized By testing the image in Figure 4a it is observed that implementing the three meta-heuristics previously introduced for the optimization of Equation (10), the correct number of thresholds needed for segmenting the image is achieved (which is 3).

The computational effort and parameters of each Gaussian distribution $(\theta_j = \{w_j, \mu_j, \sigma_j^2\})$ are summarized in Table 4. Here, the number of iterations for the algorithms were 4, i.e., the proposed method optimized Equation (10) for d = [1,2,3,4] before reaching the stopping criterion.

The behavior of the Relative Entropy function (Figure 7a) reveals its deficiency in detecting an appropriate number of distributions that will have a good description of the image histogram (Figure 8). The aforementioned is because as more distributions or thresholds are added into the mixture, it is impossible to identify a true minimum for the value of J(d) when implementing the three different meta-heuristic tools.

The additional function P(d) on the other hand shows a minimum value when a suitable number of distributions (which is the same as finding the number of thresholds) is found (Figure 7b), since its value shows an increasing pattern when more distributions are added to the mixture model. The combination of these two functions J(d) and P(d) shows that the optimal value for Θ(d) (Figure 7c) will be when the number of thresholds is three as P(d) suggested. Note that the purpose of J(d) is to find the best possible fitting with the suitable number of distributions (thresholds), and this is observed at Figure 8 where the fitted model (dotted line) provides a very good description of the original histogram (solid line) given by the image. The vertical dashed lines in Figure 8 are the values of the threshold found. In addition to the thresholding result, it was observed that VOA provides both, the smallest CPU time as well as the best objective function value among the three algorithms.

Table 4. Thresholding results over 50 runs for the test image 1.

Algorithm	Number of Thresholds	Objective Function	Threshold Values	Means	Variances	Weights	CPU Time per Iteration (s)	Total CPU Time for the Proposed Approach (s)
VOA	1	2.410	244	66.808	3836.212	0.637	0.026	0.507
				251.391	37.709	0.363		
	2	0.976	62, 244	33.209	12.892	0.483	0.071	
				175.849	1298.019	0.167		
				252.452	0.616	0.350		
	3	0.826	44, 145, 246	33.123	11.272	0.480	0.117	
				78.128	390.079	0.021		
				187.900	186.987	0.149		
				252.469	0.482	0.350		
	4	0.896	55, 97, 163, 246	33.178	12.123	0.482	0.294	
				75.522	119.910	0.017		
				128.441	258.119	0.003		
				185.922	82.920	0.142		
				252.158	6.177	0.356		
GA	1	2.410	244	69.838	4226.048	0.650	0.106	0.658
				252.452	0.616	0.350		
	2	0.994	65, 242	33.239	13.806	0.483	0.147	
				176.018	1256.142	0.166		
				252.440	0.736	0.351		
	3	0.857	45, 111, 242	33.131	11.350	0.481	0.195	
				74.412	187.027	0.019		
				186.725	207.543	0.150		
				252.440	0.736	0.351		
	4	0.882	44, 143, 155, 247	33.123	11.272	0.480	0.211	
				77.924	377.966	0.021		
				150.013	12.387	0.001		
				188.193	191.885	0.149		
				252.477	0.431	0.349		
PSO	1	2.384	247	70.158	4274.185	0.651	0.083	0.584
				252.477	0.431	0.349		
	2	0.955	65, 247	33.239	13.806	0.483	0.118	
				176.672	1288.421	0.167		
				252.477	0.431	0.349		
	3	0.857	45, 111, 242	33.131	11.350	0.481	0.172	
				74.412	187.027	0.019		
				186.725	207.543	0.150		
				252.440	0.736	0.351		
	4	0.880	42, 135, 154, 247	33.097	11.070	0.479	0.211	
				74.922	381.817	0.022		
				144.535	35.001	0.001		
				188.161	192.778	0.149		
				252.477	0.431	0.349		

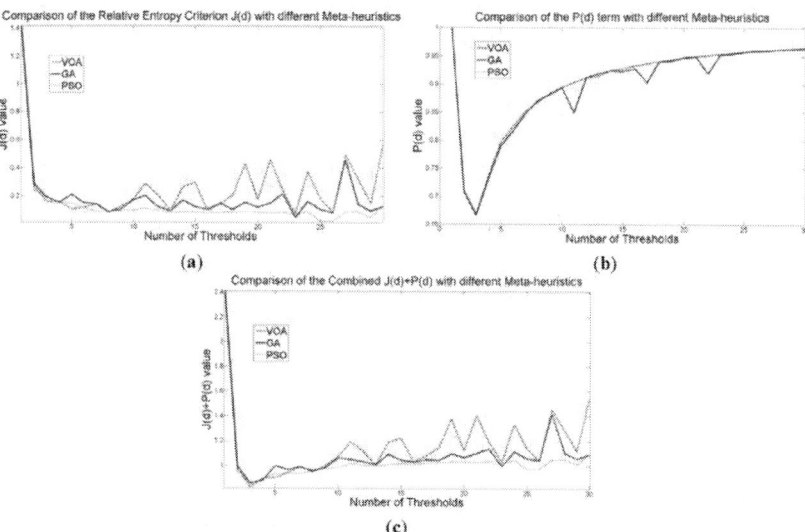

Figure 7. Behavior of (**a**) Relative Entropy function J(d), (**b**) P(d), and (**c**) Objective function Θ(d) over different numbers of thresholds with different meta-heuristics VOA, GA and PSO on test image 1.

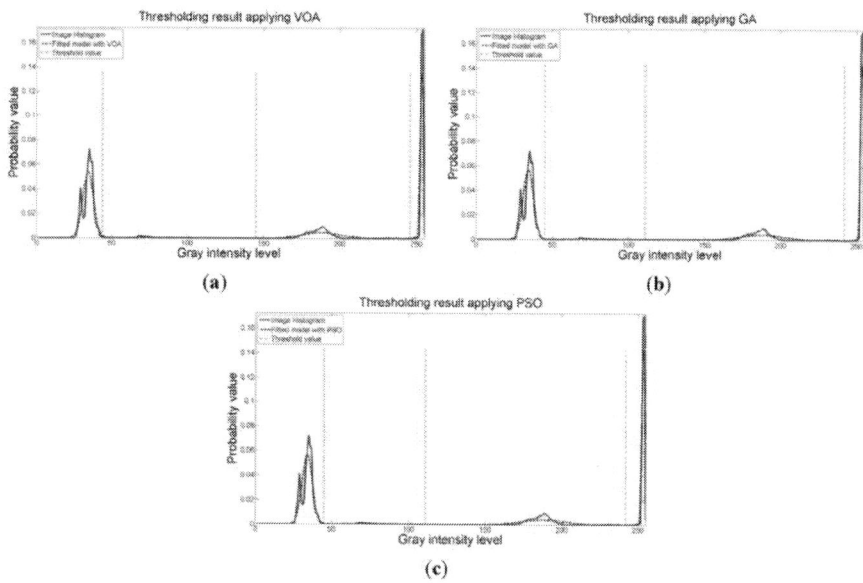

Figure 8. Fitting of the histogram of the test image 1 implementing (**a**) VOA, (**b**) GA, and (**c**) PSO.

When implementing Otsu's method, it is rather impressive to observe that the output image delivered by the algorithms when optimizing Equation (10) resembles the one given by Otsu's method (Figure 4e). The aforementioned, confirms the competitiveness of the proposed idea in segmenting a gray scale image given a number of distributions in the mixture model. Additionally, the main contribution is that we do not need to look at the histogram to determine how many thresholds will provide a good segmentation, and by implementing optimization tools such as the ones presented in this study, we can provide satisfactory results in a short period of time, where methods such as Otsu's would take too long.

Table 5. Thresholding results over 50 runs for the test image 2.

Algorithm	Number of Thresholds	Objective Function	Threshold Values	Means	Variances	Weights	CPU Time per Iteration (s)	Total CPU Time for the Proposed Approach (s)
VOA	1	1.051	143	87.326	1015.901	0.127	0.057	
				203.855	603.462	0.873		
	2	0.585	133, 205	81.590	808.054	0.114	0.089	
				182.352	280.586	0.448		0.237
				223.915	170.129	0.438		
	3	0.670	104, 174, 209	67.393	370.007	0.082	0.091	
				150.866	385.486	0.153		
				192.994	90.775	0.378		
				226.231	146.602	0.386		
GA	1	1.051	142	86.658	991.336	0.126	0.109	
				203.746	609.117	0.874		
	2	0.593	132, 204	81.075	790.076	0.113	0.145	
				181.537	279.992	0.436		0.480
				223.310	176.642	0.451		
	3	0.659	142, 149, 206	86.658	991.336	0.126	0.226	
				145.121	3.989	0.011		
				185.213	205.948	0.440		
				224.558	163.332	0.423		
PSO	1	1.051	143	87.326	1015.901	0.127	0.072	
				203.855	603.462	0.873		
	2	0.573	156, 206	97.751	1389.414	0.153	0.141	
				186.458	170.789	0.424		0.453
				224.558	163.332	0.423		
	3	0.619	101, 180, 211	66.048	335.944	0.079	0.240	
				155.144	453.146	0.195		
				195.932	76.337	0.366		
				227.489	134.942	0.359		

When an image containing text on a wrinkled paper (Figure 5a) is tested, two thresholds (or three Gaussian distributions) give the best objective function value for Equation (10) as observed in Figure 9c. Table 5 summarizes the thresholding results, i.e., Θ(d), computational effort and Gaussian parameters, for the three meta-heuristics implemented. As for the fitting result, Figure 10 shows that even though three Gaussian distributions do not provide an exact description of the image histogram, it is good enough to recognize all the characters on the thresholded image (Figure 5b–d).

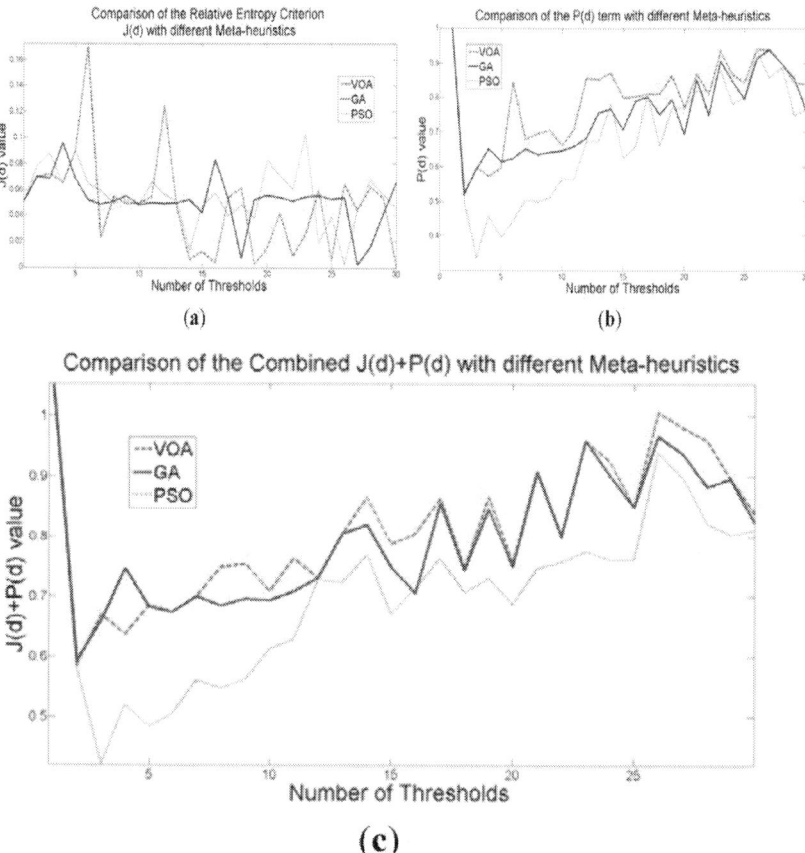

Figure 9. Behavior of (**a**) Relative Entropy function J(d), (**b**) P(d), and (**c**) Objective function Θ(d) over different numbers of thresholds with different meta-heuristics VOA, GA and PSO on test image 2.

The outstanding performance of the three meta-heuristics is observed once again when comparing with the Otsu's method (Figure 5e), where the computational effort shows the feasibility of optimizing the proposed mathematical model with heuristic optimization algorithms.

The thresholding results of the Lena image (Figure 6a) shows that four thresholds (five Gaussian distributions) have the best objective function value, which is detailed in Table 6. Visually, the thresholded images (Figure 6b–d) obtain most of the details from the original one, and in terms of objective function behavior (Figure 11) there is no need to add more distributions into the mixture model (i.e., more thresholds) because they do not provide a better objective function value.

Table 6. Thresholding results over 50 runs for the test image 3.

Algorithm	Number of Thresholds	Objective Function	Threshold Values	Means	Variances	Weights	CPU Time per Iteration (s)	Total CPU Time for the Proposed Approach (s)
VOA	1	1.043	63	48.917	51.398	0.159	0.046	1.009
				138.048	1442.095	0.841		
	2	0.575	59, 138	47.598	40.100	0.143	0.128	
				104.785	484.361	0.430		
				168.523	538.213	0.428		
	3	0.454	99, 140, 183	65.002	369.233	0.302	0.198	
				120.137	147.621	0.287		
				157.234	130.155	0.295		
				201.595	109.646	0.116		
	4	0.367	86, 126, 154, 199	57.243	192.433	0.236	0.210	
				106.550	126.938	0.236		
				139.629	67.666	0.241		
				170.694	173.733	0.218		
				208.851	42.459	0.068		
	5	0.466	68. 104, 134, 156, 190	50.380	68.188	0.175	0.428	
				88.813	108.622	0.165		
				120.054	77.656	0.203		
				145.072	39.845	0.192		
				168.597	90.400	0.167		
				204.364	77.722	0.098		
GA	1	1.044	66	49.794	60.874	0.169	0.103	0.904
				138.924	1393.456	0.831		
	2	0.606	69, 139	50.671	72.095	0.178	0.136	
				109.090	368.075	0.402		
				169.106	530.358	0.419		
	3	0.504	96, 140, 179	62.664	315.826	0.282	0.182	
				118.590	171.180	0.307		
				156.168	110.243	0.282		
				199.453	139.510	0.129		
	4	0.474	46, 99, 136, 169	40.698	14.583	0.051	0.225	
				69.954	296.623	0.251		
				118.032	124.129	0.256		
				151.053	77.603	0.267		
				192.624	235.736	0.175		
	5	0.584	65, 87, 128, 152, 205	49.504	57.550	0.166	0.258	
				76.023	40.415	0.074		
				108.258	138.292	0.250		
				139.533	49.522	0.205		
				171.707	244.613	0.257		
				211.824	28.018	0.049		
PSO	1	1.044	64	49.212	54.387	0.163	0.096	0.887
				138.353	1424.978	0.837		
	2	0.575	68, 140	50.380	68.188	0.175	0.128	
				109.403	387.984	0.414		
				169.718	522.330	0.411		
	3	0.446	97, 141, 191	63.440	333.741	0.288	0.170	
				119.639	170.199	0.309		
				159.625	168.784	0.308		
				204.790	73.728	0.095		
	4	0.337	79, 121, 148, 184	54.216	131.179	0.210	0.220	
				101.036	126.541	0.225		
				134.327	60.474	0.225		
				161.866	95.355	0.227		
				202.037	103.844	0.113		
	5	0.478	60, 115, 117, 147, 186	47.926	42.582	0.147	0.273	
				90.827	237.679	0.255		
				115.514	0.250	0.010		
				132.322	71.331	0.240		
				161.755	107.986	0.240		
				202.782	94.639	0.108		

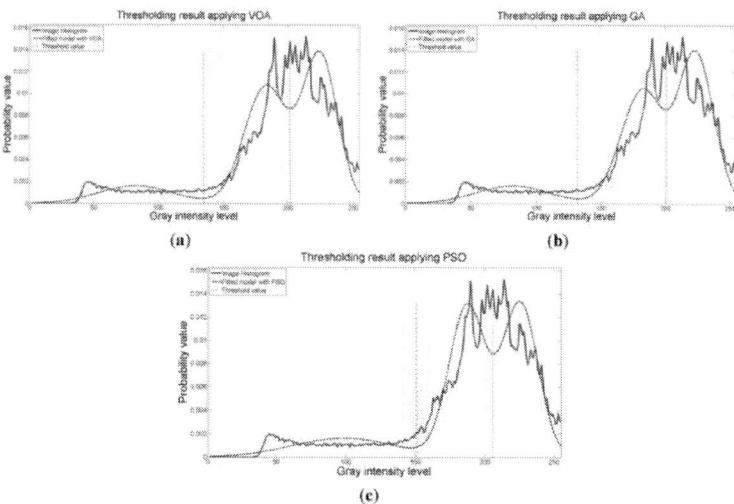

Figure 10. Fitting of the histogram of the test image 2 implementing (**a**) VOA, (**b**) GA, and (**c**) PSO.

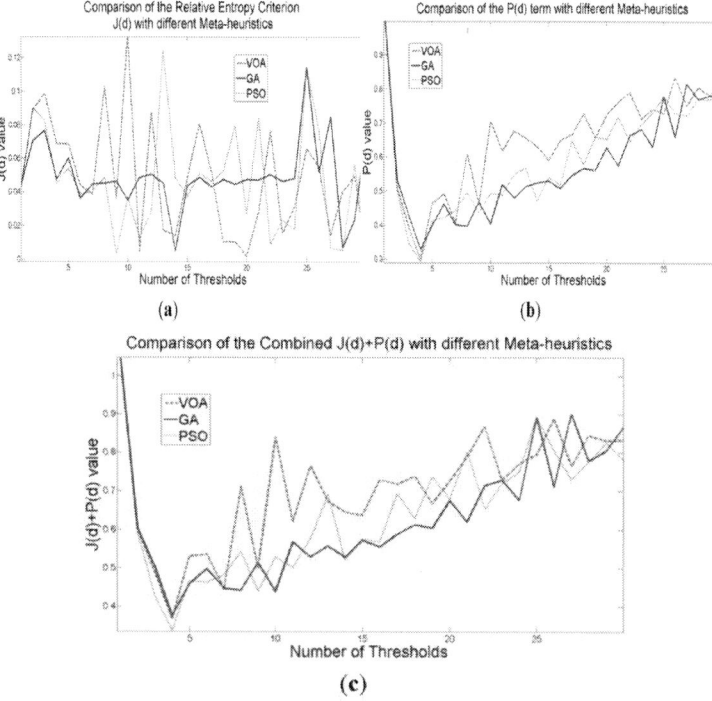

Figure 11. Behavior of (**a**) Relative Entropy function J(d), (**b**) P(d), and (**c**) Objective function Θ(d) over different numbers of thresholds with different meta-heuristics VOA, GA and PSO on test image 3.

Once again, the fitting provided by the mixture model (Figure 12) might not be the best; however, it is good enough to provide most of the details from the original image. It is interesting to observe that all the algorithmic tools are able to find satisfactory results in no more than 1.009 seconds in the case of VOA which is the slowest one, even though the algorithmic tools had to optimize Equation (10) for d = [1,2,3,4].

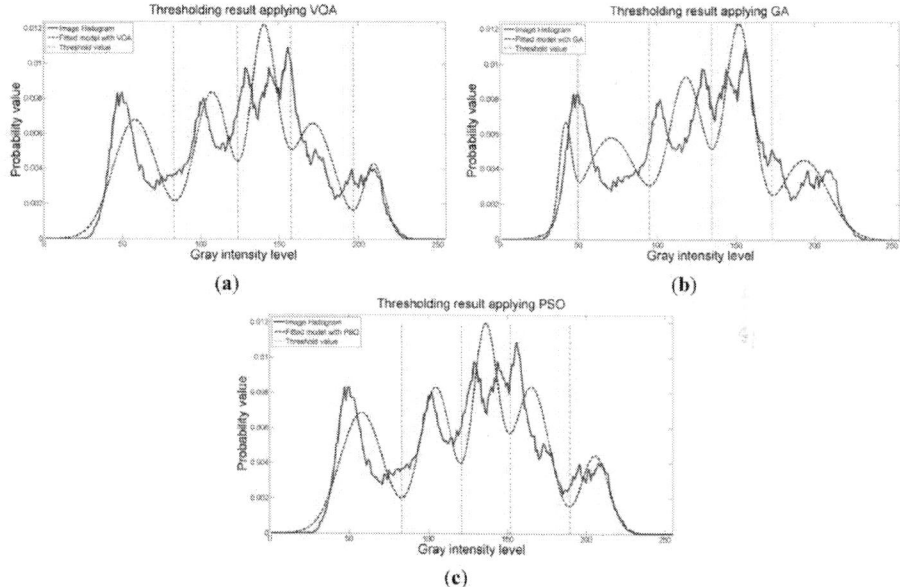

Figure 12. Fitting of the histogram of the test image 3 implementing **(a)** VOA, **(b)** GA, and **(c)** PSO.

To further test the proposed method an image with low contrast is used as illustrated in Figure 13a. It is observed that all the algorithmic tools are able to segment the image providing the correct number of thresholds which is three. Additionally, when comparing the output image given by Otsu's method and the idea proposed in this study, we are able to observe that despite its novelty the proposed method provides stable and satisfactory results for a low contrast image.

Three thresholds are suggested after the optimization of Equation (10) is performed, which is good enough for successfully segmenting the image under study (Figure 13b–d), though the fitting of the image histogram was not a perfect one (Figure 15). More importantly, the addition of more than four distributions (i.e., three thresholds) to the mathematical model Equation (10) does not achieve a better result according to Figure 14c; therefore, the power of the proposed approach is shown once again with this instance. As for the Otsu's method, even though the correct number of thresholds is provided, the low contrast causes defect in the output image (seen at the light gray region in the middle of Figure 13e.

Table 7. Thresholding result over 50 runs for the test image 4.

Algorithm	Number of Thresholds	Objective Function	Threshold Values	Means	Variances	Weights	CPU Time per Iteration (s)	Total CPU Time for the Proposed Approach (s)
VOA	1	2.112	98	102.027 174.622	504.595 36.165	0.612 0.388	0.045	0.213
	2	1.014	105, 169	91.007 146.853 176.415	2.994 157.462 0.749	0.486 0.160 0.354	0.051	
	3	0.848	101, 133, 169	90.917 109.438 150.667 176.415	1.888 46.637 17.258 0.749	0.482 0.019 0.145 0.354	0.054	
	4	0.919	103, 117, 140, 169	90.929 107.796 126.818 150.202 176.271	2.009 17.853 25.714 9.141 2.326	0.483 0.016 0.002 0.139 0.359	0.064	
GA	1	2.112	98	90.906 166.703	1.803 268.234	0.482 0.518	0.134	0.692
	2	1.051	100, 166	90.914 145.534 176.349	1.861 191.624 1.354	0.482 0.161 0.357	0.153	
	3	0.903	99, 122, 167	90.910 107.759 150.171 176.364	1.832 23.640 19.452 1.203	0.482 0.018 0.144 0.356	0.179	
	4	0.998	109, 115, 139, 177	91.270 111.849 123.655 164.654 177.000	6.829 2.575 52.305 163.335 1.455	0.494 0.003 0.005 0.323 0.176	0.225	
PSO	1	2.112	98	90.906 166.703	1.803 268.234	0.482 0.518	0.027	0.171
	2	1.005	97, 168	90.902 145.505 176.393	1.780 204.582 0.931	0.481 0.163 0.355	0.035	
	3	0.873	98, 116, 168	90.906 106.393 149.899 176.393	1.803 12.956 33.945 0.931	0.482 0.016 0.147 0.355	0.045	
	4	0.911	98, 130, 138, 169	90.906 108.777 133.110 150.704 176.415	1.803 42.833 4.122 16.707 0.749	0.482 0.019 0.001 0.144 0.354	0.065	

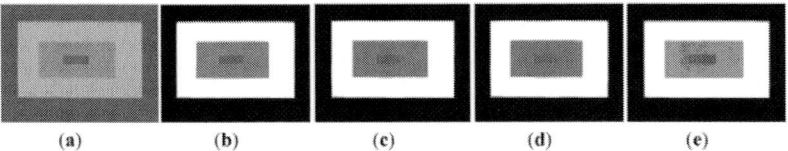

(a)   (b)   (c)   (d)   (e)

Figure 13. Test image 4: **(a)** Original image; Thresholded image implementing **(b)** VOA, **(c)** GA, **(d)** PSO, and **(e)** Otsu's method.

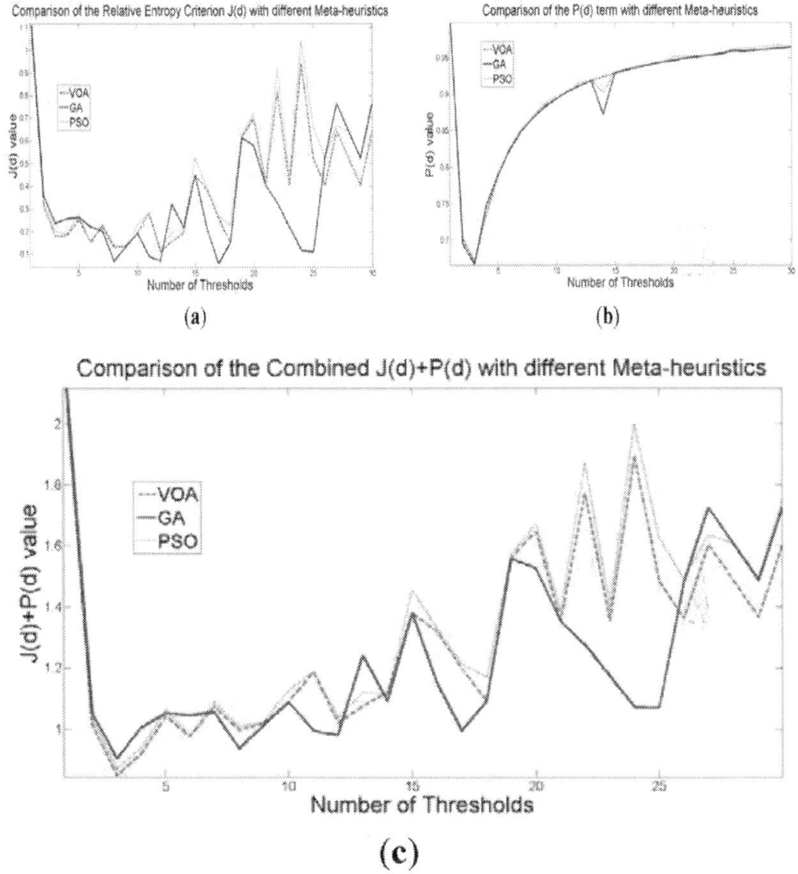

Figure 14. Behavior of **(a)** Relative Entropy function J(d), **(b)** P(d), and **(c)** Objective function Θ(d) over different numbers of thresholds with different meta-heuristics VOA, GA, and PSO on test image 4.

Table 8. Thresholding result over 50 runs for the test image 5.

Algorithm	Number of Thresholds	Objective Function	Threshold Values	Means	Variances	Weights	CPU Time per Iteration (s)	Total CPU Time for the Proposed Approach (s)
VOA	1	1.023	67	48.328	119.083	0.172	0.060	0.435
				138.301	1428.555	0.828		
	2	0.556	63, 137	46.555	100.292	0.155	0.075	
				104.962	436.410	0.421		
				168.426	564.154	0.424		
	3	0.504	88, 140, 206	57.730	278.208	0.254	0.082	
				115.780	221.287	0.346		
				165.005	334.790	0.358		
				214.735	50.956	0.042		
	4	0.481	87, 130, 151, 197	57.218	267.832	0.249	0.094	
				109.591	151.059	0.268		
				140.153	36.619	0.178		
				168.811	169.302	0.232		
				209.138	78.820	0.071		
	5	0.495	77, 116, 143, 158, 214, 255	52.347	176.710	0.208	0.124	
				97.798	120.079	0.210		
				129.636	59.464	0.208		
				149.842	18.471	0.128		
				180.649	270.308	0.226		
				220.581	34.873	0.020		
GA	1	1.024	71	49.927	139.317	0.187	0.139	1.204
				139.577	1363.879	0.813		
	2	0.669	76, 137	51.913	169.499	0.204	0.198	
				109.784	294.414	0.372		
				168.426	564.154	0.424		
	3	0.586	84, 138, 227	55.674	237.243	0.236	0.209	
				113.180	235.410	0.347		
				168.529	529.908	0.413		
				231.261	20.782	0.003		
	4	0.568	60, 103, 146, 211	45.138	87.633	0.141	0.259	
				82.720	165.607	0.195		
				125.553	152.188	0.316		
				170.644	343.107	0.320		
				218.253	40.308	0.028		
	5	0.573	57, 90, 130, 156, 212	43.635	76.301	0.126	0.400	
				72.881	99.772	0.138		
				110.800	131.973	0.254		
				142.590	55.391	0.220		
				178.408	270.837	0.237		
				219.035	38.368	0.025		
PSO	1	1.023	66	47.900	114.220	0.168	0.034	0.373
				137.948	1446.983	0.832		
	2	0.564	64, 137	47.023	104.922	0.159	0.060	
				105.419	421.751	0.416		
				168.426	564.154	0.424		
	3	0.413	97, 140, 182	63.047	390.706	0.300	0.073	
				119.448	154.217	0.300		
				157.206	130.606	0.281		
				201.012	152.773	0.119		
	4	0.410	28, 89, 133, 169	22.825	17.056	0.007	0.091	
				59.246	263.407	0.252		
				112.179	160.043	0.284		
				149.441	99.432	0.282		
				192.596	259.838	0.175		
	5	0.576	40, 70, 125, 159, 230	32.580	32.714	0.037	0.115	
				53.830	68.546	0.146		
				100.401	235.202	0.297		
				141.671	92.691	0.281		
				184.398	346.042	0.237		
				234.122	19.797	0.002		

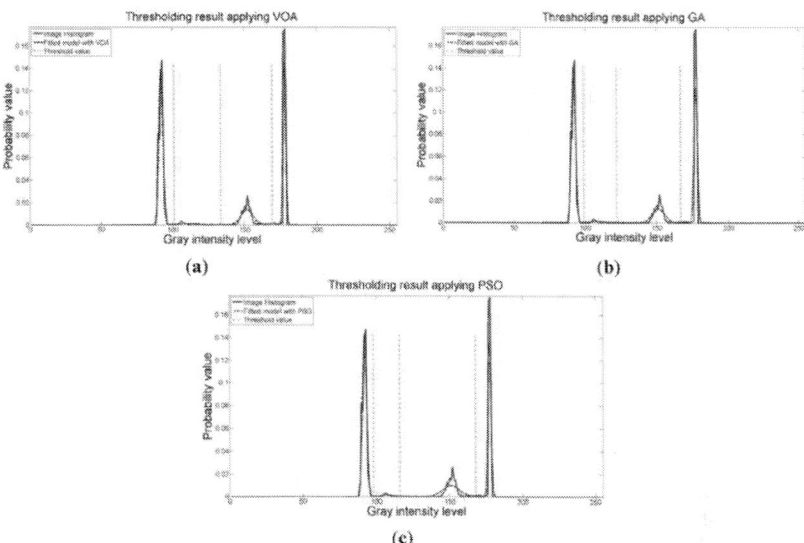

Figure 15. Fitting of the histogram of the test image 4 implementing **(a)** VOA, **(b)** GA, and **(c)** PSO.

The Lena image in which a random noise is generated will be our last test instance (Figure 16a), from this it is expected to provide clear evidence concerning robustness of the proposed method, where all the parameter values and computational results are summarized on Table 8. By observing the thresholded images when implementing the proposed approach (Figure 16b–d), we are able to conclude that random noise does not represent a major issue, even though different optimization tools are used.

The objective function behavior (Figure 17c) proved once again that when a suitable number of thresholds is achieved, the addition of more distributions into the mixture model is not necessary, since it will always achieve a larger objective function value compared with the one given by having four thresholds (or five Gaussians). Additionally, the fitting of the histogram (Figure 18) given by the image, even though is not a perfect one, is proved to be good enough to keep most of the relevant details from the original test instance.

Most of the relevant details from the original instance are kept. On the other hand, Otsu's method (Figure 16e) is not able to provide an output image as clear as the ones given by the proposed approach when implementing the meta-heuristic tools.

Figure 16. Test image 5: **(a)** Original image; Thresholded image implementing **(b)** VOA, **(c)** GA, **(d)** PSO, and **(e)** Otsu's method with 4 thresholds.

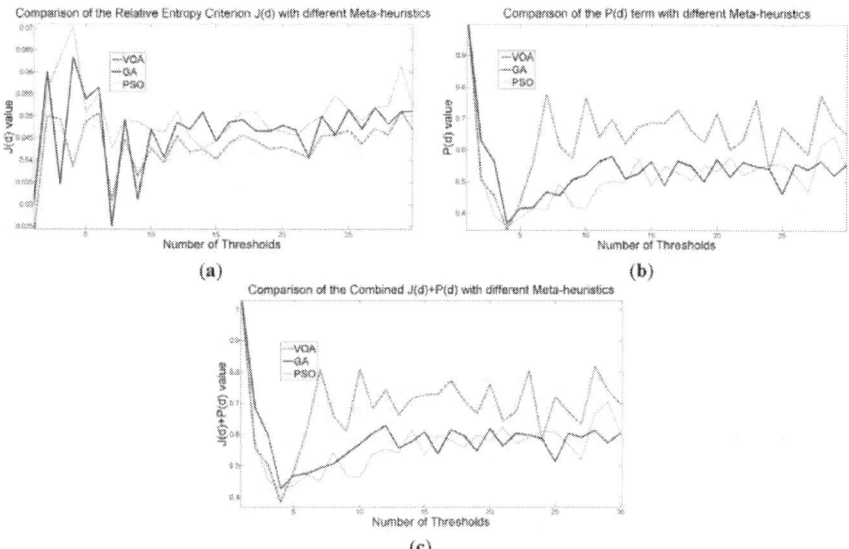

Figure 17. Behavior of **(a)** Relative Entropy function J(d), **(b)** P(d), and **(c)** Objective function $\Theta(d)$ over different numbers of thresholds with different meta-heuristics VOA, GA and PSO on test image 5.

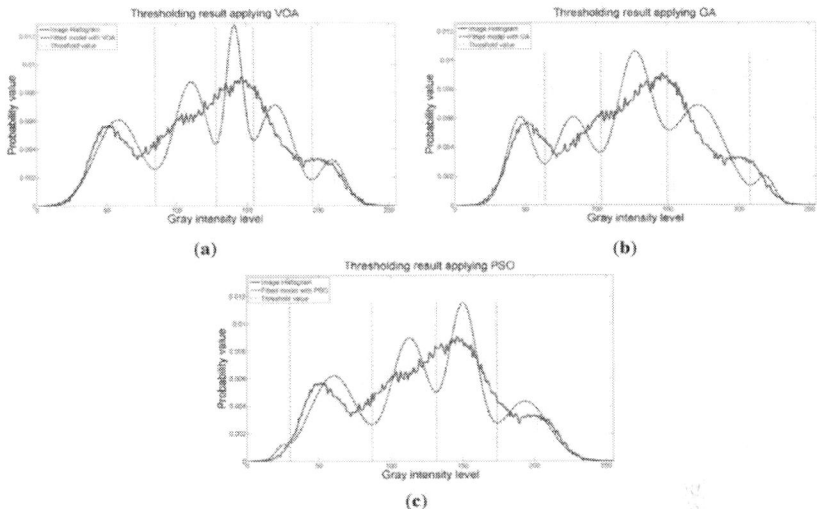

Figure 18. Fitting of the histogram of the test image 5 implementing (a) VOA, (b) GA, and (c) PSO.

4. CONCLUSIONS

In this study a new approach to automatically assess the number of components "d" in a mixture of Gaussians distributions has been introduced. The proposed method is based on the Relative Entropy Criterion (Kullback-Leibler information distance) where an additional term is added to the function and helps to determine a suitable number of distributions. Finding the appropriate number of distributions is the same as determining the number of thresholds for segmenting an image, and this study has further shown that the method proposed in [8] is powerful enough in finding a suitable number of distributions (thresholds) in a short period of time.

The novelty of the approach is that, not only an appropriate number of distributions determined by P(d) is achieved, but also a good fitting of the image histogram is obtained by the Relative Entropy function J(d). The optimization of Equation (10) was performed implementing the Virus Optimization Algorithm, Genetic Algorithm, Particle Swarm Optimization, and the output images are compared to that given by a well-known segmentation approach Otsu's method. The objective function behavior shows that the proposed model achieves a suitable number of thresholds when its minimum value is achieved, and the addition of more distributions (thresholds) into the model will cause an increasing trend of the model in Equation (10).

Comparing the proposed method with Otsu's method provided clear evidence of the effectiveness and efficiency of the approach where the algorithmic tools are used in order to reduce the computational effort when optimizing Equation (10). Additionally, the proposed method proved to reach the same value for the number of thresholds needed for the segmentation of the images tested in this study, even though different optimization algorithms were implemented.

It is worth mentioning that the proposed method proved to work remarkably well under test images with low contrast and random noise. A suitable number of thresholds and an outstanding result in the output thresholded images were obtained. Whereas for the Otsu's method, the output image showed some defects once the segmentation was performed.

The fitting result coming from the proposed approach might not be the best; however, when segmenting an image what matters the most is the fidelity in which most of the details are kept from the original picture. This is what makes the difference between a good and poor segmentation result. Future directions point toward testing the proposed method with more meta-heuristic algorithms, as well as a wider range of images to evaluate the robustness of the approach.

ACKNOWLEDGMENTS

This work was partially supported by National Science Council in Taiwan (NSC-100-2628-E-155-004-MY3).

REFERENCES

1. Shapiro, L.G.; Stockman, C. Computer Vision; Prentice Hall: Upper Saddle River, NJ, USA, 2002.
2. Jiao, L.C.; Gong, M.G.; Wang, S.; Hou, B.; Zheng, Z.; Wu, Q.D. Natural and remote sensing image segmentation using memetic computing. IEEE Comput. Intell. Mag. **2010**, 5, 78–91.
3. Zhang, H.; Fritts, J.E.; Goldman, S.A. Image segmentation evaluation: A survey of unsupervised methods. Comput. Vis. Image Underst. **2008**, 110, 260–280.
4. Ashburner, J.; Friston, K.L. Image Segmentation. In Human Brain Function, 2nd. ed.; Academic Press: Waltham, MA, USA, 2004; Chapter 35; pp. 695–706. (free version of the chapter can be found at: http://www.fil.ion.ucl.ac.uk/spm/doc/books/hbf2/pdfs/Ch5.pdf).
5. Qin, K.; Li, D.; Wu, T.; Liu, Y.C.; Chen, G.S.; Cao, B.H. A comparative study of type-2 fuzzy sets and cloud model. Int. J. Comput. Intell. Syst. **2010**, 3, 61–73.

6. Glover, F.K. Handbook of Metaheuristics; Springer: New York, NY, USA, 2003.
7. Safabakhsh, R.; Hosseini, H.S. Automatic multilevel thresholding for image segmentation by the growing time adaptive self-organizing map. IEEE Trans. Pattern Anal. Mach. Intell. **2002**, 24, 1388–1393.
8. Liang, Y.C.; Cuevas, J.R. Multilevel image thresholding using relative entropy and virus optimization algorithm. In Proceedings of the 2012 IEEE World Congress on Computational Intelligence (WCCI2012), Brisbane, Australia, 10–15 June 2012; pp. 1–8.
9. Kullback, S.; Leibler, R.A. On information and sufficiency. Ann. Math. Stat. **1951**, 22, 79–86.
10. McLachlan, G.; Peel, D. Finite Mixture Models; John Wiley & Sons: Hoboken, NJ, USA, 2001. [Google Scholar]
11. Rice, J. Mathematical Statistics and Data Analysis, 2nd ed.; Duxbury Press: Pacific Grove, CA, USA, 1995.
12. Arora, S.; Acharya, K.; Verma, A.; Panigrahi, P.K. Multilevel thresholding for image segmentation through a fast statistical recursive algorithm. Pattern Recogn. Lett. **2006**, 29, 119–125.
13. Chao, R.M.; Wu, H.C.; Chen, Z.C. Image segmentation by automatic histogram thresholding. In Proceedings of the 2nd International Conference on Interaction Sciences: Information Technology, Culture and Human, Seoul, Korea, 24–26 November 2009; pp. 136–141.
14. Hammouche, K.; Diaf, M.; Siarry, P. A comparative study of various meta-heuristic techniques applied to the multilevel thresholding problem. Eng. Appl. Artif. Intel. **2010**, 23, 676–688.
15. Yen, J.C.; Chang, F.J.; Chang, S. A new criterion for automatic multilevel thresholding. IEEE Trans. Image Process. **1995**, 4, 370–378.
16. Wang, H.J.; Cuevas, J.R.; Lai, Y.C.; Liang, Y.C. Virus Optimization Algorithm (VOA): A novel metaheuristic for solving continuous optimization problems. In Proceedings of 10th Asia Pacific Industrial Engineering & Management System Conference (APIEMS), Kitakyushu, Japan, 14–16 December 2009; pp. 2166–2174.
17. Holland, J.H. Adaptation in Neural and Artificial Systems; University of Michigan Press: Ann Arbor, MI, USA, 1975.
18. Kennedy, J.; Eberhart, R. Particle swarm optimization. In Proceedings of IEEE International Conference on Neural Networks, Perth, Australia, 27 November–1 December 1995; pp. 1942–1948.
19. Box, G.E.; Hunter, W.G.; Hunter, J.S. Statistics for Experimenters: Design, Innovation, and Discovery; John Wiley &Sons: New York, NY, USA, 2005.

20. Myers, R.H.; Montgomery, D.C.; Cook, C.M. Response Surface Methodology: Process and Product Optimization Using Designed Experiments, 3rd ed.; Wiley Series in Probability and Statistics; John Wiley & Sons: Hoboken, NJ, USA, 2009.

Experimental Matching of Instances to Heuristics for Constraint Satisfaction Problems

Jorge Humberto Moreno-Scott, José Carlos Ortiz-Bayliss, Hugo Terashima-Marín, and Santiago Enrique Conant-Pablos

National School of Engineering and Sciences, Tecnológico de Monterrey, Avenida Eugenio Garza Sada 2501 Sur, Colonia Tecnológico, 64849 Monterrey, NL, Mexico

ABSTRACT

Constraint satisfaction problems are of special interest for the artificial intelligence and operations research community due to their many applications. Although heuristics involved in solving these problems have largely been studied in the past, little is known about the relation between instances and the respective performance of the heuristics used to solve them. This paper focuses on both the exploration of the instance space to identify relations between instances and good performing heuristics and how to use such relations to improve the search. Firstly, the document describes a methodology to explore the instance space of constraint satisfaction problems and evaluate the corresponding performance of six variable ordering heuristics for such instances in order to find regions on the instance space where some heuristics outperform the others. Analyzing such regions favors the understanding of how these heuristics work and contribute to their improvement. Secondly, we use the information gathered from the first stage to predict the most suitable heuristic to use according to the features of the instance currently being solved. This approach proved to be competitive when compared against the heuristics applied in isolation on both randomly generated and structured instances of constraint satisfaction problems.

1. INTRODUCTION

Combinatorial problems are recurrent in artificial intelligence and related areas. The current literature contains a significant amount of work that has focused on designing and implementing methods that successfully solve these problems by combining the strengths of existing algorithms to improve the performance. Examples of these methods include dynamic algorithm portfolios [1–3], selection hyperheuristics [4–6], and instance specific algorithm configuration (ISAC) [7]. In general, all these methods manage a set of algorithms (solvers, heuristics, or strategies) and apply one that is suitable to the current problem state of the instance being solved. Although different names have been used in the literature, from this point on, we will refer to these methods as algorithm selectors.

Algorithm selectors relate instances to one suitable strategy to be used during the search, based on its historical performance on similar instances. These methods have proven to be reliable for solving a much wider set of instances than the algorithms they select from. These strategies keep a record of the historical performance of different algorithms on a set of solved instances in order to estimate, based on the similarity of the instances, the expected performance of such algorithms on unseen instances. To estimate how similar two instances are, the algorithm selectors compare the values of a set of features that characterize the instances. With this, algorithm selection strategies define and maintain a relation of instances to algorithms that is used to determine a suitable algorithm to be used when a new instance is presented to the system. Unfortunately, the internal representations of the relation between instances and algorithms are usually hardly interpretable by humans, making it difficult to understand how the algorithm selectors make their decisions.

In this investigation, we focus on analyzing the relation between instances and heuristics for constraint satisfaction problem (CSP), which is one of the most studied combinatorial problems in the literature. A CSP consists of a set of variables that contains the variables that need to be assigned a value from a corresponding domain , and there exists a set of constraints that restricts the values a subset of variables can simultaneously take. The importance of CSPs lies in the fact that many combinatorial problems such as scheduling [8], radio link frequency assignment [9], and microcontroller selection/pin assignment [10] can be formulated as CSPs.

CSPs are usually solved by using backtracking-based algorithms [11]. Backtracking-based algorithms explore the space of solutions by using depth-first search, where every node in the search tree represents an assignment. The process starts with an empty variable assignment that is iteratively extended until obtaining a complete assignment that satisfies all the constraints or the instance is proven to be unsatisfiable [12]. These algorithms rely on a constructive approach that takes one variable at the time and consider only one value for it. Heuristics are usually applied to decide the next variable to instantiate and which value to use. These heuristics are commonly referred to as variable and value ordering heuristics, respectively. Once a variable is assigned a value, the search evaluates whether the assignment breaks one or more constraints. If that is the case, another value must be tried for such variable. If during the search any variable runs out of values, the algorithm backtracks and assigns a new value to the variable located at the backtracking position. Thus, the algorithm tries to undo the path once a failure has been detected by going back to upper levels until it gets to a variable where it can change its value and continue the search from that point. In general, the better the decisions of the heuristics, the smaller the cost of the search. But heuristics are problem dependent and then their performance may significantly vary from one instance to another.

This study analyzes the behaviour of six variable ordering heuristics to identify the most suitable ones for specific regions of the CSP instance space and also which ones should not be used on certain areas of the space. The hypothesis is that we can use information from the individual performance of various variable ordering heuristics on a set of instances to produce easy-to-interpret rules to predict the performance of such heuristics on unseen instances. The overall goal of this investigation is to use the information collected about heuristics and their performance on various instances to produce an algorithm selector that recommends the most suitable heuristic to use according to the features of the instance at hand in order to minimize the cost of the search.

This paper provides insights into how to answer two important questions for the community: (1) given a heuristic, on which instances is it likely to perform well? (2) Given an instance, which heuristics are likely to perform well? These questions are carefully addressed by an extensive experimental setup that includes the analysis of six variable ordering heuristics on more than two hundred thousand instances. Finally, derived from this analysis, we proposed a simple but useful algorithm selector that exploits the strengths of six different heuristics to improve the search. In summary, the main contributions of this paper are as follows:(i)The analysis of the performance of six variable ordering heuristics on a large set of CSP instances by pointing out their strengths and limitations.(ii)A methodology to identify suitable and unsuitable regions of the CSP instance space for specific heuristics.(iii)Two algorithm selection strategies with an internal representation of the relation between instances and heuristics which are simple to interpret by humans. These algorithm selectors choose one suitable heuristic according to the features of the instance being solved and the information obtained from the historical performance of the heuristics on similar instances.(iv)The empirical evidence that including more than one heuristic when solving problem instances is not always beneficial for a heuristic selection strategy, as some heuristics may cancel the progress of others if used to solve the same problem at different stages of the search.

This paper is organized as follows. Section 2 introduces relevant works on the analysis of algorithms and algorithm selection through an exploration of the instance space of various combinatorial problems. A detailed description of the heuristics considered for this investigation is provided in Section 3. The analysis of the performance of variable ordering heuristics on the CSP instance space is outlined in Section 4. Section 5 describes the algorithm selection strategies proposed, the results obtained, and the discussion of these results. Finally, we present the conclusion and discuss some ideas for future work in Section 6.

2. BACKGROUND AND RELATED WORK

In general, the task of selecting the most suitable algorithm for a particular problem is referred to as the algorithm selection problem [13] and this concept has been applied to various problems in the past few years. Stützle and Fernandes [14] collected a large amount of quadratic assignment problem (QAP) instances to conduct a systematic study of the performance of some algorithms according to the features of the instances. Smith-Miles [15] proposed a framework for analyzing the performance of various algorithms for QAP instances to get insights into the relationship between instance space features and the performance of the algorithms evaluated. In a subsequent study, Smith-Miles et al. analyzed the performance of heuristics for the scheduling problem by using a decision tree [16]. To conduct the analysis, 75000 scheduling instances were generated and solved by using two common scheduling heuristics. The authors used a self-organizing map to visualize the feature space and the corresponding performance of the heuristics, in order to get insights into the heuristic performance. More recently, Smith-Miles et al. compared the strengths and weaknesses of different optimization algorithms on a broad range of graph coloring instances [17].

Bischl et al. tackled the algorithm selection problem as a classification task based on an exploratory landscape analysis [18]. The authors used systematic sampling of the instances to collect a set of features and used those features to predict a well-performing algorithm (in terms of expected runtime) out of a given portfolio. One-sided support vector regression was used to solve the resulting learning problem. López-Camacho et al. [19] applied principal component analysis as a knowledge discovery method to gain understanding on the structure of bin packing problems and how it relates to the performance of various heuristics for this problem.

With regard to CSPs, Tsang and Kwan [20] introduced the idea of systematically relating instances to suitable algorithms, based on the features of those instances. In that study, the authors presented a survey of algorithms for solving CSPs and established the first ideas that suggested that it was possible to relate the formulation of a CSP to one adequate solving algorithm for that formulation. This idea supports more recent algorithm selection approaches, like the ones described in the following lines. Ortiz-Bayliss et al. [21] studied the performance of two variable ordering heuristics on a large set of CSP instances. In their analysis, the authors found preliminary evidence that supports the idea that some heuristics for CSP can indeed be used in collaborative fashion to improve the search.

A preliminary idea of this investigation was presented by Moreno-Scott et al. [22], where three heuristics were analyzed at a much smaller detail than the one presented in this document. This investigation extends the previous study by including three more variable ordering heuristics into the analysis, formalizing the instance space characterization to help us identify regions of difficult and easy instances, and describing a simple but useful way to use the information from the analysis to predict a suitable heuristic to improve the search.

There is also a growing interest in the generation of particularly difficult or easy instances for testing algorithms, in order to understand when they are preferable to others. Usually, the generation of such instances is done by using evolutionary computation. The idea is to construct generation models that provide a more direct method for studying the relative strengths and weaknesses of each algorithm. Smith-Miles et al. proposed the use of an evolutionary algorithm to produce distinct classes of traveling salesman problem (TSP) instances that are intentionally easy or hard for certain algorithms [23, 24]. In their analysis, a comprehensive set of features is used to characterize the instances. By using the information gathered from the performance of these algorithms on the set of instances, the authors proposed a prediction algorithm that presents high accuracy on unseen instances for predicting search effort as well as identifying the algorithm likely to perform best.

For CSPs, van Hemert [25, 26] proposed a genetic algorithm to produce instances that are difficult to solve. van Hemert's model maintains a population of binary CSPs of which it changes the structure over time. Its genetic operators modify the conflicts between the pairs of variables. Under this approach, the set of variables and their domains are kept unchanged during the whole process. Thus, only the ratio of forbidden pairs of values can vary as a result of the evolutionary process. As part of the generation process, the algorithm requires solving the instances to evaluate their fitness. Moreno-Scott et al. [22] used van Hemert's model to generate extremely hard instances for specific variable ordering heuristics. Among their main findings, the authors confirmed that instances that are hard to solve for some heuristics may not be hard for others.

3. VARIABLE ORDERING HEURISTICS

Six dynamic variable ordering heuristics were considered for this investigation due to their performance in previous studies [5, 27–29]. Each heuristic assigns a score to the variables in the instance being solved, based on a specific criterion as the search progresses. According to its particular strategy, every time a variable is to be selected for instantiation the heuristic sorts the variables by their score in ascending or descending order, and the first variable in the sorted list is selected for instantiation. In all cases, ties among variables are broken by using the lexical order of the names of the variables. Regarding the order in which the values of the selected variables are tried, values are always tried in the default order in which they appear in the domain of the variable to instantiate.

Because instantiating a variable changes the problem state (as domains and constraints are updated), the scores given by any of the heuristics to the remaining variables are likely to change at different stages of the search. For this reason, the ordering of the variables is dynamic, deciding which variable to instantiate considering the current problem state and the current scores given by a particular heuristic.

I.   The following lines describe the variable ordering heuristics used in this work: (i) Minimum Domain (DOM). DOM [30] instantiates first the variable with the fewest values in its domain. Then, DOM selects the variable that minimizes among all the variables, where is the current domain size of variable.

II. (ii)Maximum Degree (DEG). This heuristic considers the degree of the variables to decide which one to instantiate before the others. The degree of a variable is calculated as the current number of constraints where the variable is involved. Thus, DEG instantiates first the variable with the largest [31].

III. (iii)Minimum Domain over Maximum Degree (DOMDEG). DOMDEG tries first the variable that minimizes the quotient among the remaining variables in the instance [32].

IV. (iv)Minimum Solution Density (RHO). This heuristic is based on the approximated calculation of the solution density of the CSP instance, [5, 28]. Let indicate the constraints in which variable is involved. Then, RHO will instantiate first the variable that minimizes $P_v = \prod_{c \in C_v}(1 - p_c)$, where is the current fraction of forbidden pairs of values among constraint .

V.   (v)Minimum Expected Solutions (SOL). SOL instantiates the variables in such a way that the resulting subproblem contains the maximum number of expected solutions [5, 28]. To do so, the search branches on the variable that minimizes.

VI.   (vi)Maximum Conflicts (MXC). This heuristic prefers the variable that maximizes the number of conflicts where it is currently involved [33]. A conflict represents a pair of values that is not allowed for two variables at the same time. A constraint between two variables and may contain zero or more conflicts (up to). The larger the number of conflicts in a constraint is, the more difficult it is to satisfy. MXC will select first the variable that maximizes.

To help clarify how these heuristics make their decisions, a simple CSP instance is depicted in Figure 1 and analyzed by using each heuristic. Table 1 presents the scores given to the variables according to each heuristic. As the reader may observe, there are cases where different heuristics select the same variable. In this example, DOMDEG, RHO, and SOL will instantiate first, while DOM, DEG, and MXC will prefer , , and , respectively.

Table 1: Scores given to the variables in the CSP instance depicted in Figure 1 by each one of the six heuristics. Values in bold indicate the scores preferred by each heuristic and ties are broken by using the lexical ordering on the names of the variables.

	DOM d_v	DEG $\deg_v$	DOMDEG $d_v/\deg_v$	RHO ρ_v	SOL sol_v	MXC cf_v
v_0	2	1	2	0.5	1	3
v_1	3	2	1.5	0.1296	0.3888	7
v_2	3	2	1.5	0.2962	0.8886	6
v_3	2	2	1	**0.1111**	**0.2222**	6
v_4	3	2	1.5	0.2469	0.7407	9
v_5	3	3	1	0.2160	0.6480	9
Selected	v_0	v_5	v_3	v_3	v_3	v_4

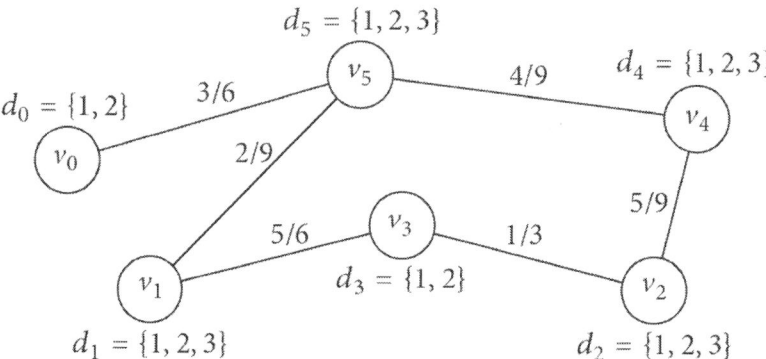

Figure 1: A CSP instance with six variables and domains of two and three values. Nodes represent variables and edges represent constraints. The values on edges indicate the fraction of forbidden pairs of values per constraint.

4. AN EXPERIMENTAL EVALUATION OF VARIABLE ORDERING HEURISTICS

In this section, we explored the CSP instance space by generating and solving a vast set of randomly generated instances. The instances were solved by using a backtracking-based algorithm where the ordering of the variables was defined dynamically by using the heuristics described in Section 3. The CSP solver used in all the experiments in this investigation was fully implemented in Java. To speed up the search, the solver incorporates backjumping [34] and constraint propagation [35]. All the experiments were conducted on an AMD 4.0 GHz 8-Core Windows machine with 32 GB of memory.

In this work, we have focused exclusively on binary CSP instances with constraints defined in extension (the forbidden pairs of values in the constraints are explicitly listed). Although the ideal case would be to support instances with constraints of -arity (for), at this point, it is difficult to define features to characterize such instances. Every instance in this investigation is characterized by using two well-studied binary CSP features: the constraint density () and the constraint tightness (). The constraint density of a CSP instance is calculated as

$$p_1 = \frac{2\,|C|}{|V|\,(|V| - 1)},$$

(1)

where and represent the number of constraints and variables in the instance, respectively. The constraint tightness of an instance is calculated as

$$p_2 = \frac{1}{|C|} \sum_{c \in C} p_c,$$

(2)

where is the fraction of forbidden pairs of values in constraint . For example, the constraint density and tightness of the example CSP shown in Figure 1 are 0.4 and 0.4815, respectively. Although other features to describe binary CSPs are available in the literature, we have decided to work with and because they have been widely used in the past and proven to provide an easy-to-interpret description of the instances.

Two main sources of CSP instances exist: randomly generated instances and structured ones from real-world applications. Structured instances from real-world applications are usually the best source but unfortunately are usually short in supply. Thus, instance generators provide a good additional source for testing algorithms. These generators have the advantage of providing a precise control over the problem features, such as size and expected hardness [36, 37], and facilitate the systematic analysis of algorithms [38].

In this investigation, we systematically explored the CSP instance space by using the instance generation model B [39] and analyzed the performance of the six heuristics on the instances produced. Model B divides the generation process into two stages (see Algorithm 1): the generation of the constraints between the variables and the generation of the forbidden pairs of values among the variables linked by a constraint. Although other generation models exist [25, 40–42], we have opted for model B for its simplicity and accuracy in producing instances with the precise values of density and tightness we required for detailed sampling of the CSP instance space.

We produced a grid of 50 × 50 sampling points distributed on the instance space . The points in this grid are separated by steps of 0.02 in each axis, resulting in a grid of 2500 uniformly distributed sampling points. For each point in the grid, we generated 50 instances of 20 variables and 10 values in their domains, with the values of and of the corresponding sampling point.

Algorithm 1: Instance generation model B.

```
(1) procedure MODELB(|V|, d, p₁, p₂)
(2)    C ← []
(3)    G ← GENERATECONSTRAINTGRAPH(|V|, p₁);
(4)    for each edge in G do
(5)        C ← C ∪ GENERATECONFLICTS(edge.from, edge.to, d, p₂)
(6)    end for
(7) end procedure
(8) procedure GENERATECONSTRAINTGRAPH(|V|, p₁)
(9)    G ← ∅
(10)    for i = 0 to n do
(11)       for j = i + 1 to |V| do
(12)          G ← G ∪ CREATEEDGE(from = i, to = j)
(13)       end for
(14)    end for
```

$$(15) \quad G \leftarrow \text{RANDOMSUBSET}\left(G, p_1 \times \frac{|V|(|V|-1)}{2}\right)$$

```
(16) end procedure
(17) procedure GENERATECONFLICTS(d, p₂)
(18)    S ← ∅
(19)    for each x in d do
(20)       for each y in d do
(21)          S ← S ∪ (x, y)
(22)       end for
(23)    end for
(24)    S ← RANDOMSUBSET(S, p₂ × d²)
(25) end procedure
```

Once the grid was produced, we used the variable ordering heuristics described in Section 3 to solve all the instances. For the purpose of this investigation, an instance is considered solved when the first solution is found or the instance is proven unsatisfiable. Because of this, any solution is equally desirable. To estimate the cost of the search, we used the number of consistency checks executed during the search. Every time a constraint was revised, the counter for consistency checks increased by one. Thus, the more the revisions of the constraints, the higher the cost of the search. In total, 125,000 instances were generated to construct the grid of instances and each one of these instances was solved six times, one time per heuristic analyzed. At this point, it is important to stress the difference between the average cost per sampling point and cost of solving a specific instance. The cost of solving an instance is the number of consistency checks required to solve an instance by using one particular heuristic. Consequently, the cost of a sampling point is the average cost of one particular heuristic on the 50 instances generated for such sampling point.

Figure 2 shows the landscape of the instance space according to the average consistency checks required by SOL to solve the 50 instances in every sampling point in the grid. The surfaces obtained for the rest of the heuristics are very similar in shape to Figure 2, but with differences in the number of consistency checks required to solve the instances. Independently of the heuristic used to solve the instances, a phase transition phenomenon is observed [43–45]. This phase transition region is where instances abruptly stop being satisfiable and contains, in average, the most difficult to solve instances, regardless of the heuristic used. Figure 2 offers a view of the instance space that illustrates the phase transition region for the grid of instances produced. In practice, the phase transition region is depicted as a narrow strip in which the maximum uncertainty about the satisfiability of the instances is reached (Figure 3).

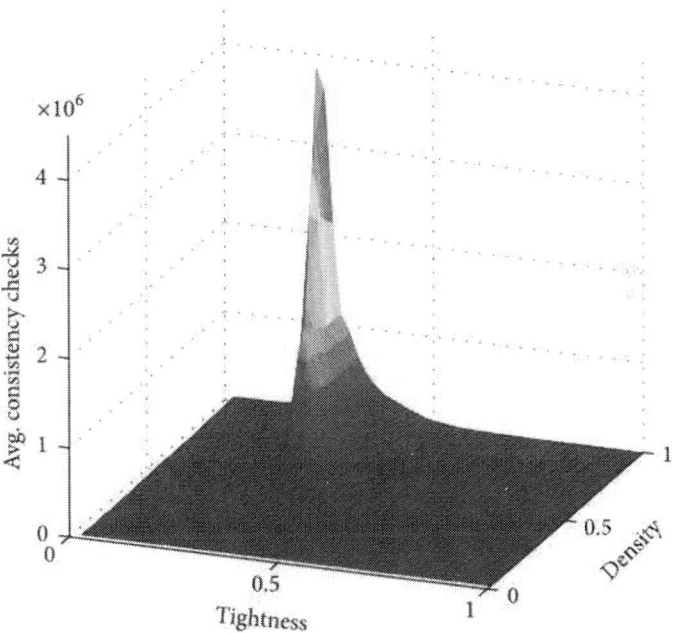

Figure 2: Average consistency checks per sampling point in the grid required by SOL.

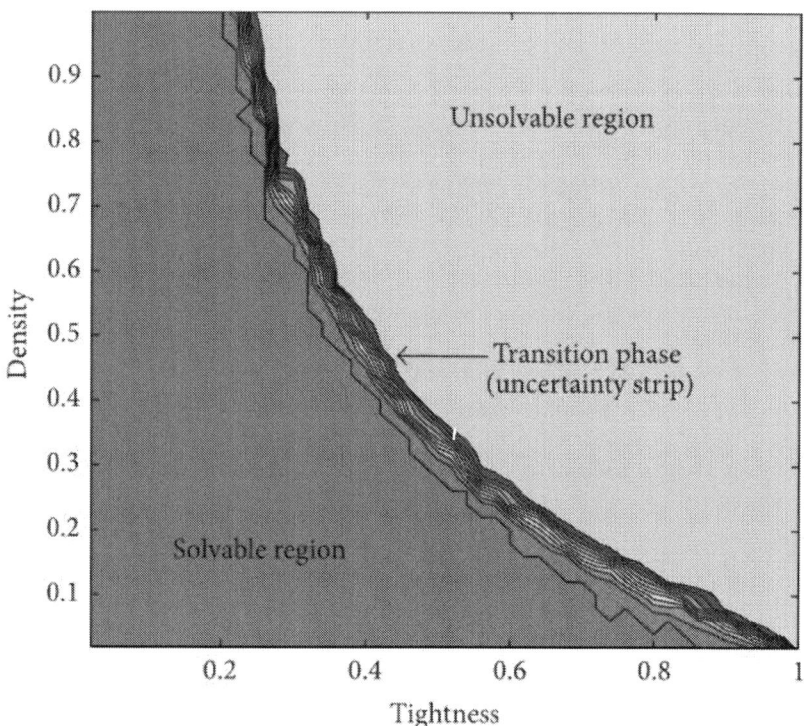

Figure 3: Phase transition for the grid of instances with 20 variables and 10 values per domain.

Table 2 identifies the values of and of the sampling points for which each of the heuristics required the maximum average consistency checks. Unsurprisingly, the highest average costs of the six heuristics are located at the same point (and). These values of and correspond to the peak in average consistency checks on the phase transition region shown in Figure 3. This observation is not new and only confirms what we already know about the phase transition: in average, it contains the hardest-to-solve instances for any backtracking-based method. But there is more to learn from these results. For example, there is a large difference between the consistency checks required by each heuristic for this particularly hard sampling point. Despite this point being difficult for all the heuristics, one of them is better than the others, showing that there is something that can be learnt to improve the search (even for points in the most difficult to solve region).

Table 2: Values of and of the points in the grid that, in average, required the largest number of consistency checks for each specific heuristic.

Heuristic	p_1	p_2	Consistency checks (millions)		
			(Avg.)	**(Max.)**	**(Min.)**
SOL	1.00	0.22	3.87	8.44	0.13
DEG	1.00	0.22	11.17	25.89	0.26
DOM	1.00	0.22	4.76	10.33	0.16
MXC	1.00	0.22	16.14	37.27	0.04
RHO	1.00	0.22	4.86	9.99	0.20
DOMDEG	1.00	0.22	11.17	25.89	0.26

In this sampling point (and in general for all the points in the phase transition region), SOL proved to be a very competent heuristic. The average cost per point on the phase transition region is usually smaller for SOL than for the other heuristics. This means that for the 50 instances in each point on the hardest-to-solve region SOL required, in average, the fewest consistency checks to solve those specific instances. The fact that SOL was the best average heuristic for the specific region of the instance space where the most difficult instances take place must not be interpreted as a proof that SOL is the best absolute heuristic over the whole space. As we will show in Section 4.1, heuristics are specialists for specific regions in the instance space and outside those regions their performance decreases significantly.

Table 2 also includes the maximum and minimum consistency checks required by each heuristic in the point (,). It is interesting to observe that the difference between the hardest and easiest instance per heuristic in such point is huge. The former indicates that it is possible to find a mixture of hard and easy instances, even for points located inside the phase transition region.

4.1. Matching Instances to Variable Ordering Heuristics

We collected all the information about the average performance of each of the six heuristics on every point in the grid. Then, we compared their results to produce mapping between regions of the instance space and the expected performance of the heuristics. This information is summarized graphically in Figure 4(a), where the best performer per point in the grid among the six heuristics is shown. The smaller the average cost per point, the better the performance of the heuristic. Although there is no dominant heuristic for all the instances in the grid, some regions seem more suitable for certain heuristics than for others (a heuristic is said to dominate heuristic on point if the average cost of is lower than the average cost of for point). The regions of dominance seem to somehow follow the shape of the phase transition curve. It is interesting to observe that SOL clearly dominates the other heuristics on the phase transition region. But this situation dramatically changes as we move away from this region. As we approach the region where all the instances are satisfiable (on the left side of the phase transition region), some strips of dominance become visible: regions where RHO and MXC obtain the best performance (from right to left). On the right side of the phase transition, where most of the instances are unsatisfiable, there is small dominance of DOM and DOMDEG, which indicates that these heuristics are useful when the instances are unsatisfiable. However, in most of the unsatisfiable region of the instance space, two or more heuristics obtain the best performance (there is no single dominant heuristic). This is illustrated in Figure 4(a) with the label ND for "no dominance." Similarly, on the bottom left corner of the graph, where the problems are easy and satisfiable, there is no dominance of one heuristic over the others because most of the instances are trivially solved. Note that heuristic DEG does not appear in the figure, since it is never the dominant heuristic for any of the points in the grid.

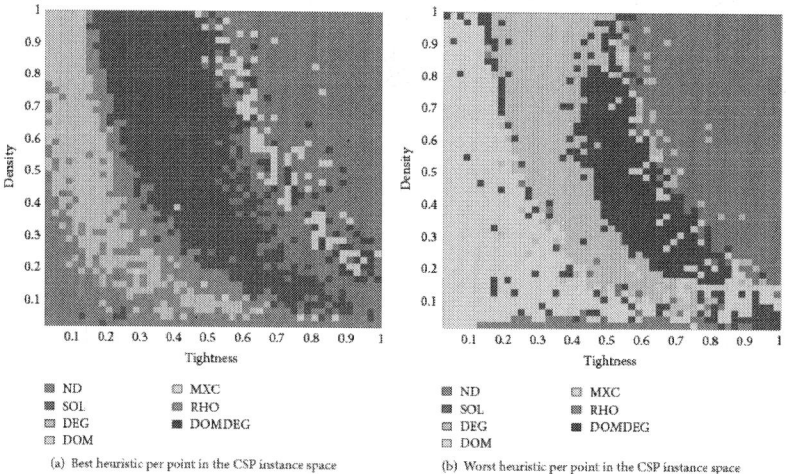

(a) Best heuristic per point in the CSP instance space

(b) Worst heuristic per point in the CSP instance space

Figure 4: Best and worst heuristics in the CSP instance space.

We have identified the most suitable heuristics for specific regions of the instance space, but it is also relevant to identify the regions where it may not be wise to apply a particular heuristic. For this reason, Figure 4(b) shows a similar analysis to the previous one but focused on the worst heuristic per point. As in the previous case, some patterns of dominance are clearly visible. At the region where the problems are satisfiable, DOM is the worst heuristic, as it is dominated by the other five heuristics. This result makes sense, as DOM proved to be a specialist only for some unsatisfiable instances (see Figure 4(a)). MXC is the worst choice for points close to the phase transition region. There is a narrow strip that follows the phase transition curve, on the unsatisfiable region where DOMDEG presents its worst performance. Finally, on the right side of the space where the instances are unsatisfiable, no heuristic is always dominated by the others. This means that at least two heuristics showed the same poor performance.

We have obtained information about the average performance of the heuristics on the instance space. This allowed us to identify regions where certain heuristics should be used and which ones should not. But what is the performance of the heuristics on the 50 instances of each particular point in the grid? For example, Table 3 presents the results obtained by each of the six heuristics on the 50 instances contained in the point (,), where the highest average cost in the grid was achieved. The best average heuristic for this point, SOL, obtained the lowest cost in 42 instances. MXC, as expected, obtained the worst costs in 46 instances. If we consider only the average performance of these heuristics, it might seem obvious that MXC should never be used for instances with these values of and . However, MXC obtained the best result in one isolated case (instance 27) with an impressive performance three times better than SOL. This indicates that even though we can estimate the average performance of the heuristics on the instances based on their values of and , there is no guarantee that we will always select the best option for all instances as there are cases where an apparently suboptimal performer may be a really useful solving option. An important consequence of this result is that although we can use and to properly estimate the average expected performance of the heuristics, more features are needed to fully distinguish extremely strange cases where one usually bad heuristic should be preferred. Including additional features represents an important step towards improving the mapping between instances and heuristics in the future.

Table 3: Consistency checks required by each heuristic at point $(p_1 = 1.0, p_2 = 0.22)$. The best and worst result per instance are indicated with symbols and respectively.

Instance ID	Consistency checks (millions)					
	SOL	DEG	DOM	MXC	RHO	DOMDEG
0	↓ 2.41	6.02	2.61	↑ 6.10	3.40	6.02
1	↓ 6.91	17.65	8.23	↑ 27.99	9.43	17.65
2	↓ 7.06	23.22	8.07	↑ 26.42	8.54	23.22
3	↓ 6.98	19.95	9.21	↑ 30.94	9.50	19.95
4	↓ 4.64	15.14	6.02	↑ 16.80	7.14	15.14
5	↓ 6.91	18.43	8.56	↑ 30.52	9.03	18.43
6	↓ 1.14	↑ 5.09	2.33	4.17	1.22	↑ 5.09
7	↓ 7.56	24.24	9.43	↑ 37.27	9.42	24.24
8	1.64	↑ 3.96	↓ 1.45	3.76	1.93	↑ 3.96
9	↓ 2.05	4.62	2.24	↑ 5.49	2.63	4.62
10	↓ 6.64	19.08	8.16	↑ 29.92	8.27	19.08
11	↓ 6.19	15.26	7.70	↑ 25.76	7.68	15.26
12	↓ 6.06	19.97	10.33	↑ 31.11	9.13	19.97
13	↓ 7.11	25.89	8.62	↑ 32.52	8.77	25.89
14	↓ 1.44	3.91	1.93	↑ 7.25	1.57	3.91
15	1.68	3.30	1.63	↑ 4.70	↓ 1.41	3.30
16	5.89	13.46	6.83	↑ 22.24	↓ 5.78	13.46
17	↓ 1.40	3.64	1.71	↑ 5.02	1.64	3.64
18	↓ 1.88	8.57	2.81	↑ 8.64	2.41	8.57
19	↓ 4.98	↑ 17.18	7.27	16.75	5.89	↑ 17.18
20	↓ 4.46	10.48	5.58	↑ 16.08	5.17	10.48
21	↓ 1.07	2.52	1.35	↑ 2.93	1.23	2.52
22	↓ 7.34	20.13	8.88	↑ 29.32	8.44	20.13
23	↓ 2.12	6.46	2.65	↑ 9.77	2.58	6.46
24	↓ 5.76	14.57	7.01	↑ 20.10	6.91	14.57
25	↓ 1.26	3.99	1.55	↑ 4.22	1.93	3.99
26	↓ 0.48	2.42	0.62	↑ 5.21	0.74	2.42
27	0.13	↑ 0.26	0.16	↓ 0.04	0.20	↑ 0.26
28	↓ 8.44	20.38	10.03	↑ 27.53	8.50	20.38
29	↓ 7.69	24.05	8.52	↑ 31.94	9.99	24.05
30	2.32	4.95	↓ 2.14	↑ 7.97	2.90	4.95
31	↓ 3.60	9.53	4.45	↑ 14.10	4.92	9.53

32	1.48	2.51	1.67	↑ 4.93	↓ 1.36	2.51
33	↓ 3.76	11.13	5.86	↑ 18.12	4.69	11.13
34	↓ 1.62	4.38	1.80	↑ 5.52	1.83	4.38
35	↓ 3.35	8.65	3.65	↑ 15.53	4.33	8.65
36	↓ 7.18	20.15	7.84	↑ 29.46	8.97	20.15
37	↓ 5.45	15.53	6.42	↑ 25.83	6.85	15.53
38	1.68	5.00	2.26	↑ 7.05	↓ 1.64	5.00
39	↓ 1.99	4.26	2.77	↑ 8.67	2.80	4.26
40	↓ 6.27	20.40	7.51	↑ 24.96	7.28	20.40
41	↓ 2.23	5.92	2.72	↑ 11.06	3.18	5.92
42	↓ 0.34	1.18	0.36	↑ 1.79	0.82	1.18
43	↓ 5.85	17.98	6.85	↑ 30.57	8.08	17.98
44	↓ 1.15	2.13	1.39	↑ 4.25	1.22	2.13
45	↓ 4.83	15.43	6.48	↑ 24.17	6.64	15.43
46	↓ 1.25	4.46	1.41	↑ 5.51	1.55	4.46
47	↓ 6.94	21.70	8.35	↑ 32.63	9.01	21.70
48	1.93	4.36	↓ 1.65	↑ 8.87	2.40	4.36
49	↓ 0.73	4.96	1.14	↑ 5.35	1.84	4.96

5. USING THE MATCHING OF INSTANCES TO HEURISTICS TO IMPROVE THE SEARCH: A HEURISTIC SELECTION APPROACH

In previous sections, we described the mapping of instances to heuristics obtained by the systematic exploration of the instance space. We have found that some heuristics are, in average, better than others for certain regions of the instance space. In this part of the investigation, we were interested in using the information gathered from the previous experiments to see whether it is possible to use such information to improve the search when tested on a wider set of instances, on both randomly generated and structured ones. The assumption is that, by selecting the right heuristic according to the initial problem state, we can reduce the cost of the search and show a better performance than with the variable ordering heuristics applied in the traditional fashion.

We proposed two heuristic selection strategies for this experiment. These strategies consider the current conditions of the problem, described exclusively by the values of and of the instance to solve, and apply the best heuristic for those conditions based on the patterns depicted in Figure 4(a). The heuristic selection strategies consist of a grid of easy-to-interpret rules, one rule per cell of the mapped instance space in the form (,) heuristic. The rule with the smallest Euclidean distance from its condition to the current values of and of the instance to be solved is the one that fires and determines the heuristic to be applied. In other words, the values of and of the instance to solve are used to place the instance on the grid shown in Figure 4(a) and the best average heuristic for that point is used. The first heuristic selector (SHS) is static and it only evaluates the state of the instances at the beginning of the search. This results in a selection strategy that only makes one decision per instance and once the decision is made, the same heuristic is used to solve the whole instance, from the beginning to the end. The second heuristic selector (DHS) is dynamic. DHS evaluates the problem state every time a variable is to be assigned and then different heuristics are applied as part of the solving process. Although both heuristic selectors use the same information to make their decisions, they use such information differently. Two consequences of the difference in the use of the information are observable. First, the time for making the decisions is slightly shorter for SHS, as it only evaluates the problem state once. In this work, the additional time required by DHS to compute the problem state was not an issue, as the set of features to characterize the instances is small and the features are easy to compute. But in other cases, it may represent significant delays if more hard-to-compute features are considered. The second aspect to consider is the interaction of heuristics during the search. While some heuristics may contribute to others and improve the search, there is also a risk that some heuristics cancel the effect of others, taking the search into unpromising areas of exploration. Thus, we are likely to observe higher variance in the results if DHS is used.

We do not claim that the heuristic selection strategies described in this document are the best models among the heuristic selection methods described in the literature. In fact, a comparison between models would be a good idea for future work. At this point, our only idea is to show that there is actually a practical use of the information obtained from the analysis of the relation between instances and the performance of heuristics to solve unseen CSP instances.

5.1. TESTING THE HEURISTIC SELECTORS ON RANDOMLY GENERATED INSTANCES

The heuristic selectors described before were tested on three additional grids of instances generated exclusively for testing purposes. Each one of these new grids contains 2500 sampling points (as the one used for exploring the performance of the heuristics) with 10 instances per point. The instances in these grids contain 20 variables and 10 values in their domains. In total, 75,000 additional instances were generated for the testing phase. The analysis of the heuristic selector is based on two criteria: the percentage of points where the heuristic selector is better than each heuristic among all the points per grid and the actual reduction in consistency checks obtained by using the heuristic selector with respect to the heuristics among all the instances in each test grid. The first criterion estimates how stable the heuristic selector is, while the second one provides an idea of the benefit of using this method over the single heuristics.

Tables 4 and 5 present the percentage of sampling points in the instance space in which the heuristic selectors SHS and DHS performed better than each particular heuristic. Except for DOM (which was the worst average performer for the test grids), SHS always obtained better results than DHS in the head-to-head comparison against the heuristics. Regardless of the differences between SHS and DHS, it is interesting to note the high percentage of instances where the heuristic selection strategies dominate SOL (which was the best performer for the instances within the phase transition region). Although SOL is a good heuristic for hard-to-solve instances, the percentage of instances where the selectors are better than SOL is larger than the percentage obtained for the other heuristics. The reason for this is that SOL is not a competent heuristic for solving not-so-hard instances (which cover around 60% of the instance space).

Table 4: Head-to-head comparison of SHS and each heuristic. The results indicate the percentage of points per test grid where SHS dominated each particular heuristic.

Grid	SOL	DEG	DOM	MXC	RHO	DOMDEG
Test grid I	84.72%	68.60%	67.48%	68.64%	71.00%	68.12%
Test grid II	84.08%	68.92%	67.64%	68.76%	70.08%	68.08%
Test grid III	83.40%	68.68%	66.80%	69.32%	69.80%	67.64%

Table 5: Head-to-head comparison of DHS and each heuristic. The results indicate the percentage of points per test grid where DHS dominated each particular heuristic.

Grid	SOL	DEG	DOM	MXC	RHO	DOMDEG
Test grid I	66.48%	64.84%	59.32%	71.48%	61.15%	60.8%
Test grid II	65.80%	64.84%	59.64%	70.64%	60.56%	60.8%
Test grid III	66.40%	63.84%	58.32%	71.08%	60.92%	59.72%

Tables 6 and 7 complement the information about the performance of SHS and DHS. In these tables, we indicate the percentage of consistency checks saved by using SHS and DHS with respect to the heuristics applied in isolation. The reason for this comparison is that it is not enough to know that the heuristic selectors reduced the number of consistency checks required by a specific variable ordering heuristic; we are also interested in knowing how many consistency checks we can save by using these strategies. From Tables 6 and 7, we can clearly observe what we discussed before about SOL and its performance inside the phase transition region. SOL is a very reliable heuristic for the region where the hardest instances occur and as a consequence, the differences between the best and worst performer are huge, stressing the consequences of bad choices. Outside that region, the number of consistency checks is significantly lower for all the heuristics, resulting in smaller differences between the best and the worst performer (which reduces the impact of bad choices). Although SOL is a specialist for the region where the phase transition occurs, SHS is able to reduce the number of consistency checks of this heuristic in more than 1.5% for all the test instances. The performance of SHS, when compared to the other heuristics, is outstanding, but it is mainly because of the good performance of this selector on the hardest-to-solve instances. On the other hand, the dynamic selection of heuristics by DHS was not as effective as the strategy of SHS. Although DHS is able to obtain better results than SOL in around 65% of the test instances, the cost of using DHS to solve all the instances increases the number of consistency checks with respect to SOL in around 9.16% consistency checks.

Table 6: Head-to-head comparison of SHS and each heuristic. The results indicate the percentage of saved consistency checks per test grid by using SHS with respect to each particular heuristic.

Grid	SOL	DEG	DOM	MXC	RHO	DOMDEG
Test grid I	1.89%	70.34%	42.36%	60.29%	14.23%	56.87%
Test grid II	1.79%	64.80%	35.90%	59.38%	13.56%	51.47%
Test grid III	1.76%	67.83%	41.93%	60.39%	14.20%	56.62%

Table 7: Head-to-head comparison of DHS and each heuristic. The results indicate the percentage of saved consistency checks per test grid by using DHS with respect to each particular heuristic (negative numbers indicate increases in the percentage of consistency checks).

Grid	SOL	DEG	DOM	MXC	RHO	DOMDEG
Test grid I	−9.43%	66.9%	35.74%	55.73%	4.38%	51.9%
Test grid II	−9.51%	60.75%	28.51%	54.70%	3.60%	45.87%
Test grid III	−8.55%	64.45%	35.83%	56.24%	5.2%	52.07%

In average, SHS performed better than DHS among the test grids used in this investigation. The results confirm that, for the instances used in this investigation, changing to a different heuristic once the search has started is not as beneficial as staying with a suitable initial choice of heuristic (choice based on what we know about the instances and the heuristics). The reason for this is that the patterns shown in Figure 4(a) estimate the best performer per instance assuming that the same heuristic is used from start to end. Because the patterns do not capture any information about the changes of heuristics as the search progresses, the dynamic selector is working on a pattern that may not be valid for making its decisions in the best way.

5.2. Testing SHS on Structured Instances

We have confirmed the idea that, by using the methodology proposed, it is possible to accurately predict a suitable heuristic for solving instances similar to the ones used for finding the relation between instances and heuristics. The historical information about the heuristics was collected from randomly generated instances to produce one static heuristic selection strategy that, according to the initial features of the instance to solve, decided the most suitable heuristic to apply. Although the results seem encouraging at this point, it is difficult, based only on the results obtained for randomly generated instances, to visualize how well the relation between instances and heuristics obtained could scale to larger structured instances.

For this reason, we applied the best heuristic selector from the ones described before to solve a set of 250 instances with very different properties to the ones used for obtaining the patterns of use. Thus, the static heuristic selector, SHS, was used for the rest of the experiments. Contrary to the instances used so far in this investigation, the constraints in the instances used for this experiment contain some structure. For this investigation, we took and combined six files from a public repository (the public repository can be accessed at http://www.cril.univ-artois.fr/~lecoutre/benchmarks.html): composed-25-10-20.tgz, composed-75-1-80.tgz, ehi-85.tgz, geom.tgz, QCP-10.tgz, and QCP-15.tgz. Files QCP-10.tgz and QCP-15.tgz contain 15 instances each, with 100 and 225 variables, respectively, and nonuniform domains. The 20 instances from files composed-25-10-20.tgz and composed-75-1-80.tgz are composed of a main underconstrained fragment and some auxiliary fragments [46] and contain 105 and 83 variables, respectively, and 10 values in the domain of each variable. File ehi-85.tgz contains 100 3-SAT unsatisfiable instances represented as binary CSPs. The equivalent binary instances contain 297 variables and eight values in their domains. File geom.tgz contains 100 geometrical instances (instead of a density parameter, a distance parameter is used to define the distribution of the constraints among the instance) with 50 variables and 20 values in their domains. Figure 5 presents the distribution of these structured instances on the instance space. In this case, we do not have the flexibility of the random generation model and as a result, most of the instances are distributed over some specific regions in the instance space.

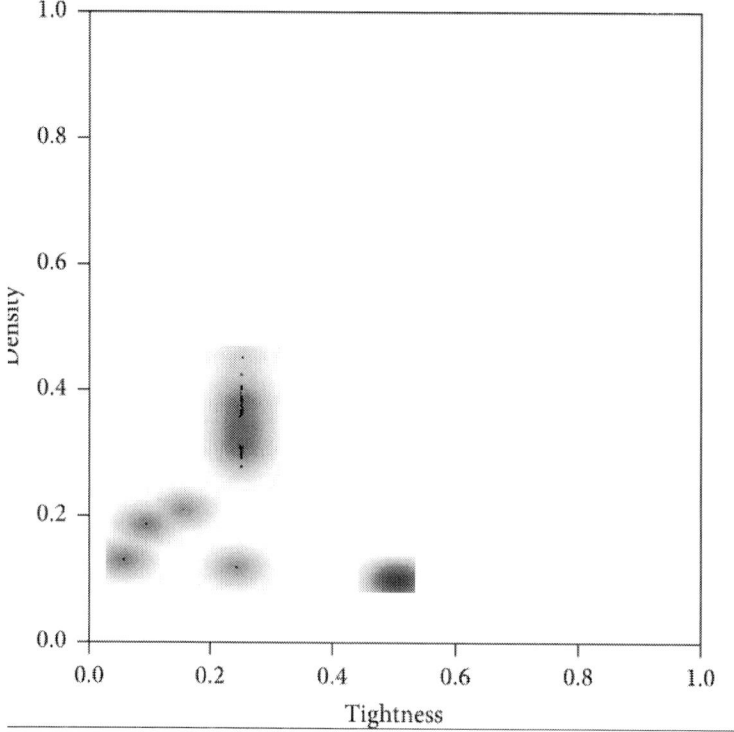

Figure 5: Density scatter plot for the structured instances used in this investigation (darker regions indicate a higher concentration of instances).

When we compared the performance of SHS against each heuristic on this set of structured instances, we observed that the good performance shown on randomly generated instances is also presented in the set of structured ones. SHS dominates SOL in 51.83% of the instances and saves 19.22% consistency checks with respect to this heuristic. The results are better for the other heuristics. For example, in the case of DEG and MXC, the heuristic selection strategy dominates these heuristics in 76.73% and 91.84% of the instances, respectively, and reduces the number of consistency checks by 71.62% and 74.82%, with respect to each of these heuristics.

For this experiment, we were also interested in a more challenging comparison. For this reason, we tested SHS against the best possible result obtained from the six heuristics. Figure 6 presents the percentage of instances where each method obtained the minimum number of consistency checks per instance. For some instances, two or more methods are tied with the minimum number of consistency checks. For this reason, when we sum up the percentages where each method required the minimum consistency checks, the result is greater than 100%.

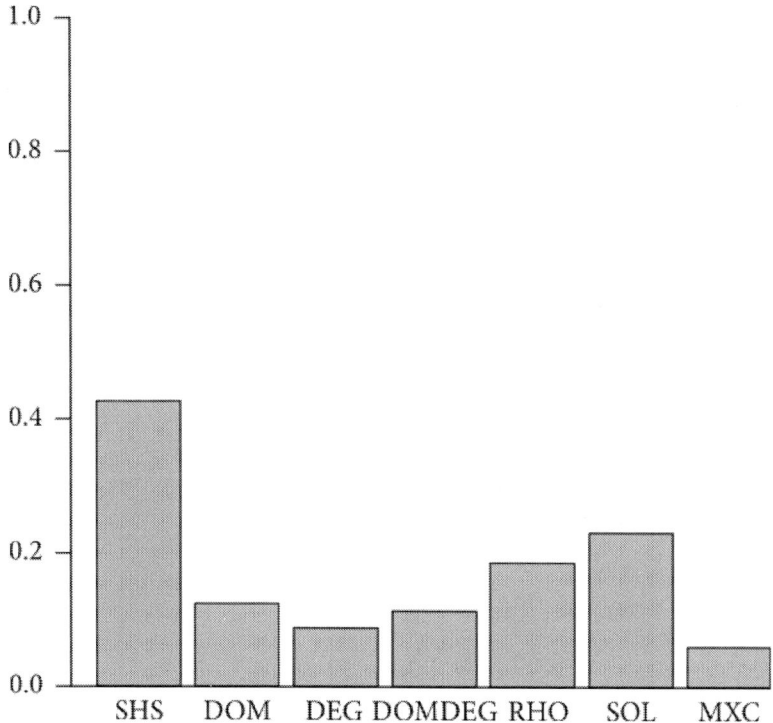

Figure 6: Percentage of instances on the set of structured instances where each method required the minimum number of consistency checks.

The results depicted in Figure 6 remark the potential of the heuristic selection strategy proposed. By using SHS, we can, in more than 40% of the instances, obtain a cost comparable to the best possible one obtained with any of the six heuristics. This is more than we can achieve by using any of the heuristics applied in isolation (SOL is the closest heuristic in performance, achieving the minimum cost in 23% of the instances).

Predicting the best heuristic only in around 40% is sufficient to overcome the performance of the six heuristics evaluated in this investigation but gives plenty of room for improvement in the future. The small percentage of instances where the prediction is accurate is a consequence of the change in the type of instances used. As heuristics may present a different behaviour on randomly generated instances with respect to structured ones, the rules obtained for randomly generated instances may not be accurate for structured instances. We are aware of this situation and recognize that, for properly solving structured instances, the selectors should use information from similar instances. But with this experiment, we have proven that it is possible to improve the search by using the information about the historical performance of heuristics on instances that may not correspond to the same types used for extracting the information and producing the rules of application.

5.3. Comparison of Performance with Other Dynamic Heuristics

We have discussed the performance of SHS on different structured instances with respect to the heuristics available for the heuristic selector. In this experiment, we compare the performance of SHS versus the other two reliable dynamic ordering heuristics that are not available for the heuristic selector: activity-based search (ABS) [47] and weighted degree (WDEG) [48].

The performance of the three methods is depicted in Figure 7. From this figure, we can observe that the median of SHS in the set of structured instances is lower than the median of ABS, but very similar to the one of WDEG. But the variance is higher for WDEG than for SHS and ABS.

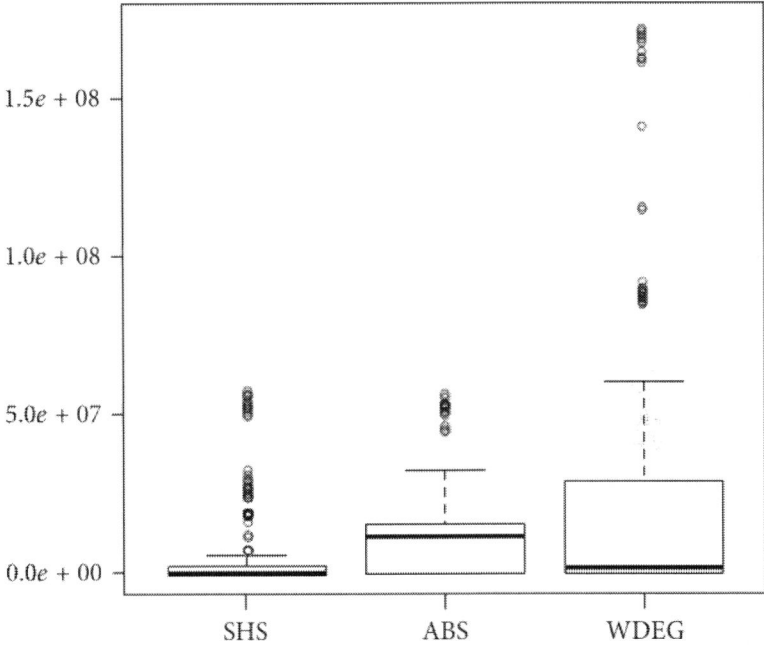

Figure 7: Performance of SHS, ABS, and WDEG on the set of structured instances.

Although the statistical evidence is insufficient to claim that SHS is better than ABS and WDEG (by using a unilateral paired -test with 5% of significance), there are important savings in the cost of the search by using SHS that are worth discussing. By using SHS, we require, in average, 44.18% less consistency checks than ABS and 66.86% less consistency checks than WDEG on the set of structured instances. We are aware that the performance of the methods discussed may change if other sets of instances are used, but at this point, the results confirm that SHS is capable of competing against other reliable heuristics that were not part of the selection process.

6. CONCLUSION

This paper described a methodology to characterize the CSP instance space in order to analyze the performance of different variable ordering heuristics. This analysis allowed us to locate regions where some heuristics are better than others and also regions where some of these heuristics should not be used. The results confirmed that a large fraction of hard-to-solve instances are located on the same region of the instance space regardless of the heuristic used. But we also found evidence that, even for those regions, some heuristics are more desirable than others and that we can use such evidence to improve the search.

We identified regions in the instance space (characterized by the constraint density and tightness) where one heuristic dominates the others in average performance, but there is no absolute dominance among all the instances generated for specific values of density and tightness. For example, the best average heuristic for a set of instances can sometimes be defeated by an apparently weaker heuristic for some exact region of the instance space. We think that more features are required to identify such unusual situations and properly characterize those cases.

The instance space characterization provided an opportunity for understanding how heuristics work since it allowed us to identify where a heuristic should be used or avoided. Among the heuristics studied, SOL proved to be the most competent one for the region where the hardest average instances occur. An explanation for this good performance relies on the way different aspects of the problem are analyzed by the heuristic. While heuristics such as DOM, DEG, and MXC focus only on one aspect of the problem to decide the next variable to instantiate (the domain size, the degree, and the number of conflicts, resp.), RHO considers a mixture of two of them (the degree and the proportion of conflicts). For this reason, RHO is, in general, better than DOM, DEG, and MXC. But including more information may not always result in a better average performance. For example, DOM/DEG also combines information about the domain size and the degree of the variables but it does not seem to perform as well as RHO. Then, the way the different aspects of the problem are combined is also related to the performance of the heuristics. When we look at SOL, we observe that it is basically a revision of RHO, which is by itself a competent heuristic that considers two aspects of the problem. SOL extends RHO and improves its performance by including information about the domain size.

Although this information is by itself relevant and useful, we wanted to show that it can indeed be used to improve the search. By using the information obtained from the analysis of the heuristics and the instance space, we produced matching from instances to heuristics that was transparently translated into a heuristic selection strategy. This strategy was implemented by using the patterns obtained from the exploration of the performance of the heuristics on the CSP instance space. It is important to stress that the information the selector uses to make its decisions is easily interpretable by humans. This has various advantages, being among the most important ones that it is possible to visualize the relative strengths and weaknesses of these heuristics, allowing a more reliable heuristic performance prediction.

Among the two heuristic selectors proposed, the static one proved to be reliable and competitive for solving instances, both randomly generated and structured ones, as it improved the results obtained by any of the variable ordering heuristics when applied in isolation. The results confirmed our initial idea that the information obtained from the historical performance of heuristics on a set of instances can be used to improve the search (even when using such a simple strategy like the one described in this document). Now that we have confirmed our initial ideas, we can consider extending the matching process by including more features and heuristics and designing a more robust heuristic selection strategy.

As future work, we are interested in extending our investigation to explore other problem domains. Although the methodology described in this paper is focused on exploring the CSP instance space to predict the performance of some heuristics on such instances, it is not limited to this particular problem domain. The same model can be used on other domains provided that a systematic way to produce instances (to generate instances in a vast region of the instance space) and a suitable representation exist. Finally, we would like to use the information gathered from the mapping of heuristics to instances to produce more robust heuristic selectors that make better use of the information on the performance of the heuristics.

ACKNOWLEDGMENTS

This research was supported in part by ITESM Strategic Project PRY075, ITESM Research Group with Strategic Focus in Intelligent Systems, and CONACyT Basic Science Projects under Grants 99695 and 241461.

REFERENCES

1. S. L. Epstein, E. C. Freuder, R. Wallace, A. Morozov, and B. Samuels, "The adaptive constraint engine," in Principles and Practice of Constraint Programming—CP 2002: 8th International Conference, CP 2002 Ithaca, NY, USA, September 9–13, 2002 Proceedings, vol. 2470 of Lecture Notes in Computer Science, pp. 525–542, Springer, Berlin, Germany, 2002.
2. E. O'Mahony, E. Hebrard, A. Holland, C. Nugent, and B. O'Sullivan, "Using case-based reasoning in an algorithm portfolio for constraint solving," in Proceedings of the 19th Irish Conference on Artificial Intelligence and Cognitive Science, Cork, Ireland, August 2008.
3. S. Petrovic and R. Qu, "Case-based reasoning as a heuristic selector in a hyper-heuristic for course timetabling problems," in Proceedings of the 6th International Conference on Knowledge-Based Intelligent

Information Engineering Systems and Applied Technologies (KES '02), vol. 82, pp. 336–340, September 2002.

4. B. Crawford, R. Soto, C. Castro, and E. Monfroy, "A hyperheuristic approach for dynamic enumeration strategy selection in constraint satisfaction," in New Challenges on Bioinspired Applications, vol. 6687 of Lecture Notes in Computer Science, pp. 295–304, Springer, Berlin, Germany, 2011.

5. J. C. Ortiz-Bayliss, H. Terashima-Marín, and S. E. Conant-Pablos, "Learning vector quantization for variable ordering in constraint satisfaction problems," Pattern Recognition Letters, vol. 34, no. 4, pp. 423–432, 2013.

6. R. Soto, B. Crawford, E. Monfroy, and V. Bustos, "Using autonomous search for generating good enumeration strategy blends in constraint programming," in Computational Science and Its Applications—ICCSA 2012: 12th International Conference, Salvador de Bahia, Brazil, June 18–21, 2012, Proceedings, Part III, vol. 7335 of Lecture Notes in Computer Science, pp. 607–617, Springer, Berlin, Germany, 2012.

7. Y. Malitsky, "Evolving instance-specific algorithm configuration," in Instance-Specific Algorithm Configuration, pp. 93–105, Springer, Basel, Switzerland, 2014.

8. P. Hell and J. Nešetřil, "Colouring, constraint satisfaction, and complexity," Computer Science Review, vol. 2, no. 3, pp. 143–163, 2008.

9. N. Dunkin and S. Allen, "Frequency assignment problems: representations and solutions," Tech. Rep. CSD-TR-97-14, University of London, 1997.

10. J. A. Berlier and J. M. McCollum, "A constraint satisfaction algorithm for microcontroller selection and pin assignment," in Proceedings of the IEEE SoutheastCon, pp. 348–351, IEEE, Concord, NC, USA, March 2010.

11. J. R. Bitner and E. M. Reingold, "Backtrack programming techniques," Communications of the ACM, vol. 18, no. 11, pp. 651–656, 1975.

12. N. Jussien and O. Lhomme, "Local search with constraint propagation and conflict-based heuristics," in Proceedings of the 17th National Conference on Artificial Intelligence and 12th Conference on Innovative Applications of Artificial Intelligence, pp. 169–174, AAAI Press, MIT Press, Austin, Tex, USA, July-August 2000.

13. J. R. Rice, "The algorithm selection problem," Advances in Computers, vol. 15, pp. 65–118, 1976.

14. T. Stützle and S. Fernandes, "New benchmark instances for the qap and the experimental analysis of algorithms," in Evolutionary Computation in Combinatorial Optimization, J. Gottlieb and G. Raidl, Eds., vol. 3004 of

Lecture Notes in Computer Science, pp. 199–209, Springer, Berlin, Germany, 2004.

15. K. A. Smith-Miles, "Towards insightful algorithm selection for optimisation using meta-learning concepts," in Proceedings of the IEEE International Joint Conference on Neural Networks (IJCNN '08), pp. 4118–4124, IEEE, Hong Kong, June 2008.

16. K. A. Smith-Miles, R. J. James, J. W. Giffin, and Y. Tu, "A knowledge discovery approach to understanding relationships between scheduling problem structure and heuristic performance," in Learning and Intelligent Optimization, T. Stützle, Ed., vol. 5851 of Lecture Notes in Computer Science, pp. 89–103, Springer, Berlin, Germany, 2009.

17. K. Smith-Miles, D. Baatar, B. Wreford, and R. Lewis, "Towards objective measures of algorithm performance across instance space," Computers and Operations Research, vol. 45, pp. 12–24, 2014.

18. B. Bischl, O. Mersmann, H. Trautmann, and M. Preuß, "Algorithm selection based on exploratory landscape analysis and cost-sensitive learning," in Proceedings of the 14th Annual Conference on Genetic and Evolutionary Computation (GECCO '12), pp. 313–320, ACM, Philadelphia, Pa, USA, July 2012.

19. E. López-Camacho, H. Terashima-Marín, G. Ochoa, and S. E. Conant-Pablos, "Understanding the structure of bin packing problems through principal component analysis," International Journal of Production Economics, vol. 145, no. 2, pp. 488–499, 2013.

20. E. Tsang and A. Kwan, "Mapping constraint satisfaction problems to algorithms and heuristics," Tech. Rep. CSM-198, Department of Computer Sciences, University of Essex, 1993.

21. J. C. Ortiz-Bayliss, H. Terashima-Marin, P. Ross, J. I. Fuentes-Rosado, and M. Valenzuela-Rend, "A neuro-evolutionary approach to produce general hyper-heuristics for the dynamic variable ordering in hard binary constraint satisfaction problems," in Proceedings of the 11th Annual Conference on Genetic and Evolutionary Computation (GECCO '09), pp. 1811–1812, Montreal, Canada, July 2009.

22. J. H. Moreno-Scott, J. C. Ortiz-Bayliss, H. Terashima-Marín, and S. E. Conant-Pablos, "Challenging heuristics: evolving binary constraint satisfaction problems," in Proceedings of the 14th Annual Conference on Genetic and Evolutionary Computation (GECCO '12), pp. 409–416, ACM, New York, NY, USA, July 2012.

23. K. Smith-Miles, J. van Hemert, and X. Y. Lim, "Understanding TSP difficulty by learning from evolved instances," in Learning and Intelligent Optimization, C. Blum and R. Battiti, Eds., vol. 6073 of

Lecture Notes in Computer Science, pp. 266–280, Springer, Berlin, Germany, 2010.

24. K. Smith-Miles and J. van Hemert, "Discovering the suitability of optimisation algorithms by learning from evolved instances," Annals of Mathematics and Artificial Intelligence, vol. 61, no. 2, pp. 87–104, 2011.

25. J. I. van Hemert, "Evolving binary constraint satisfaction problem instances that are difficult to solve," in Proceedings of the IEEE Congress on Evolutionary Computation (CEC '03), vol. 2, pp. 1267–1273, IEEE, December 2003.

26. J. I. Van Hemert, "Evolving combinatorial problem instances that are difficult to solve," Evolutionary Computation, vol. 14, no. 4, pp. 433–462, 2006.

27. F. Boussemart, F. Hemery, and C. Lecoutre, "Revision ordering heuristics for the constraint satisfaction problem," in Proceedings of the 1st International Workshop on Constraint Propagation and Implementation (CPAI '04), pp. 9–43, Toronto, Canada, September 2004.

28. P. Gent, E. MacIntyre, P. Prosser, B. M. Smith, and T. Walsh, "An empirical study of dynamic variable ordering heuristics for the constraint satisfaction problem," in Principles and Practice of Constraint Programming—CP96: Second International Conference, CP96 Cambridge, MA, USA, August 19–22, 1996 Proceedings, vol. 1118 of Lecture Notes in Computer Science, pp. 179–193, Springer, Berlin, Germany, 1996.

29. R. J. Wallace, "Analysis of heuristic synergies," in Recent Advances in Constraints, B. Hnich, M. Carlsson, F. Fages, and F. Rossi, Eds., vol. 3978 of Lecture Notes in Computer Science, pp. 73–87, Springer, Berlin, Germany, 2006.

30. R. M. Haralick and G. L. Elliott, "Increasing tree search efficiency for constraint satisfaction problems," in Proceedings of the 6th International Joint Conference on Artificial Intelligence, vol. 1, pp. 356–364, Morgan Kaufmann, San Francisco, Calif, USA, 1979.

31. R. Dechter and I. Meiri, "Experimental evaluation of preprocessing algorithms for constraint satisfaction problems," Artificial Intelligence, vol. 68, no. 2, pp. 211–241, 1994.

32. D. Brélaz, "New methods to color the vertices of a graph," Communications of the ACM, vol. 22, no. 4, pp. 251–256, 1979.

33. H. Terashima-Marín, J. C. Ortiz-Bayliss, P. Ross, and M. Valenzuela-Rendón, "Hyper-heuristics for the dynamic variable ordering in constraint satisfaction problems," in Proceedings of the 10th Annual

Conference on Genetic and Evolutionary Computation (GECCO '08), pp. 571–578, ACM, Atlanta, Ga, USA, July 2008. Vie

34. J. G. Gaschnig, "A general backtrack algorithm that eliminates most redundant tests," in Proceedings of the 5th International Joint Conference on Artificial Intelligence (IJCAI '77), vol. 1, p. 457, Morgan Kaufmann, Cambridge, Mass, USA, August 1977.

35. E. C. Freuder, "Synthesizing constraint expressions," Communications of the ACM, vol. 21, no. 11, pp. 958–966, 1978. View at Publisher · View at Google Scholar · View at MathSciNet ·

36. D. Achlioptas, C. Gomes, H. Kautz, and B. Selman, "Generating satisfiable problem instances," in Proceedings of the 17th National Conference on Artificial Intelligence (AAAI '00), pp. 256–301, Austin, Tex, USA, July-August 2000.

37. J. Culberson, "Hidden solutions, tell-tales, heuristics and anti-heuristics," in Proceedings of the Workshop on Empirical Methods in Artificial Intelligence (IJCAI '01), H. Hoos and T. Stuëtzle, Eds., pp. 9–14, 2001.

38. Rish and D. Frost, "Statistical analysis of backtracking on inconsistent CSPs," in Principles and Practice of Constraint Programming—CP97, G. Smolka, Ed., vol. 1330 of Lecture Notes in Computer Science, pp. 150–162, Springer, Berlin, Germany, 1997.

39. B. M. Smith, "Locating the phase transition in binary constraint satisfaction problems," Artificial Intelligence, vol. 81, no. 1-2, pp. 155–181, 1996.

40. D. Achlioptas, M. S. O. Molloy, L. M. Kirousis, Y. C. Stamatiou, E. Kranakis, and D. Krizanc, "Random constraint satisfaction: a more accurate picture," Constraints, vol. 6, no. 4, pp. 329–344, 2001.

41. E. MacIntyre, P. Prosser, B. M. Smith, and E. MacIntyre, "Random constraint satisfaction: theory meets practice," in Principles and Practice of Constraint Programming—CP98: 4th International Conference, CP98 Pisa, Italy, October 26–30, 1998 Proceedings, vol. 1520 of Lecture Notes in Computer Science, pp. 325–339, Springer, Berlin, Germany, 1998.

42. K. Xu and W. Li, "Many hard examples in exact phase transitions," Theoretical Computer Science, vol. 355, no. 3, pp. 291–302, 2006. View at Publisher · View at Google Scholar · View at Zentralblatt MATH ·

43. P. Prosser, "Hybrid algorithms for the constraint satisfaction problem," Computational Intelligence, vol. 9, no. 3, pp. 268–299, 1993.

44. B. M. Smith, "Constructing an asymptotic phase transition in random binary constraint satisfaction problems," Theoretical Computer Science, vol. 265, no. 1-2, pp. 265–283, 2001.

45. Y. Fan and J. Shen, "On the phase transitions of random k-constraint satisfaction problems," Artificial Intelligence, vol. 175, no. 3-4, pp. 914–927

46. C. Lecoutre, F. Boussemart, and F. Hemery, "Backjump-based techniques versus conflict-directed heuristics," in Proceedings of the 16th IEEE International Conference on Tools with Artificial Intelligence (ICTAI '04), pp. 549–557, Boca Raton, Fla, USA, November 2004.

47. L. Michel and P. Van Hentenryck, "Activity-based search for black-box constraint programming solvers," in Integration of AI and OR Techniques in Contraint Programming for Combinatorial Optimzation Problems, N. Beldiceanu, N. Jussien, and É. Pinson, Eds., vol. 7298 of Lecture Notes in Computer Science, pp. 228–243, Springer, Berlin, Germany, 2012.

48. F. Boussemart, F. Hemery, C. Lecoutre, and L. Sais, "Boosting systematic search by weighting constraints," in Proceedings of the European Conference on Artificial Intelligence (ECAI '04), pp. 146–150, IOS Press, 2004.

A Novel Spatial-Temporal Voronoi Diagram-Based Heuristic Approach for Large-Scale Vehicle Routing Optimization with Time Constraints

Wei Tu [1,2,3,*]**, Qingquan Li** [2,3,4,*]**, Zhixiang Fang** [4] **and Baoding Zhou** [2,3]

[1] *College of Information Engineering, Shenzhen University, Shenzhen 518060, China*

[2] *Shenzhen Key Laboratory of Spatial Information Smart Sensing and Services, College of Civil Engineering, Shenzhen University, Shenzhen 518060, China*

[3] *Key Laboratory for Geo-Environmental Monitoring of Coastal Zone of the National Administration of Surveying, Mapping and GeoInformation, Shenzhen University, Shenzhen 518060, China*

[4] *State Key Laboratory of Information Engineering in Surveying, Mapping, and Remote Sensing, Wuhan University, Wuhan 430079, China*

ABSTRACT

Vehicle routing optimization (VRO) designs the best routes to reduce travel cost, energy consumption, and carbon emission. Due to non-deterministic polynomial-time hard (NP-hard) complexity, many VROs involved in real-world applications require too much computing effort. Shortening computing time for VRO is a great challenge for state-of-the-art spatial optimization algorithms. From a spatial-temporal perspective, this paper presents a spatial-temporal Voronoi diagram-based heuristic approach for large-scale vehicle routing problems with time windows (VRPTW). Considering time constraints, a spatial-temporal Voronoi distance is derived from the spatial-temporal Voronoi diagram to find near neighbors in the space-time searching context. A Voronoi distance decay strategy that integrates a time warp operation is proposed to accelerate local search procedures. A spatial-temporal feature-guided search is developed to improve unpromising micro route structures. Experiments on VRPTW benchmarks and real-world instances are conducted to verify performance. The results demonstrate that the proposed approach is competitive with state-of-the-art heuristics and achieves high-quality solutions for large-scale instances of VRPTWs in a short time. This novel approach will contribute to spatial decision support community by developing an effective vehicle routing optimization method for large transportation applications in both public and private sectors.

Keywords: spatial decision support system; vehicle routing problem; Voronoi diagram; heuristics; local search; distance decay

1. INTRODUCTION

Flexible transportation service not only reduces travel cost, but also alleviates related energy and environmental concerns, such as traffic congestion, energy consumption, and carbon emission [1,2]. With the advancement of geographic information science (GIS), space-related problem decision-making has been embedded in spatial decision support systems (SDSSs) to provide flexible transportation service in both public and private sectors [3], such as the design of school bus routes [4], collection of solid waste [5,6], distribution of goods in chain business [7], design of police patrol routes [8], and planning express deliveries [9]. Spatial function tools are used for intelligent transportation operation in many services including management of geo-referenced transport network data, geocoding massive human demands, positioning moving vehicles, and navigating routes for drivers [5,6,7,8,9].

Vehicle routing optimization (VRO) aims to design least cost routes to satisfy geographically-distributed human demand through spatial intelligence of SDSSs in the field of transportation [10], including the shortest path problem (SPP) [11,12,13], traveling salesman problem (TSP) [14], and the vehicle routing problem (VRP) [5,6,7]. One type of VRO is modeled as the vehicle routing problem with time window (VRPTW), which imposes time constraints on customers, vehicles, and depots to accept or distribute high quality service [4,6,14]. It specifies that a customer only accepts certain transportation-related service in a given time interval

[15,16,17]. The VRPTW is computationally intractable in many real-life applications [18]. The heuristic algorithm is the only justified approach as it can find good solutions in a reasonable time [16,17]. There are two types of heuristic algorithms for the VRPTW, which include local search and evolutionary optimization. Local search iteratively modifies local route structures to improve solution quality and follows a single trajectory to explore the solution space [3,6,7]. Evolutionary optimization simultaneously searches multiple parts of the whole solution space to converge on the best solution [5,19,20]. Recently, state-of-the-art heuristic algorithms have been reported to yield a high quality solution of the VRPTW with hundreds of customers [5,9,18]. More efficient heuristics are needed for even larger VRPTW applications.

Spatial principle has exhibited good performance in space-related problem decision-making [21,22,23] in problems such as bus stop location [3], hazmat route planning [20], and path covering [24]. To efficiently solve VRO, from a spatial perspective, a few speed-up strategies have been developed to handle spatial issues and extend solvable VRO instances up to thousands of customers, which include the granular neighborhood [25], the kth nearest neighbors [26], and the k-ring-shaped Voronoi neighbors [27]. The common denominator of these effective strategies is to search only a limited number of spatial neighbors to save computing efforts [28]. However, such useful strategies become ineffective for the VRPTW since time constraint imposes an exceptional influence on the proximity of consecutively served customers in a route. A long wait will occur if two spatially-close customers with quite different time windows (for example, one is 8:00–9:00 a.m. and the other is 3:00–4:00 p.m.) are visited consecutively in a route, which challenges the motivation of spatial proximity-based heuristics approaches. Hence, spatial proximity measures must be extended to address the additional temporal constraints [18]. Moreover, how to accelerate the VRO with the help of spatial-temporal thinking should be investigated.

Voronoi diagrams describe both spatial proximity and topological proximity between spatial objects [29,30,31,32]. By extending it to the space-time context, a spatial-temporal Voronoi diagram (STVD) is well suited for measuring the proximity of customers with time constraints. Therefore, this article proposes a spatial-temporal Voronoi diagram-based heuristic approach to solve very large-scale VRPTWs. A derived spatial-temporal Voronoi neighbors from the Voronoi diagram is used to select promising candidates for a local search. Furthermore, motivated by the truth that nearer neighbors have a higher possibility to be consecutively served in a route [28], a Voronoi distance decay strategy is proposed to assign reasonable search efforts on neighbors with different Voronoi distances. A spatial-temporal feature guided search is used to reconstruct unpromising micro route structures. Intensive experiments on benchmark datasets (25–1000 customers) and real-world large scale VRPTWs and its multi-depot variants (2000–10,000 customers) have been implemented to assess the performance of the proposed approach. This efficient and effective approach enables current local search heuristics to solve super large-scale VRPTWs and will contribute to improving spatial intelligence for transportation-oriented SDSSs.

The remainder of this paper is organized as follows. Section 2 reviews related literature. Section 3 introduces the VRPTW, the STVD, the derived spatial-temporal Voronoi distance and neighbors. The proposed spatial-temporal Voronoi diagram-based heuristic for the VRPTW and its multi-depot variants are described in Section 4, and details of the experiments, results, and comparisons are reported in Section 5. Section 6 discusses the impact of the STVD and spatial-temporal proximity behind the best found solutions. The last section draws the conclusions.

2. LITERATURE REVIEW

This section reviews heuristic algorithms for the VRPTW and the integration of vehicle routing optimization and GIS.

2.1. Heuristic Algorithms for the VRPTW

Heuristic algorithms for VRPTWs are divided into two categories: local search and evolutionary optimization. Starting from an initial solution, local search shifts from a current solution to a better neighborhood solution with a simple change of local routes [33]. The change is made iteratively by moving nodes or exchanging arcs within a route or between routes such as 1–0 exchange, 1–1 exchange, 2-Opt, and λ-opt. Usually, the change process becomes trapped in a local optimum. To overcome unpromising results, many intelligent strategies that accept some worse solutions have been developed to further improve the solution quality, such as simulated annealing [34], iterated local search [35], large neighborhood search [36], variable neighborhood search [37], and tabu search [38]. Numerous tests on small- and medium-size VRPTWs with 25–100 customers have demonstrated good performance of local search heuristics [34,36,37].

Evolutionary optimization concurrently improves many solutions and results in high quality solutions. Four major steps are involved: representation, selection, combination, and mutation. Some of the most successful evolutionary optimization algorithms for the VRPTW are provided by Repoussis et al. [39] and Vidal et al. [40]. Mester and Bräysy [41] used the standard evolutionary optimization framework to guide exploration in the VRPTW solution space with solution initialization and evolution. Gehring and Homberger [42] parallelized the genetic algorithm and first solved VRPTW instances up to 1000 customers. Usually, local search is used as a component in the evolutionary optimization approach. Fast and simplified node or arc exchange-based local search have been used for offspring education in the genetic algorithm [41]. Therefore, more efficient local search heuristics also help to improve evolutionary optimization.

For larger VRPTW instances, recent advances have proposed several accelerating techniques to save computing efforts while still achieving high quality solutions. Following the local search principle, taking spatial distance into account, Cordeau [38] limited the local search candidates with kth nearest neighbors to shorten computing time. Toth and Vigo [25] selected the local search evaluated nodes with a fixed distance threshold. Considering time constraints, Ibaraki [43] proposed a time-oriented neighbor list-based strategy to reduce the local search evaluation. Some unpromising search candidates had been excluded with the unmatched time windows. But the spatial proximity has been ignored. Nagata [44] developed a time warp operation for customers with a late service to avoid complex time window analysis. An additional penalty function was added to the original objectives to justify the local search heuristics. However, the spatial dimension and the temporal dimension are still separated in these local search heuristics.

Another effective accelerating approach is to decompose large-scale VRPTW instances into many smaller subproblems. Using spatial clustering, Dondo and Cerdá [45] divided the large scale VRPTW into a set of small-sized problems. Routes were generated for each small problem to provide a high quality initial solution. Qi et al. [18] developed an effective spatial-temporal distance measure to deal with both space and time issues. A large-scale VRPTW was partitioned with a k-medoid clustering method to generate a good initial solution. A standard genetic algorithm was used to improve the initial solution. Good performance of these effective approaches has been verified by VRPTW instances with no more than 1000 customers.

With regard to the Voronoi diagram, the only work has been provided by Milthers [46] who used two-dimensional Voronoi diagrams to split the VRPTW into many subproblems and then solve them with large neighborhood search heuristics. It showed the effectiveness of Voronoi diagrams in guiding the search process. However, this study dealt only with decomposition of the VRPTW in the solution construction stage. Use of the spatial-temporal Voronoi diagram in speeding up the local search for the VRPTW should still be further considered.

2.2. Integration of Vehicle Routing Optimization and GIS

Effective approaches to find high quality solutions for VRPTWs have been embedded in a GIS-based SDSS to deal with many real world applications. Generally, the integration between GIS and vehicle routing optimization is tightly coupled. Spatial data management, processing, and visualization tools are used to collect customer orders, georeference related data, activate the solving process, and display routes for VROs, as in GIS software such as ArcGIS and TransCAD. In line with local search, Weigel and Cao [7] first reported their efforts in the implementation of a tabu heuristics approach in a GIS environment to deal with VRO for a major American retailer. Following the evolutionary optimization line, Mendoza et al. [5] integrated a customized routing module, which improved solution quality with commercial solutions such as SAP/R3 and ArcGIS to cope with VROs in public utilities. Experiments on a real-world case in Bogotá, Colombia with 323–601 customers verified the effectiveness of evolutionary optimization and GIS software.

Recently, progress of GIS related to cloud computing transfers the integration of GIS and VRO to a loose coupling way. Using Google maps, Santos et al. [47] developed user-friendly web-based SDSSs that embed VRO to deal with urban trash collection in Coimbra, Portugal. TU et al. [48] presented a cloud GIS-based spatial decision support framework with variable neighborhood search heuristics for dynamic vehicle routing using historical traffic information. These practices have proven the dominance of VROs in real-life transportation applications. However, they also indicate that current spatial intelligence should be improved to enhance the ability of SDSSs in transportation departments to cope with increasing numbers of customers.

In summary, state-of-the-art heuristics, including local search and evolutionary optimization, solve small- and medium-size VRPTWs in a reasonable time. They have been verified to be effective in a GIS environment for real-life transportation cases with no more than 1000 customers. For larger-size real-life applications, more efficient heuristic algorithms are needed to reduce computing time. Using the spatial-temporal Voronoi diagram, this paper proposes an efficient local search-based approach to address VRPTWs up to 10,000 customers in a reasonable time. Differing from the neighborhood in the spatial context [25,26,27,48,49], the spatial-temporal Voronoi neighborhood considering both spatial distance and time windows is proposed to limit the search space. A Voronoi distance decay strategy is developed to assign more searching efforts to nearer neighbors. The spatial-temporal feature guided search is used to escape from local minima. Details of the STVD, Voronoi distance, Voronoi neighbors, and the proposed approach for the VRPTW are described in the next sections.

3. THE VRPTW AND THE SPATIAL-TEMPORAL VORONOI DIAGRAM

This section introduces the VRPTW, the STVD, and the spatial-temporal Voronoi distance.

3.1. The VRPTW

The VRPTW has five components: customers, depot, vehicle, routes, and the solution.

Customers: a set of customers $Vc=\{v1,v2,...,vn\}$ expects to be served. A customer v_i ($I>0$) located at (x_i, y_i) has a unique demand, q_i, a service duration, s_i, and a service time window $<e_i, l_i>$. The service is required to arrive after start time, e_i, and before end time, l_i. Such a time window is termed "soft" when it could be violated with a penalty cost, and termed "hard" when no violation is permitted.

Depot: The depot $v0$ located at (x_0, y_0) provides service in a time window $<e_0, l_0>$.

Vehicles: A fleet of K identical vehicles with the capacity Q is located at the depot. Each vehicle departs from the depot, serves multiple customers, and returns to the depot. Travel distance and time between nodes (includes customers and the depot) i, j are denoted as d_{ij} and t_{ij}, respectively, where $i \neq j$.

Routes: A route $r=\langle v0,v1,v2,..., vR, vR+1\rangle$ is formed by a sequence of visited customers, which represents the service process of a vehicle, where R denotes the number of visited customers. $vR+1=v0$, so that the vehicle must return to the depot. The earliest departure time at the depot $v0$ is a_r, the earliest service time at a customer vi is a_{vi}, and the earliest return time at the depot is a_{vR+1}, which are defined recursively as follows:

$$a_{v_0} = a_r$$
$$\left\{ a_{v_i} = \max\{a_{v_{i-1}} + s_{v_{i-1}} + t_{i-1,i}, e_{v_i}\}, (i = 1, ..., R + 1) \right. \tag{1}$$

A route is specified following two restrictions: (1) The route duration time $a_{vR+1}-a_r$ should be no more than the maximum duration T_{max}; (2) The total served demand $a_{v_{R+1}} - a_r$ should be no more than the vehicle capacity Q. The total travel distance of the route is defined as $\sum_{i=0}^{R} d_{v_i v_{i+1}}$. A route is defined as feasible if the customer's service time window, the route duration time, and the vehicle capacity are satisfied.

The VRPTW solution: A solution for a VRPTW instance is a set of feasible routes $r1,r2,r3,..., rK$ such that a single vehicle visits each customer once. A hierarchy of objectives of the VRPTW is followed in this paper. The number of used vehicles is first minimized and then the total travel distance is reduced.

3.2. The Spatial-Temporal Voronoi Diagram for VRPTWs

Components of the VRPTW, such as customers, depot, vehicles, and routes, have both spatial and temporal attributes. Spatial attributes refer to customers' locations, the depot location, and the vehicle trajectories.

Temporal attributes refer to the time window, service time, travel time, and route duration time. To represent them simultaneously, a spatial-temporal coordinate system is built by taking the temporal dimension to be perpendicular to the planar XY plane. Therefore, time windows of the customers and the depot are modeled as vertical lines rooted in their locations.

Figure 1 illustrates a small-size VRPTW with five customers (v_1 to v_5), one depot (v_0), and two routes (R1, R2). Customers (v_1-v_5) and the depot (v_0) are modeled as vertical lines from its start time point to end time point, as line segments L_1-L_5 and L_0. Projections on the XY plane denote their spatial locations, whereas projections on the time dimension denote their time windows and actual service time. The transportation between customers denotes the service process in the routes (R1 or R2).

Distance in this coordinate system is defined as Equation (2), where (x_i, y_i) and (x_j, y_j) denote the spatial location of i and j, and t_i and t_j denote the time at i and j, $\bar{v}$ is the mean travel speed:

$$d(v_i, v_j) = \sqrt{(x_i - x_j)^2 + (y_i - y_j)^2 + (\bar{v}(t_i - t_j))^2}$$

(2)

Voronoi diagrams partition space into adjacent Voronoi cells with given points [30]. Following this principle, a spatial-temporal Voronoi diagram divides the spatial-temporal space into many cells. To generate a Voronoi diagram for the VRPTW, the spatial-temporal central point of a customer's spatial temporal line, (x_i, y_i, ($e_i + l_i$)/2.0), which is the red point in Figure 1, is used as the seed point. The STVD for VRPTWs is built by the fast quickhull algorithm [50].

In the STVD, a pair of points whose cells share a boundary (Voronoi face or Voronoi edge) are Voronoi neighbors [30]. The spatial-temporal Voronoi distance from a given point i to the point j, vd(i, j), is defined as the minimum number of crossings of the spatial-temporal Voronoi boundary in any route from i to j [27,31]. Figure 2 provides an example that illustrates the Voronoi distance. The Voronoi distance from P1 to P3 is 2 for the two crossings of Voronoi boundaries (C1–C2). The Voronoi distance from P1 to P13 is also 2 for the crossings C3–C4. The Voronoi distance from P1 to P12 is 3 for the crossing C5–C7. Thus, Voronoi neighbors with different spatial-temporal proximity can be measured. As both spatial distance and time windows are considered, it is well suited for accelerating local search for the solving of VRPTW. Using these useful definitions, this paper proposes an efficient local search heuristic to address large-scale VRPTWs.

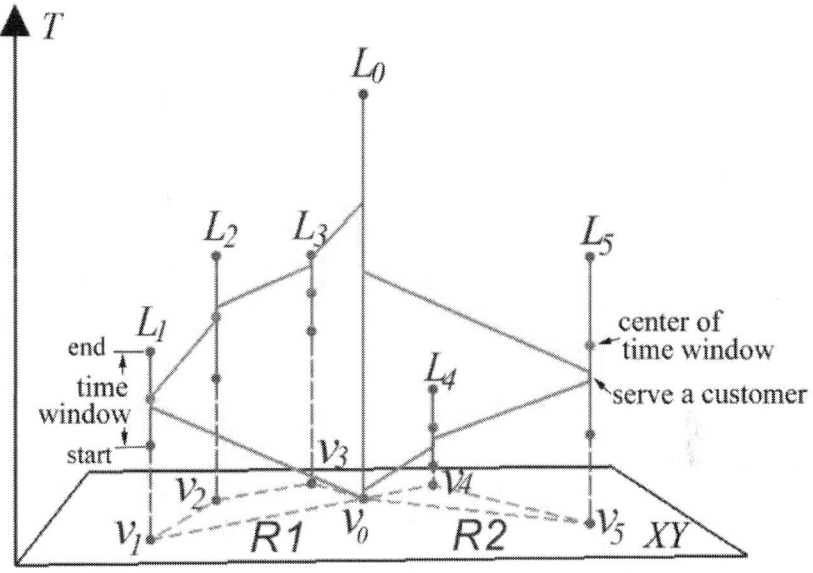

Figure 1. Spatial-temporal representation of VRPTW.

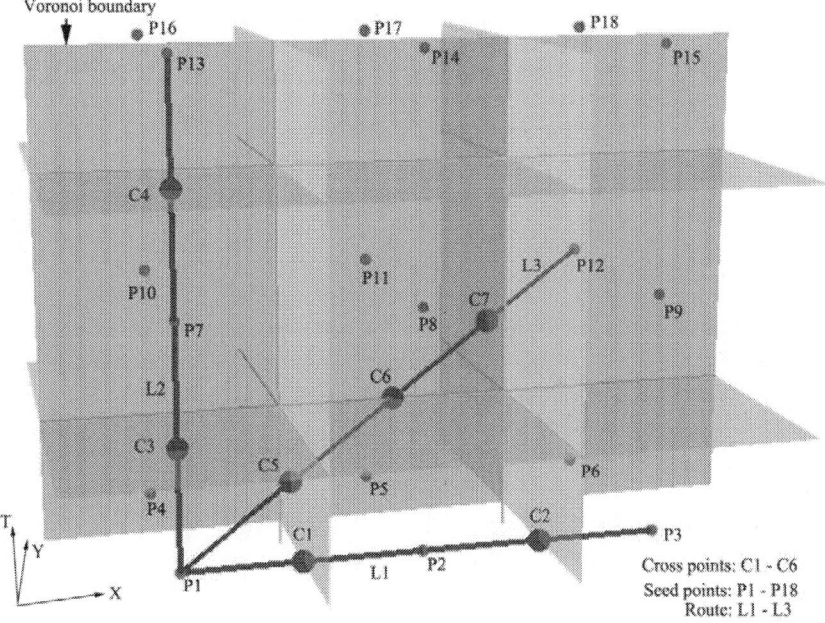

Figure 2. The spatial-temporal Voronoi diagram and the spatial-temporal Voronoi distance.

4. THE SPATIAL-TEMPORAL VORONOI DIAGRAM-BASED HEURISTIC

This section introduces the spatial-temporal Voronoi diagram-based heuristic (STVDH) for large-scale VRPTWs. The principle of the STVDH is to speed up the local search procedure by using useful spatial thinking, including the spatial-temporal Voronoi neighbors, distance decay search strategy, and local route features. The overall framework of the proposed heuristic is summarized in Figure 3. After the construction of an initial solution (step 1-1), a local search heuristic based on spatial-temporal Voronoi neighborhood is followed to improve solution quality (step 1-1 to step 1-4). A spatial-temporal Voronoi distance decay strategy is developed to assign reasonable searching efforts on neighbors with different distances (step 1-2). Spatial-temporal route features are identified and guides the searching process escape from local minima (step 1-3). The stopping condition is a limitation on the number of iterations (step 1-4). Finally, the best found solution is reported (step 1-5).

[The spatial-temporal Voronoi diagram based heuristic (STVDH)]

Input: VRPTW instance,

  parameters setting:

  N_{max}: maximum number of iterations

  Z_{max} : maximum number of non-improved iterations to start spatial-temporal guided local search

  k_{max} :maximum local search Voronoi distance

  α : parameter controlling the distance decay speed.

Output: the found best solution S_{best}

Step 1-0: Read VRPTW instance and build the spatial-temporal Voronoi diagram with customer nodes.

Step 1-1: Generate an initial solution S_0 with the construction algorithm in section 4.1. Set S_{best} = S_0.

Step 1-2: Explore the solution neighborhood by the local search in section 4.2. Update S_{best} if a better solution is found. If the solution quality isn't improved in Z_{max} iterations, record the local minima S_{local}, go to step 1-3; else, repeat this step.

Step 1-3: Implement the spatial-temporal feature guided local search in section 4.3 to disturb S_{local}. Spatial-temporal features in S_{local} are identified and a modification process is followed.

Step 1-4: Check the stopping criteria with N_{max}. If the stopping criterion is accepted, go to step 1-5. Otherwise, go to step 1-2.

Step 1-5: report S_{best} and exit.

Figure 3. The framework of the spatial-temporal Voronoi diagram-based heuristic algorithm.

4.1. The Construction Algorithm

The construction algorithm iteratively inserts an unrouted node into a proper position until all customers are routed. During the insertion, instead of evaluating all unrouted nodes, neighbor customers within a given Voronoi distance are selected as candidates of the next inserted nodes. The details of the construction algorithm are described in Figure 4. First, customers are sorted by the start time window and stored in a queue L (step 2-0). Then, L's front unrouted node is popped out as the operated node μ (step 2-1). An empty route r is initialized with u (step 2-2). Another unrouted customer v is randomly selected from μ' Voronoi neighbors. Node v will be inserted before or after a routed node i in the route r (step 2-3). If the insertion is performed, u is updated with v. The best insertion place is defined as the one that minimizes the added total travel length and the total value of the end time window violation as Equation (3), where j denotes the next visited node of i. λ is a random value in [0.5, 2]. It should be noted that the arrival time at v, j should not be before the start time, as Equations (4) and (5):

$$\Delta F = \left(d_{iv} + d_{vj} - \lambda d_{ij}\right) + \bar{v}(a_i + s_i + t_{iv} - l_v) \quad (3)$$

$$a_i + s_i + t_{iv} > e_v \quad (4)$$

$$a_i + s_i + t_{iv} + s_v + t_{vj} > e_j \quad (5)$$

[Construction algorithm]

Step 2-0: Label all customers as unrouted. Sort customers with start time window and store them in a queue L. Initialize an empty route r.

Step 2-1: If L is empty, go to step 2-4;

　　Else,

　　　　pop L's front node u.

　　　　If u is unrouted, go to step 2-2;

　　　　Else, repeat step 2-1.

Step 2-2: Insert node u into an empty route r and label u as routed.

Step 2-3: Select an unrouted node v randomly from Voronoi neighbors of the last inserted node u. Find the best insertion place i, insert v before or after i, set u=v, and label v as routed. Repeat this step until there is no place for insertion, go to step 2-1.

Step 2-4: If all customers are routed, report solution and exit.

Figure 4. The construction algorithm.

During the insertion, several constraints including the start time window, route duration time, vehicle capacity, and the maximum route length should be satisfied. If no place is found for insertion, we pop L's front node as u, insert it into a new route, and label it as routed (step 2-2 and step 2-3). The insertion will end if all customers have been labeled as routed (step 2-5). Finally, obtaining all routes is the initial solution.

4.2. Local Search

Local search iteratively explores the solution space to improve solution quality. Usually, the exploration removes or adds edges from the current solution by using local search operators. Three typical local search operators, including the 1-0 exchange move, 1-1 exchange move, and 2-Opt move [51], are used to explore the neighborhood solution in this article.

- 1-0 exchange move injects a node from the original place and inserts it after another. As Figure 5a shows, node b is removed from its current position and inserted before node i.

- 1-1 exchange move swaps positions of a pair of nodes. As Figure 5b shows, the places of nodes b and i are swapped.

- 2-Opt move swaps the ends of two routes after the positions of a pair of nodes. As Figure 5c illustrates, the partial routes after b and i are swapped.

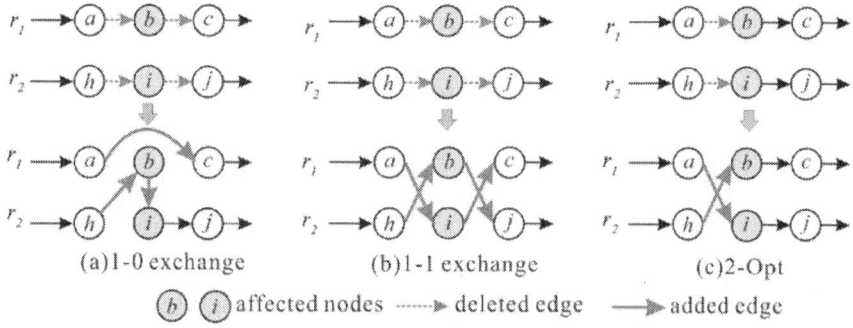

(a)1-0 exchange   (b)1-1 exchange   (c)2-Opt

b i affected nodes ------▶ deleted edge ——▶ added edge

Figure 5. Local search operators **(a)** 1-0 exchange; **(b)** 1-1 exchange; and **(c)** 2-Opt.

The computing complexity of these three operators depends on two aspects: operated nodes and local search evaluation. Two operated nodes, named b and i, are involved as shown in Figure 5. Therefore, the number of operated candidates is $n(n - 1)$, where n is the number of customers. The time complexity is $O(n2)$. For each candidate pair, local search evaluation checks the time window for each customer involved, total duration time, and vehicle capacity for each route. Due to the cascade effect of the arrival time, it will, on average, take $O(n/2R)$ time, where R is the number of routes. To utilize limited computing effort more efficiently, we propose a spatial-temporal Voronoi distance-decay strategy to assign searching efforts on selected candidates. A time warp operation is also used to shorten computing time for each candidate.

4.2.1. The Spatial-Temporal Distance Decay Strategy

Differing from the equal searching intensity of the k nearest neighbors [26] and the k-ring Voronoi neighbors [27,48], the spatial-temporal Voronoi distance decay strategy searches more on nearer neighbors than further neighbors to balance the solution quality and computing time. For spatial-temporal Voronoi neighbors with Voronoi distance k, the searching probability P_k is defined as Equation (6). As the distance k increases, searching probability P_k will decrease. Therefore, most computing effort will focus on near neighbors, but some necessary efforts are left for further neighbors.

$$P_k = \frac{\alpha^{-k}}{\sum_{k=1}^{k_{max}} \alpha^{-k}}$$

(6)

where $k = 1, ..., k_{max}$, and k_{max} is the maximum searchable Voronoi distance, $\alpha \in [1, \infty)$ is a value that controls the spatial-temporal distance decay speed. As the value of α increases, the decay speed becomes faster.

The probability range $[TP_{k-1}, TP_k]$ $(k = 1, 2, ..., k_{max})$ for neighbors no farther than Voronoi distance k is calculated as Equation (7), where $TP_0 = 0$.

$$TP_k = \sum_{i=1}^{k} P_i$$

(7)

To implement the Voronoi distance decay strategy, local search randomly generates a value of δ between (0, 1), finds the probability range in which δ is located, and then selects Voronoi neighbors with the corresponding distance k as candidates to evaluate. As it avoids evaluating all neighbors, this useful strategy puts more searching efforts on closer nodes and therefore significantly accelerates the local search. As the number of Voronoi neighbors is fixed, the computing complexity reduces from $O(n^2)$ to $O(n)$.

4.2.2. Time Warp Operation

The local search evaluates distance and time constraints for customers and routes, which cost the most computing effort [43]. Nagata [45] proposed a new penalty function in which change can be computed in $O(1)$ time for the VRPTW. When a later service happens for a customer, a time warp operation is performed such that the arrival time is adjusted to the end of the time window, and a penalty is added to the original objective. Figure 6 shows an example of the time warp. The additional penalized cost is defined as the time warp used in Equations (8) and (9), where tw_i is the cost value at customer i, and $(a_i - e_i)^+$ is a non-negative value of the service late time at i. Therefore, the objective is changed to minimize the total route number, total route length, and total penalized cost TW(s). For O(1) time consumed, the time warp operation is very fast.

$$TW(s) = \sum_{i=1}^{n} tw_i \tag{8}$$

$$tw_i = (a_i - e_i)^+ \tag{9}$$

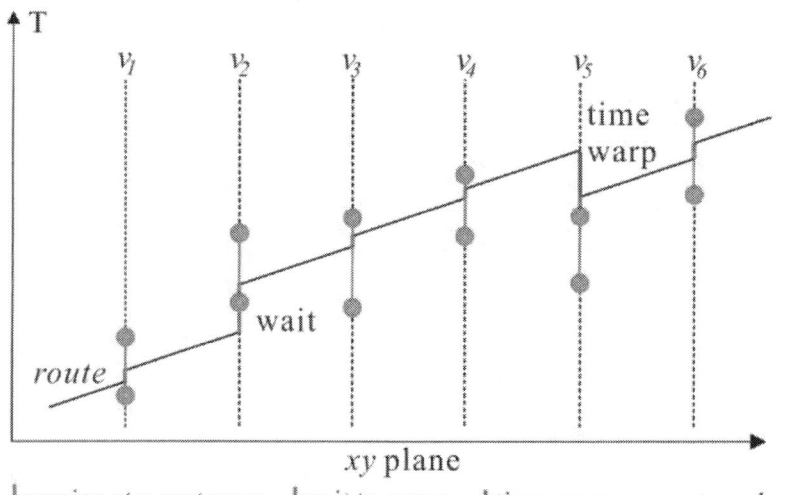

Figure 6. Wait time and time warp at a customer.

4.2.3. Acceptance Criterion

Local search will naturally drop into local minima. A threshold acceptance criterion is used to escape. If the found solution is not 0.5% worse than the current solution, it will be accepted as the new solution. Otherwise, it will be rejected. If the best solution does not improve in Z_{max} iterations, the spatial-temporal features-guided search will be started.

4.3. Spatial-Temporal Features-Guided Search

Due to the used spatial-temporal Voronoi neighbors, local search explores a relatively small proportion of the neighborhood solutions. Some potential neighborhood solutions will be rejected due to spatial and temporal constraints, such as route duration, time window, and vehicle capacity. Different from traditional random perturbations [37], spatial-temporal features guided search destroys unpromising structures defined by spatial-temporal rules and rebuilds a part of the current solution to escape local minima. It also increases the searchable probability for farther neighborhood solutions. Three components are the spatial-temporal features, the removal algorithm, and the reinsertion algorithm.

4.3.1. Spatial-Temporal Features

Three types of spatial-temporal features are defined in this paper including time window violation nodes φt, longest distance nodes φl, and smallest route's nodes φs. Their definitions are described below.

- Time-window violation nodes

  The time warp operation will introduce time window lateness in which the arrival time at a customer may be later than the end time of its time window. However, this route feature should not belong to the best solution. We remove such unpromising nodes from the current solution and reinsert them later. Formally, time window violation nodes φt are defined as Equation (10):

$$tw_i = (a_i - e_i)^+$$

$$(10)$$

- Longest-distance nodes

  To cooperate with the time window, some long segments may be kept in routes and difficult to replace in local search, which leads to the increase of total route length [27]. Longest-distance nodes φl are end points of the long segment, defined as Equation (11), where S denotes the current solution, e_{ij} or e_{ji} denotes a segment in S. e^- denotes the mean spatial distance of the Voronoi neighbors with Voronoi distance 1. The length of the long segment is more than three times e^-.

$$\varphi_l = \{v_i | d_{ij} > 3\bar{e}, e_{ij} \in S || e_{ji} \in S, v_i \in V_c\}$$

$$(11)$$

- The smallest route's nodes

  In some solutions, small routes serve only one, two, or three customers near the depot, which requires more vehicles [48]. Usually, this feature is kept for the suboptimal property. The small route's nodes are defined as φs, as Equation (12), where $|r|$ denotes the number of visited customers in the route r:

$$\varphi_l = \{v_i | v_i \in r, |r| \leq 3, v_i \in V_c\}$$

$$(12)$$

4.3.2. Removal Algorithm

To increase diversification of the guided search, more nodes are removed from the current solution in this study. Additional nodes φa are Voronoi neighbors of defined features, just as:

$$\varphi_a = \{v_i | vd(v_i, v_j) = 1, v_j \in (\varphi_t \cup \varphi_l \cup \varphi_s), v_i \notin (\varphi_t \cup \varphi_l \cup \varphi_s), v_i \in V_c\}$$

(13)

Therefore, removed nodes φ are spatial-temporal features nodes and their Voronoi neighborsφa, as in Equation (14). The removal algorithm sequentially ejects all nodes belonging to φ and leaves a partial solution for the insertion.

$$\varphi_a = \{v_i | v_i \in (\varphi_t \cup \varphi_l \cup \varphi_s \cup \varphi_a), v_i \in V_c\}$$

(14)

4.3.3. Insertion Algorithm

This algorithm inserts all removed nodes into new places in the partial solution. First, all removed nodes are randomly pushed into a queue L. Then, the front node of L is popped out and inserted at the best place before or after its Voronoi neighbors to keep spatial-temporal proximity, which minimizes insertion cost as Equation (3), under the limitation of total duration time, time window, and vehicle capacity. It is of note that the end time windows should not be violated. If there is no proper place, a new route is created, and the front node of L is popped out and initialized as the first-served customer. The insertion operation is repeated until all removed nodes are located in proper places. Due to the best choice at each insertion, the final solution is up to the node sequence in the queue. In this article, the sequence is randomly arranged 50 times to obtain different rebuild solutions. Finally, the obtained result of the permutation that minimizes the total objective is accepted as the final solution.

4.4. An Extension of the Solving Algorithm for MDVRPTW

MDVRPTW is a variant of VRPTW, which has more than one depot to serve geographically scattered customers. Compared with the VRPTW, MDVRPTW has an additional problem of how to allocate customers to the right depot. Voronoi partition that divides a study area into many regions with given objects is another typical function of the Voronoi diagram. For the effectiveness, it has been widely used in service area analysis [32]. By using Voronoi partition characteristics of the depots' spatial-temporal Voronoi diagram, we assign customers to be served by the depot of its located Voronoi regions. By that, the presented STVDH is easy to be extended to deal with a large-scale MDVRPTW by decomposing it to many smaller VRPTWs.

Figure 7 illustrates the workflow of the STVDH's adaption for the MDVRPTW. The depots' spatial-temporal Voronoi diagram is built to allocate customers to the located depot so that the MDVRPTW is divided into several VRPTWs, each of which has a depot and many assigned customers. Routes for each VRPTW are first initialized using the construction algorithm in Section 4.1 and then separately improved by the spatial-temporal Voronoi diagram-based local search heuristic in Section 4.2 and Section 4.3. To overcome the border effect of Voronoi diagram, an additional improvement between depots is used to move or exchange nodes between routes served by different depots. This is done by using 1-0 and 1-1 exchanges to move nodes between routes served by different depots after the spatial-temporal features guided search. Details of this process are referred to TU et al. [49]. The stopping condition is the same as the STDVH. Finally, the best found solution is reported.

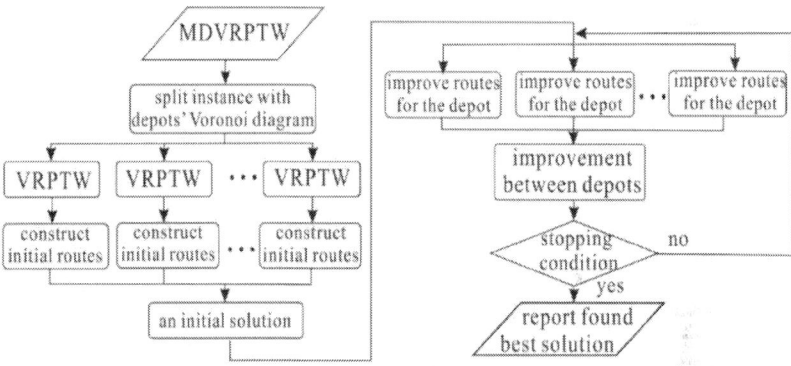

Figure 7. The adaption of the STVDH for the MDVRPTW.

5. EXPERIMENT AND COMPARISON

To assess the performance of the STVDH, experiments with benchmark problems and large-scale real-world VRPTW and MDVRPTW problems in Shanghai, China were conducted. The STDVH algorithm was implemented with C++, running on a Windows 7 64-bit system with an Intel Core i7-3.4G processor and 16 GB memory. This section reports test datasets, parameter tuning, computing results, and comparisons with other heuristic algorithms. Computing times of compared algorithms are transformed to the computing platform of STVDH with Dongarra factor [52].

5.1. Test VRPTW and MDVRPTW Problem Dataset

Two types of VRPTW and MDVRPTW problem datasets are tested. The first dataset includes the VRPTW benchmarks of Solomon [15] and Gehring and Homberger [53]. The benchmarks of Solomon have 56 problems with 100 customers. They are divided into six groups with eight to twelve problems, named R1, R2, C1, C2, RC1, and RC2. In the Euclidean plane, customers are randomly distributed (R1 and R2), cluster distributed (C1 and C2) or semi-cluster distributed (RC1 and RC2). Every instance has a single depot within the spatial domain of the customers. The travel time between nodes is equal to the corresponding Euclidean distance. R1, C1, and RC1 have a short time scheduling horizon, whereas R2, C2 and RC2 have a long time scheduling horizon. Similar to Solomon's VRPTW problems, the VRPTW benchmarks of Gehring and Homberger [53] have 300 VRPTW instances with 200, 400, 600, 800 or 1000 customers [54]. Customers have the same distributions.

The second dataset is a large-scale VRPTW and MDVRPTW problem dataset in Shanghai, China [55]. Five VRPTW instances (Sh1a–Sh5a) and 5 MDVRPTW instances (Sh1b–Sh5b) with 2000 to 10,000 customers and one to six depots are generated to simulate the daily parcel delivery service of a logistics company in Shanghai, China. Customers are randomly selected from the commercial points of interest (POI) using a professional digital navigation map to obtain real-world spatial locations. The time window of each customer is a random value in min bounded by 30 and 60, and the service duration is a random value from 1 to 4 min. Demand is a random value in the interval [1, 100]. Depots are located in professional logistic districts in the city. All vehicles have the same capacity, 2000, and maximum route duration, 480 min. The mean travel speed is set to be 45 km/h.

5.2. Parameter Tuning

There are four parameters to be tuned in the presented STVDH algorithm as shown in Figure 3. Following Coy's calibration approach [56], a preliminary experiment is conducted on an instance in the problem dataset to find the best parameter settings. The tuning process initially identifies the first two parameters related to the stopping condition and then the last two parameters related to the Voronoi distance decay strategy. The first two parameters, the maximum number of iterations, N_{max}, and the maximum number of iterations to start the spatial-temporal features guided local search, Z_{max}, control the exit of the STVDH. After intensive experiments with different stopping conditions, N_{max} should be set to 5000N to converge on a stable high-quality solution, where N is the number of customers. Z_{max} is set to 100N to start the spatial-temporal feature guided search.

The last two parameters control the Voronoi diagram-based speedup strategy. The third parameter, $kmax$, indicates the maximum searchable Voronoi neighbors. As the value of $kmax$ increases, local search will evaluate more neighbors, which requires more computing efforts. According to the distribution of the Voronoi neighbors, $kmax$ should be no more than 5 [49]. The last parameter, α, determines the distance decaying speed. A high value indicates a tendency for fast decay. The proper value of

α is within the bound [1,4]. The value space of parameters are summarized in Table 1.

Table 1. Parameter settings of the STDVH algorithm.

Notation	Parameter	Setting
N_{max}	the maximum number of iterations	5000N
Z_{max}	the maximum number of iterations to start the spatial-temporal features guided search	100N
k_{max}	the maximum searching Voronoi neighbors	1, 2, 3, 4, 5
α	Voronoi distance decaying speed	(0, 4)

N is the number of customers

Experiments on some selected VRPTW instances are conducted to set parameter values for one type of VRPTW dataset. Taking the second VRPTW dataset as an example, we solved Sh2a with all combinations of the last two parameters. Finally, we set $kmax$ = 3 and = 2 for the best performance in the experiment. Without loss of generality, identical parameter values were set for each instance. The impact of the Voronoi diagram will be discussed in Section 6. A summary of parameter settings is listed in Table 1. Taking Sh2a as an example, Figure 8 displays the convergence of the number of routes and the total length of routes with the parameter settings used in solving Sh2a.

5.3. The Results of the Solomon [15] and Gehring and Homberger [53] VRPTW Benchmarks

The parameter setting for these VRPTW benchmarks is as: N_{max} = 5000**N**, Z_{max} = 100**N**, $kmax$ = 2 and α = 1.5. For each problem, the STDVH was run 10 times. The total objective, total routes length, and computing time were recorded. Results of three state-of-the-art methods, including the arc-guided evolutionary algorithm (AGEA) [39], the hybrid genetic algorithm with adaptive diversity management (HGASDC) [40], and the local search based heuristics VRPEJ [57], which limits the searching space with the kth nearest neighbor strategy, are compared with the STVDH.

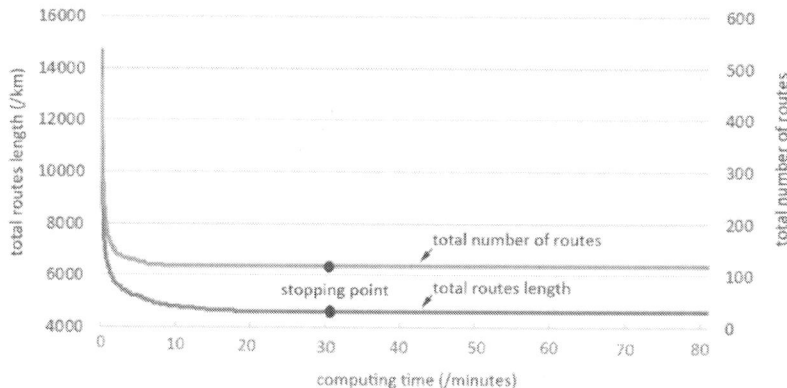

Figure 8. The convergence of the total number of routes and the total routes length (Sh2a).

Table 2 and Table 3 present the comparison of the results on the Solomom and Gehring and Homberger VRPTW benchmarks respectively. Each entry reports the average of computing results of a group (average computing time | the average total routes length). They indicate that the presented STVDH is able to achieve high quality solutions of the VRPTWs with no more than 1000 customers in a relatively short computing time. In terms of the number of routes, the STVDH achieves the fewest routes in both the Solomon (total of 405 routes) and Gehring and Homberger (total of 10296 routes) VRPTW benchmarks. In terms of the total route length, the STVDH performs better than the VRPEJ for all size VRPTW instances, only inferiorly to AGEA on Solomon benchmarks, but superiorly to AGEA on the Gehring and Homberger benchmarks. However, the STVDH's results are still worse than that of the HGSADC. For the computing efficiency, the STVDH consumes the least computing time in all six size VRPTW instances (100~1000). Therefore, the proposed STDVH exhibits a good performance on the VRPTW benchmarks of Solomon and Gehring and Homberger.

5.4. The Results of Large Scale VRPTW and MDVRPTW Problem in Shanghai, China

The parameter setting for this large scale VRPTW and MDVRPTW problem is as follow: N_{max} = 5000N, Z_{max} = 100N, $kmax$ =3 and α = 2. For each problem, the STVDH algorithm was also run 10 times. Total travel distance, number of used vehicles, and computing time were recorded. The obtained results are summarized in Table 4 (https://github.com/spatialsmart /VRPTW/tree/master/Shanghai/Solution).

Table 2. Comparison of results on the Solomon VRPTW benchmarks [15].

Instance	N	AGEA	HGSDAC	VRPEJ	STVDH	STVDH
		1 run	Best 5 run	Best 10 run	AVG 10 run	Best 10 run
R1	100	11.92|1211.43	11.92|1210.69	11.92|1214.67	11.92|1213.85	11.92|1212.65
R2	100	2.73|954.05	2.73|951.51	2.73|954.10	2.73|954.01	2.73|953.20
C1	100	10.0|828.38	10.0|828.38	10.0|828.38	10.0|828.38	10.0|828.38
C2	100	3.0|589.86	3.0|589.36	3.0|589.86	3.0|589.86	3.0|589.86
RC1	100	11.50|1384.83	11.5|1384.17	11.5|1387.81	11.5|1386.93	11.5|1386.01
RC2	100	3.25|1121.26	3.25|1119.24	3.25|1127.38	3.25|1126.49	3.25|1125.72
CNV		405	405	405	405	405
CTD		57,254.73	57,195	57,366.96	57,341.97	57,305.10
T(/min)		180	2.68	3.62	0.91	0.9
T2(/min)		52.3	1.56	3.62	0.91	0.9

Instance: group name. N: number of customers. CNV: cumulative number of vehicles. CND: cumulative travel distance. T: time reported by metaheuristics. T2: time on the computing platform of the STVDH.

Table 3. Comparison of results on the Gehring and Homberger VRPTW benchmarks [53].

Instance	N	AGEA	HGSDAC	VRPEJ	STVDH	STVDH
		1 run	Best 5 run	Best 10 run	AVG 10 run	Best 10 run
R1	200	18.2|3640.11	18.2|3613.16	18.2|3664.28	18.2|3653.24	18.2|3617.54
R2	200	4.0|2941.99	4.0|2929.41	4.0|2938.53	4.0|2967.87	4.0|2938.53
C1	200	18.9|2721.90	18.9|2718.41	18.9|2749.18	18.9|2758.67	18.9|2721.47
C2	200	6.0|1833.36	6.0|1831.59	6.0|1880.47	6.0|1858.34	6.0|1835.52
RC1	200	18.0|3224.63	18.0|3180.48	18.0|3205.81	18.0|3218.83	18.0|3179.89
RC2	200	4.3|2554.33	4.3|2536.20	4.3|2574.92	4.3|2584.37	4.3|2544.44
CNV		694	694	694	694	694
CTD		169,163	168,092	176,440.8	170,415.5	168,373.7
T(/min)		90	8.40	5.4	1.22	1.2
T2(/min)		26.7	4.9	5.4	1.22	1.2

R1	400	36.4|8514.11	36.4|8402.57	36.4|8615.29	36.4|8598.39	36.4|8446.36
R2	400	8.0|6258.82	8.0|6152.92	8.0|6274.20	8.0|6279.13	8.0|6160.84
C1	400	37.6|7273.90	37.6|7170.47	37.6|7339.88	37.6|7514.01	37.6|7186.10
C2	400	11.7|3941.70	11.6|3950.95	11.7|4024.82	11.7|3995.21	11.7|3951.71
RC1	400	36.0|8088.46	36.0|7907.14	36.0|8107.86	36.0|8098.32	36.0|7952.00
RC2	400	8.40|5516.59	8.5|5215.21	8.4|5394.54	8.4|5393.49	8.4|5292.92
CNV		1381	1381	1381	1381	1381
CTD		395,936	388,013	397,565.9	398,785.5	389,385.5
T(/min)		180	34.1	9.8	2.30	2.3
T2(/min)		53.3	19.8	9.8	2.30	2.3
R1	600	54.5|18,781.79	54.5|18,023.18	54.5|18,620.73	54.5|18,587.90	54.5|18,237.74
R2	600	11.0|12,804.60	11.0|12,352.38	11.0|12,615.07	11.0|12,612.60	11.0|12,343.51
C1	600	57.3|14,236.86	57.4|14,058.46	57.3|14,605.53	57.3|14,585.55	57.3|14,271.58
C2	600	17.4|7729.80	17.4|7594.41	17.4|7748.47	17.4|7728.74	17.4|7589.10
RC1	600	55.0|16,767.72	55.0|16,097.05	55.0|16,529.63	55.0|16,524.77	55.0|16,203.93
RC2	600	11.4|11,311.81	11.5|10,511.86	11.4|10,879.26	11.4|10,824.00	11.4|10,626.35
CNV		2066	2068	2068	2066	2066
CTD		816,326	786,793	809,986.9	808,635.65	792,722.1
T(/min)		270	99.4	16.2	3.91	3.9
T2(/min)		80.0	57.8	16.2	3.91	3.9
R1	800	72.8|32,734.57	72.8|31,311.38	72.8|32,281.48	72.8|32,108.01	72.8|31,540.28
R2	800	15.0|20,618.21	15.0|19,933.39	15.0|20,448.88	15.0|20,339.05	15.0|19,969.61
C1	800	75.2|25,911.44	75.4|24,876.93	75.2|26,097.53	75.1|25,972.63	75.1|25,490.85
C2	800	23.4|11,835.72	23.3|11,475.05	23.4|11,897.31	23.3|11,826.41	23.3|11,621.87
RC1	800	72.0|33,975.61	72.0|29,404.32	72.0|31,071.16	72.0|30,904.01	72.0|30,390.41
RC2	800	15.5|17,536.54	15.4|16,495.82	15.5|16,878.69	15.4|16,733.78	15.4|16,467.01
CNV		2739	2739	2739	2736	2736
CTD		1,424,321	1,334,963	1,386,750.4	1,378,838.8	1,354,800

T(/min)	360	215	24.8	5.82	5.8
T2(/min)	106.6	125.1	24.8	5.82	5.8
R1	91.9|51,414.26	91.9|47,759.66	91.9|49,741.43	91.9|49,396.91	91.9|48,523.49
R2	19.0|30,804.79	19.0|29,076.45	19.0|29,871.68	19.0|29,595.30	19.0|29,092.01
C1	94.2|43,111.60	94.1|41,572.86	94.1|43,089.45	94.1|42,682.27	94.1|41,977.06
C2	29.3|16,810.22	28.8|16,796.45	29.0|117,340.13	28.9|17,174.73	28.8|16,877.68
RC1	90.0|46,753.61	90.0|44,333.40	90.0|46,152.97	90.0|45,824.65	90.0|44,974.63
RC2	18.4|25,588.52	18.2|24,131.12	18.4|24,951.31	18.3|24,655.48	18.2|24,248.11
CNV	3428	3420	3421	3419	3419
CTD	2,144,830	2,036,700	2,111,469.7	2,093,293.5	2,056,930
T(/min)	450	349	34.5	7.75	7.7
T2(/min)	133.3	203.1	34.5	7.75	7.7

Instance: group name. N: number of customers. CNV: cumulative number of vehicles. CND: cumulative travel distance. T: time reported by metaheuristics. T2: time on the computing platform of the STVDH.

As Table 4 indicates, the STVDH algorithm reports the best found solution for large-scale VRPTWs up to 10,000 customers in 139.5 min (about 2.32 hours). As the number of customers increases from 2000 to 10,000, the computing time of the STVDH increases from 15.7 min to 120.8 min, whereas for the MDVRPTW, the time increases from 16.1 min to 139.5 min. Compared with the VRPTW, more computing efforts are required for the MDVRPTW due to additional customer allocation between depots. In terms of solution quality, the STVDH utilizes an average of 847.1 total routes that travelled 30,184.725 km to serve all customers in the five VRPTW instances. For the five MDVRPTW instances, 906.4 routes that travelled 26,353.572 km are required to satisfy customer needs. It should be noted that the performance of the STVDH is robust, as indicated by the gap between S_{avg} and S_{best} (−1.60% for the VRPTW and −1.73% for the MDVRPTW). Therefore, the proposed STVDH provides promising results for large-scale VRPTW and VRPTW within reasonable computing times.

Taking problem Sh1a as an example, Figure 9 displays the best found solution and the spatial-temporal vehicle route with ArcScene. It demonstrates the best route compromise between spatial proximity and time constraints. The same vehicle may not yet serve spatially-near customers as Figure 9b shows.

To assess the solution quality of the obtained results, we compare them with solutions reported by two heuristic algorithms. One solution is obtained by solving the simulated instances with the network analysis module in ArcGIS™ 10.2, which uses a tabu heuristics algorithm to solve complex real-world VRPTW problems. Another compared solution is calculated using VRPEJ [55]. Other metaheuristics such as AGEA and HGSDAC are not compared due to unavailability. Each algorithm was run 10 times to find the best result.

Table 4. Results from the large-scale VRPTW and MDVRPTW instances in Shanghai, China.

Instance	Type	N	M	D	Q	STVDH			
						S_{avg} 10 (/km)	T_{avg} 10 (/min)	S_{Best} 10(/km)	T_{Best} 10 (/min)
Sh1a	VRPTW	2000	1	200	2000	59.3|3024.870	15.7	58|2996.104	15.9
Sh2a	VRPTW	4000	1	300	2000	119.8|4674.122	28.3	118|4598.592	30.6
Sh3a	VRPTW	6000	1	400	2000	171.0|6145.245	60.0	169|6052.084	65.9
Sh4a	VRPTW	8000	1	500	2000	220.3|7511.361	93.6	218|7372.345	98.9
Sh5a	VRPTW	10,000	1	600	2000	276.8|8829.127	120.8	274|8681.006	128.8
CNV						847.1		837	
CTD(/km)						30,184.725		29,701.132	
Gap to S_{avg}								−1.60%	
Total Time (/min)							318.4		340.1
Sh1b	MDVRPTW	2000	2	200	2000	70.2|3045.601	16.1	69|2999.616	16.9
Sh2b	MDVRPTW	4000	3	300	2000	127.9|4265.365	29.8	126|4174.144	31.5
Sh3b	MDVRPTW	6000	4	400	2000	172.1|6247.152	62.2	169|5260.810	69.6
Sh4b	MDVRPTW	8000	5	500	2000	237.4|6849.848	98.4	235|6238.438	105.4
Sh5b	MDVRPTW	10,000	6	600	2000	298.8|6845.606	138.9	293|6784.751	139.5
CNV						906.4		892	
CTD (/km)						26,353.572		26,289.418	
Gap to S_{avg}								−1.73%	
Total Time (/min)							345.4		362.9

Instance: problem name. Type: VRPTW or MDVRPTW; N: number of customers. M: number of depots. D: number of vehicles at a depot. Q: vehicle capacity. S_{avg} 10: average result of 10 runs (number of routes\| total routes length). T_{avg} 10: average computing time. S_{Best} 10: best of 10 runs (number of routes\| total routes length). T_{Best} 10: the computing time of the best solution's run. CNV: cumulative number of vehicles. CND: cumulative travel distance.

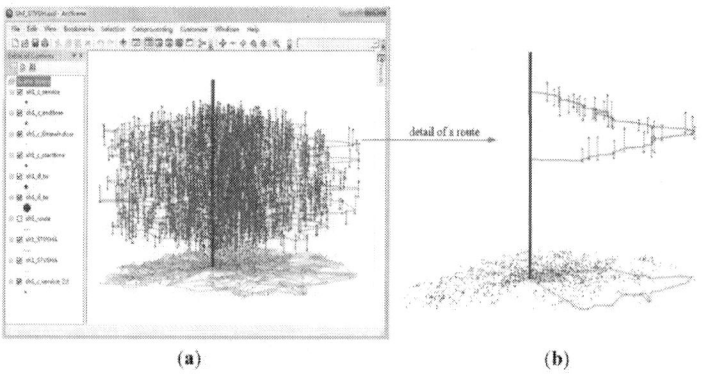

(a)   (b)

Figure 9. The result for large scale VRPTW problem Sh1a. **(a)** the final best solution; and **(b)** a spatial-temporal route.

Table 5 compares the obtained results. It indicates that the STVDH presented in this study outperforms both ArcGIS and VRPEJ. In terms of efficiency, comparison with ArcGIS (total of 2141.1 min) and VRPEJ (total of 2355.3 min) indicates that the STVDH costs the least computing effort (total of 703 min). In terms of solution quality, the STVDH requires 1729 routes travelling 55,990.55 km to serve all customers in 10 instances. ArcGIS requires more vehicles (1764 routes) that travel a greater distance (62,047.809 km) to provide the same service. The VRPEJ, which utilizes a spatially kth nearest neighbors strategy, requires the most routes (1893 routes) that travelled 63,659.365 km. Hence, it confirms that the spatial-temporal Voronoi neighbor strategy in the presented STVDH is much better than the spatial neighbor strategy for local search to solve large scale VRPTWs.

Table 5. Comparison of the results of the STVDH with other heuristic algorithms.

Instance	Type	N	M	STVDH		ArcGIS		VRPEJ	
				S_{Best} 10(/km)	T_{Best} 10 (/min)	S_{Best} 10(/km)	T_{Best} 10 (/min)	S_{Best} 10(/km)	T_{Best} 10 (/min)
Sh1a	VRPTW	1	2000	58I2996.104	15.9	62I3296.104	60.5	59I3285.404	50.8
Sh2a	VRPTW	1	4000	118I4598.592	30.6	120I4022.248	125.4	126I5234.168	122.2
Sh3a	VRPTW	1	6000	169I6052.084	65.9	173I6738.745	197.2	190I6948.372	242.7
Sh4a	VRPTW	1	8000	218I7372.345	98.9	223I8045.327	254.8	242I8438.982	290.8
Sh5a	VRPTW	1	10,000	274I8681.006	128.8	278I9874.135	407.2	312I9814.247	428.8
Sh1b	MDVRPTW	2	2000	69I2999.616	16.9	72I3308.245	60.5	74I3374.126	58.1
Sh2b	MDVRPTW	3	4000	126I4174.144	31.5	124I4610.283	125.4	132I4878.785	142.3
Sh3b	MDVRPTW	4	6000	169I5260.810	69.6	174I6645.397	197.2	180I7013.522	248.5
Sh4b	MDVRPTW	5	8000	235I6238.438	105.4	239I6982.136	254.8	260I6989.971	280.5
Sh5b	MDVRPTW	6	10,000	293I6784.751	139.5	299I7525.189	568.1	339I7881.788	490.6

6. DISCUSSION

Taking the large scale VRPTW and MDVRPTW in Shanghai, China as an example, this section discusses the impact of the spatial-temporal Voronoi diagram and spatial-temporal proximity behind the best found results.

6.1. Impact of the Spatial-Temporal Voronoi Diagram

To evaluate the impact of the Voronoi diagram, this study investigated the balance of solution quality and computing time with different values for parameters $kmax$ and α. As Figure 10 illustrates, considering the spatial-temporal Voronoi neighbors, an increase in $kmax$ improves solution quality but decreases computing efficiency because more spatial-temporal Voronoi neighbors are involved in local search procedures. With regard to the effect of the Voronoi distance decay strategy, as Figure 10a shows, both the slow decaying speed ($\alpha = 1$) and the fast decaying speed ($\alpha = 3$) generate worse results. However, as Figure 10b shows, as a fast decaying speed requires less evaluation on far Voronoi neighbors, the STVDH consumes less computing efforts as the parameter α increases.

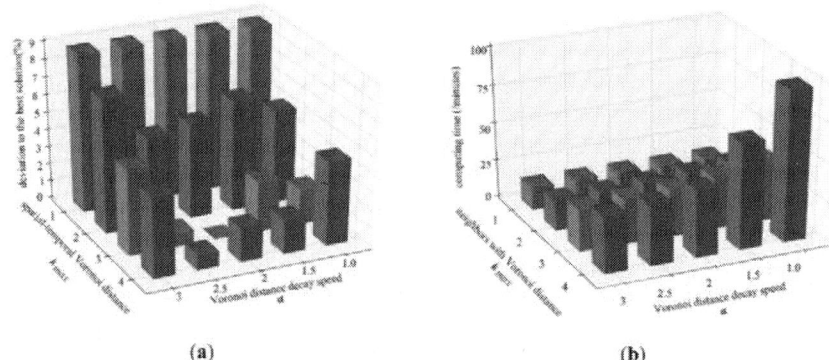

(a)   (b)

Figure 10. The impact of the spatial-temporal Voronoi diagram on the STVDH. (**a**) Deviation to the best solution; and (**b**) computing time.

6.2. Spatial-Temporal Proximity Analysis on the Best Found Results

To understand spatial-temporal proximity behind the obtained results, we conducted an analysis on the Voronoi distance between consecutively-visited customers in the route of best found solutions in Section 5.2. Figure 11 displays the average percentage of Voronoi distance in the best found solutions. Two typical features are indicated below.

- The number of larger Voronoi distances is only a small proportion of the best found results. As Figure 11 displays, the percentages of Voronoi distances greater than three in the VRPTW and MDRPTW solutions are only 1.77% and 1.58%, respectively. Such a distribution agrees with the setting of parameter *kmax* (*kmax*= 3) in Section 5.2. Therefore, the spatial-temporal Voronoi neighborhood is very typical in the found solution.
- The percentage decreases sharply as the Voronoi distance increases. For the large-scale VRPTW dataset (Sh1a–Sh5a), the percentages for Voronoi distances 1, 2, 3, and >= 4 are 59.44%, 30.13%, 8.66%, and 1.77%, respectively, which is similar to the distribution in the MDVRPTW solution. This verifies that there is a spatial-temporal local compact structure in the routes of best found solutions. Compared with the theoretical searching intensity (as Equation (7) in Section 4.2.1 where *kmax*=3, α=2) in the Voronoi distance decay strategy, these percentages systematically slightly shift to small Voronoi distances. This result demonstrates the effectiveness of the Voronoi distance-

decay strategy, which searches more on near neighbors in the local search but still spends necessary efforts on far neighbors.

In summary, analysis of best found results demonstrates reasons why the spatial-temporal Voronoi diagram is effective in the STVDH algorithm.

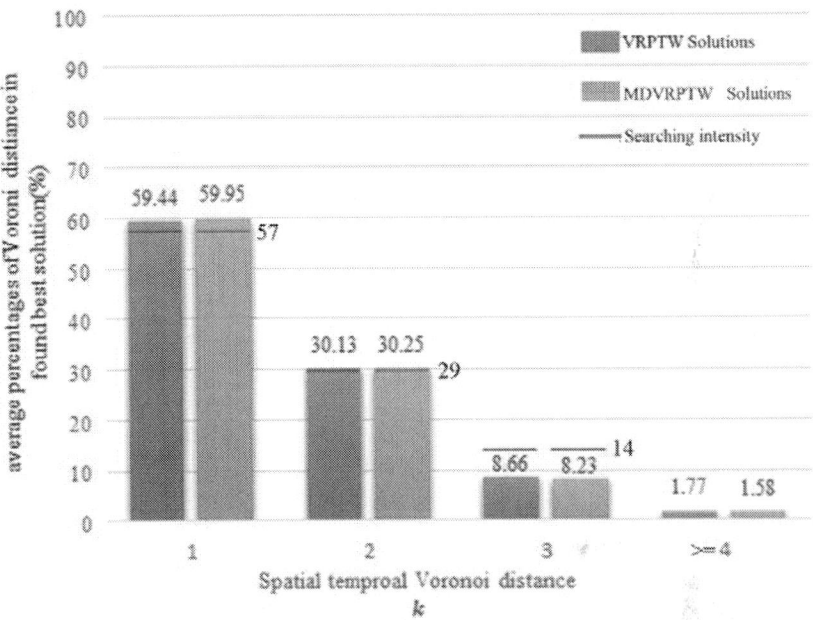

Figure 11. Average percentage of the Voronoi distance in the best found solutions.

7. CONCLUSIONS

Vehicle routing designs the least costly routes to satisfy the geographical distribution of human needs. Embedded in the GIS environment, it not only benefits transportation decision-making for both public and private sectors, but also enriches the value of geoinformation with the assistance of location-based service. Inspired by spatial-temporal proximity, this article presents a novel spatial-temporal Voronoi diagram-based heuristic approach to solve large-scale VRPTWs quickly. The derived spatial-temporal Voronoi neighborhood measures proximity considering both spatial and temporal issues. The used Voronoi neighbors limits searching space in the construction algorithm, local search, and spatial-temporal feature-guided search. Moreover, the presented Voronoi distance-decay strategy assigns more searching efforts on spatial-temporal near neighbors.

Experiments on the VRPTW benchmarks and large-scale VRPTW instances in Shanghai, China with 2000–10,000 customers have demonstrated the good performance on VRPTW and MDVRPTW problems of different sizes. The obtained results indicate that the STVDH presented in this study can provide high quality solutions for large-scale VRPTWs in a reasonable time with the help of a spatial-temporal Voronoi diagram. It also verifies that the spatial-temporal proximity is very typical in best found solutions as 98% of segments had a Voronoi distance less than three.

The main contributions of this article are twofold. First, considering both time windows and spatial locations, the spatial-temporal Voronoi diagram is proposed to accelerate the solving process for VROs. In contrast to the spatial proximity based speeding up strategies [25,26,27], it makes use of spatial-temporal Voronoi neighbor to further reduce computing effort for local search based heuristics. The idea behind the presented approach that makes uses of spatial principles to accelerate complex optimizing processes can be extended to facilitate other space-related decision-making problems such as near real-time emergency response and large-scale facility location. Second, a spatial-temporal Voronoi diagram-based heuristic was developed to solve large-scale VRPTWs. This novel approach exhibits good performance on super large-scale VRPTWs up to 10,000 customers, which can cope with challenges from many complex real-world transportation and logistics applications.

Among future developments that we intend to undertake, we plan to integrate the presented approach with outdoor/indoor ubiquitous positioning and friendly navigating technologies for vehicles with cloud GIS. The presented effective approach will be used to upgrade the network analyst module in SDSSs to provide a more flexible transportation service for daily large-scale logistics and distribution activities. It will further enhance the spatial intelligence of modern transportation and logistics applications.

REFERENCES

1. Miller, H.J.; Shaw, S.L. Geographic Information Systems for Transportation: Principles and Applications; Oxford University Press: Oxford, UK, 2001.
2. Lee, E.; Oduor, P. Using multi-attribute decision factors for a modified all-or-nothing traffic assignment. ISPRS Int. J. Geo-Inf.2015, 4, 883–899.
3. Huang, Z.; Liu, X. A hierarchical approach to optimizing bus stop distribution in large and fast developing cities. ISPRS Int. J. Geo-Inf.2014, 3, 554–564.

4. Schittekat, P.; Kinable, J.; Sörensen, K.; Sevaux, M.; Spieksma, F.; Springael, J. A metaheuristic for the school bus routing problem with bus stop selection. Eur. J. Oper. Res.**2013**, 229, 518–528.

5. Mendoza, J.E.; Medaglia, A.L.; Velasco, N. An evolutionary-based decision support system for vehicle routing: The case of a public utility. Decis. Support. Syst.**2009**, 46, 730–742.

6. Arribas, C.A.; Blazquez, C.A.; Lamas, A. Urban solid waste collection system using mathematical modelling and tools of geographic information systems. Waste. Manage. Res.**2010**, 28, 355–363.

7. Weigel, D.; Cao, B. Applying GIS and OR techniques to solve sears technician-dispatching and home-delivery problems. Interfaces**1999**, 29, 112–130.

8. Kuo, P.-F.; Lord, D.; Walden, T.D. Using geographical information systems to organize police patrol routes effectively by grouping hotspots of crash and crime data. J. Transp. Geogr.**2013**, 30, 138–148.

9. Janssens, J.; van den Bergh, J.; Sörensen, K.; Cattrysse, D. Multi-objective microzone-based vehicle routing for courier companies: From tactical to operational planning. Eur. J. Oper. Res.**2015**, 242, 222–231.

10. Thill, J.C. Geographic information systems for transportation in perspective. Transp. Res. C-Emer.**2000**, 8, 3–12.

11. Antikainen, H. Using the hierarchical pathfinding A* algorithm in GIS to find paths through rasters with nonuniform traversal cost. ISPRS. Int. J. Geo.-Inf.**2013**, 2, 996–1014.

12. Vanhove, S.; Fack, V. An effective heuristic for computing many shortest path alternatives in road networks. Int. J. Geogr. Inf. Sci.**2012**, 23, 1031–1050.

13. Li, R.; Leung, Y.; Huang, B.; Lin, H. A genetic algorithm for multiobjective dangerous goods route planning. Int. J. Geogr. Inf. Sci.**2012**, 27, 1073–1089.

14. Curtin, K.M.; Voicu, G.; Rice, M.T.; Stefanidis, A. A comparative analysis of traveling salesman solutions from geographic information systems. Trans. GIS.**2014**, 18, 286–301.

15. Solomon, M.M. Algorithms for the vehicle routing and scheduling problems with time window constraints. Oper. Res.**1987**, 35, 254–265.

16. Bräysy, O.; Gendreau, M. Vehicle routing problem with time windows, part I: Route construction and local search algorithms. Transp. Sci.**2005**, 39, 104–118.

17. Braysy, O.; Gendreau, M. Vehicle routing problem with time windows, part II: Metaheuristics. Transp. Sci.**2005**, 39, 119–139.

18. Qi, M.; Lin, W.-H.; Li, N.; Miao, L. A spatiotemporal partitioning approach for large-scale vehicle routing problems with time windows. Transp. Res. E-Log.**2012**, 48, 248–257.

19. Ray, J.J. A web-based spatial decision support system optimizes routes for oversize/overweight vehicles in delaware. Decis. Support. Syst.**2007**, 43, 1171–1185.

20. Huang, B.; Cheu, R.L.; Liew, Y.S. GIS and genetic algorithms for hazmat route planning with security considerations. Int. J. Geogr. Inf. Sci.**2004**, 18, 769–787.

21. Huang, B.; Fery, P.; Xue, L.; Wang, Y. Seeking the pareto front for multiobjective spatial optimization problems. Int. J. Geogr. Inf. Sci.**2008**, 22, 507–526.

22. Wright, D.J.; Wang, S. The emergence of spatial cyberinfrastructure. P. Natl. Acad. Sci.**2011**, 108, 5488–5491.

23. Tong, D.; Murray, A.T. Spatial optimization in geography. Ann. Amer. Geogr.**2012**, 102, 1290–1309.

24. Li, X.; He, J.; Liu, X. Ant intelligence for solving optimal path-covering problems with multi-objectives. Int. J. Geogr. Inf. Sci.**2009**, 23, 839–857.

25. Toth, P.; Vigo, D. The granular tabu search and its application to the vehicle-routing problem. Inf. J. Comput.**2003**, 15, 333–346.

26. Li, F.; Golden, B.; Wasil, E. Very large-scale vehicle routing: New test problems, algorithms, and results. Comput. Oper. Res.**2005**, 32, 1165–1179.

27. Fang, Z.; Tu, W.; Li, Q.; Shaw, S.-L.; Chen, S.; Chen, B.Y. A Voronoi neighborhood-based search heuristic for distance/capacity constrained very large vehicle routing problems. Int. J. Geogr. Inf. Sci.**2013**, 27, 741–764.

28. Tu, W.; Fang, Z.; Li, Q. The empirical Voronoi veighborhood analysis of heuristic solutions for vehicle routing problems. In Proceedings of 2013 International Symposium on Recent Advances in Transport Modelling, Goldcoast, Australia, 21–23 April 2013.

29. Okabe, A.; Boots, B.; Sugihara, K.; Chiu, S.N. Spatial Tessellations: Concepts and Applications of Voronoi Diagrams; John Wiley & Sons: Hoboken, NJ, USA, 2000.

30. Chen, J.; Zhao, R.; Li, Z. Voronoi-based k-order neighbour relations for spatial analysis. ISPRS. J. Photogram. Remote Sens.**2004**, 59, 60–72. Zhao, R.; Chen, J.; Li, Z. K-order spatial neighbours based on Voronoi diagram: Description, computation and applications. Int. Arc. Photogram. Remote Sens. Spat. Inf. Sci.**2002**, 34, 10–15.

31. Tu, W.; Li, Q.; Fang, Z. Large scale multi-depot logistics routing optimization based on network Voronoi diagram. Acta Geod. et Cartogr. Sin.**2014**, 43, 1075–1082.

32. Laporte, G. Fifty years of vehicle routing. Transp. Sci.**2009**, 43, 408–416.

33. Baños, R.; Ortega, J.; Gil, C.; Fernández, A.; de Toro, F. A simulated annealing-based parallel multi-objective approach to vehicle routing problems with time windows. Expert. Syst. Appl.**2013**, 40, 1696–1707.

34. Alabas-Uslu, C.; Dengiz, B. A self-adaptive local search algorithm for the classical vehicle routing problem. Expert. Syst. Appl.**2011**, 38, 8990–8998.

35. Ropke, S.; Pisinger, D. An adaptive large neighborhood search heuristic for the pickup and delivery problem with time windows. Transp. Sci.**2006**, 40, 455–472.

36. Polacek, M.; Benkner, S.; Doerner, K.F.; Hartl, R.F. A cooperative and adaptive variable neighborhood search for the multi depot vehicle routing problem with time windows. BuR-Business Res.**2008**, 1, 207–218.

37. Cordeau, J.F.; Laporte, G.; Mercier, A. A unified tabu search heuristic for vehicle routing problems with time windows. J. Oper. Res. Soc.**2001**, 52, 928–936.

38. Repoussis, P.P.; Tarantilis, C.D.; Ioannou, G. Arc-guided evolutionary algorithm for the vehicle routing problem with time windows. IEEE. Trans. Evol. Comput.**2009**, 13, 624–647.

39. Vidal, T.; Crainic, T.G.; Gendreau, M.; Prins, C. A hybrid genetic algorithm with adaptive diversity management for a large class of vehicle routing problems with time-windows. Comput. Oper. Res.**2012**, 40, 475–489.

40. Mester, D.; Bräysy, O. Active guided evolution strategies for large-scale vehicle routing problems with time windows. Comput. Oper. Res.**2005**, 32, 1593–1614.

41. Gehring, H.; Homberger, J. Parallelization of a two-phase metaheuristic for routing problems with time windows. J. Heuristics.**2002**, 8, 251–276.

42. Ibaraki, T.; Imahori, S.; Kubo, M.; Masuda, T.; Uno, T.; Yagiura, M. Effective local search algorithms for routing and scheduling problems with general time-window constraints. Transp. Sci.**2005**, 39, 206–232.

43. Nagata, Y.; Bräysy, O.; Dullaert, W. A penalty-based edge assembly memetic algorithm for the vehicle routing problem with time windows. Comput. Oper. Res.**2010**, 37, 724–737.

44. Dondo, R.; Cerdá, J. A cluster-based optimization approach for the multi-depot heterogeneous fleet vehicle routing problem with time windows. Eur. J. Oper. Res.**2007**, 176, 1478–1507.

45. Milthers, N.P.M. Solving VRP Using Voronoi Diagrams and Adaptive Large Neighborhood Search. Master's thesis, University of Copenhagen, Copenhagen, Denmark, 2009.

46. Santos, L.; Coutinho-Rodrigues, J.; Antunes, C.H. A web spatial decision support system for vehicle routing using google maps. Deci. Support. Syst.**2011**, 51, 1–9.

47. Tu, W.; Li, Q.; Chang, X.; Yue, Y.; Zhu, J. A Spatio-temporal decision support framework for large scale logistics distribution in Metropolitan area. In Advances in Spatial Data Handling and Anslysis; Harvey, F., Leung, Y., Eds.; Springer: Berlin, Germany, 2015; pp. 193–206.

48. Tu, W.; Fang, Z.; Li, Q.; Shaw, S.-L.; Chen, B.Y. A bi-level Voronoi diagram based heuristics for large scale mutli-depot vehicle routing problem with time window. Transp. Res. E-Log**2014**, 61, 84–97.

49. Barber, C.B.; Dobkin, D.P.; Huhdanpaa, H.T. The quickhull algorithm for convex hulls. ACM. T. Math. Softw.**1996**, 22, 469–483.

50. Zachariadis, E.E.; Kiranoudis, C.T. A strategy for reducing the computational complexity of local search-based methods for the vehicle routing problem. Comput. Oper. Res.**2010**, 37, 2089–2105.

51. Dongarra, J. Performance of Various Computers Using Standard Linear Equations Software. Available online: http://www.netlib.org/benchmark/performance.ps (accessed on 30 August 2015).

52. Gehring, H.; Homberger, J. A parallel hybrid evolutionary metaheuristic for the vehicle routing problem with time windows. In Proceedings of the EUROGEN99, Jyväskylä, Finland, 30 May–3 June 1999; pp. 57–64.

53. The VRPTW Benchmarks of GEHRING and HOMBERGER. Available Online: http://www.bernabe.dorronsoro.es/vrp/ (accessed on 10 October 2015).

54. The Large-Scale VRPTW and MDVRPTW Problem Dataset in Shanghai, China. Available Online: https://github.com/spatialsmart/VRPTW/tree/master/Shanghai/Problem (accessed on 10 October 2015).

55. Coy, S.; Golden, B.; Runger, G.; Wasil, E. Using experimental design to find effective parameter settings for heuristics. J. Heuristics.**2001**, 7, 77–97. [Google Scholar] [CrossRef]

56. Groër, C.; Golden, B.; Wasil, E. A library of local search heuristics for the vehicle routing problem. Math. Program. Comput.**2011**, 2, 79–101.

Heuristic Evaluation on Mobile Interfaces: A New Checklist

Rosa Yáñez Gómez, Daniel Cascado Caballero, and José-Luis Sevillano

*Department of Computer Technology and Architecture, ETS Ingenieria
Informatica, Universidad de Sevilla, Avenida Reina Mercedes s/n. 41012
Seville, Spain*

ABSTRACT

The rapid evolution and adoption of mobile devices raise new usability
challenges, given their limitations (in screen size, battery life, etc.) as well as
the specific requirements of this new interaction. Traditional evaluation
techniques need to be adapted in order for these requirements to be met.
Heuristic evaluation (HE), an Inspection Method based on evaluation
conducted by experts over a real system or prototype, is based on checklists
which are desktop-centred and do not adequately detect mobile-specific
usability issues. In this paper, we propose a compilation of heuristic
evaluation checklists taken from the existing bibliography but readapted to
new mobile interfaces. Selecting and rearranging these heuristic guidelines
offer a tool which works well not just for evaluation but also as a best-
practices checklist. The result is a comprehensive checklist which is
experimentally evaluated as a design tool. This experimental evaluation
involved two software engineers without any specific knowledge about
usability, a group of ten users who compared the usability of a first
prototype designed without our heuristics, and a second one after applying
the proposed checklist. The results of this experiment show the usefulness
of the proposed checklist for avoiding usability gaps even with nontrained
developers.

1. INTRODUCTION

Usability is the extent to which a product can be used with effectiveness, efficiency, and satisfaction in a specified context of use [1]. While usability evaluation of traditional browsers from pc environments—desktop or laptop—has been widely studied, mobile browsing from smartphones, touch phones, and tablets present new usability challenges [2]. Additionally, mobile browsing is becoming increasingly widespread as a way of accessing online information and communicating with other users. Specific usability evaluation techniques adapted to mobile browsing constitute an interesting and increasingly important study area.

Usability evaluation assesses the ease of use of a website's functions and how well they enable users to perform their tasks efficiently [3]. To carry out this evaluation, there are several usability evaluation techniques.

Usability evaluation techniques can be classified as shown in Figure 1 [4–8]. Over real systems or prototypes, the best alternatives are evaluations conducted by experts, also known as Inspection Methods, or evaluations involving users, which are divided into inquiry methods and testing methods depending on the methodology adopted. With a more academic focus, predictive evaluation offers some predictions over the usability of a potential and not-yet-existent prototype.

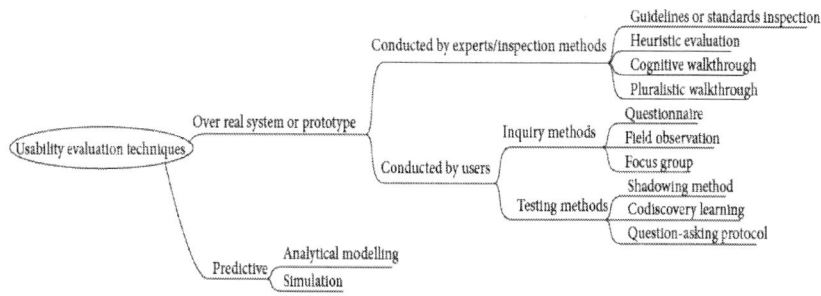

Figure 1: Classification of some usability evaluation techniques.

Heuristic evaluation (HE) is an inspection method based on evaluation over real system or prototype, conducted by experts. The term "expert" is used as opposed to "users" but in many cases evaluators do not need to be usability experts [9, 10]. In HE, experts check the accomplishment of a given heuristic checklist. Due to its nature, this inspection cannot be performed automatically.

HE, like other usability assurance techniques, has to take into account the fact that usability is not intrinsically objective in nature but is rather closely intertwined with an evaluator's personal interpretation of the artefact and his or her interaction with it [11]. But, evaluations can be designed to compensate for personal interpretation as much as possible.

Moreover, inspection methods are often criticized for only being able to detect a small number of problems in total together with a very high number of cosmetic ones [12]. But, HE presents several advantages over other techniques: its implementation is easy, fast, and cheap, and it is suitable for every life-cycle software phase and does not require previous planning [7]. Furthermore, it is not mandatory for evaluators to be usability experts [9, 10]. It is possible for engineers or technicians with basic usability knowledge to drive an evaluation. Furthermore, regarding the number of evaluators, Nielsen demonstrated empirically that between three and five experts should be enough [13].

Because of all these advantages, HE is a convenient usability evaluation method: the worst usability conflicts are detected at a low cost. But, traditional HE checklists are desktop-centred and do not properly detect mobile-specific usability issues [2].

In this study, we propose a heuristic guideline centred in mobile environments based on a review of previous literature. This mobile-specific heuristic guideline is not only an evaluation tool but also a compilation of recommended best-practices. It can guide the design of websites or applications oriented to mobile devices taking usability into account.

The following section describes the methods followed to define the mobile heuristic guideline. Then, Results and Discussion section is divided according to the steps defined in the methodology. We have included a brief discussion of the results for each task. The final sections include Conclusions and Future Work, Acknowledgments, and References.

2. METHODS

To obtain a heuristic guideline centred in mobile environments and based on a review of previous literature, we will follow a six-step process.

1) A clear definition of the problem scope is necessary as a first step to define and classify the special characteristics of mobile interaction.

2) Next, we rearrange existing and well-known heuristics into a new compilation. We can reuse heuristic guidelines from the literature and adapt them to the new mobile paradigm because heuristic checklists derive from human behaviour, not technology [14]. This heuristics is general checks that must be accomplished in order to achieve a high level of usability.

3) After building this new classification of heuristics, we will develop a compilation of different proposed subheuristics. "Heuristic" in this paper refers to a global usability issue that must be evaluated or taken into account when designing. In contrast, the term "subheuristic" refers

to specific guidelines items. The main difference between the two concepts lies in the level of expertise required of the evaluator and the abstraction level of the checklist. The resulting selection of subheuristics in this step takes into account some of the mobile devices restrictions presented in the first step. But, the result of this stage does not include many mobile specific questions, as they are not covered in traditional heuristic guidelines.

4) The fourth step in this work consists of enriching the list with mobile-specific subheuristics. This subheuristics is gleaned from mobile usability studies and best practices proposed in the literature.

5) One further step is required to homogenize the redaction and format of subheuristics in order to make it useful for nonexperts.

6) Finally, we conduct an evaluation of the usefulness of the tool as an aid in designing for mobile.

This process differs slightly from the methodology proposed by Rusu et al. [15], but we can subsume their phases when establishing new usability heuristics in our proposed method.

It is worth remarking that popular mobile operating systems are now providing usability guidelines [16, 17] which focus mainly on maintaining coherent interaction and presentation through applications over the whole platform. These guidelines could in some cases enrich certain aspects of our proposal, although we have opted to keep it essentially agnostic of specific platforms aesthetics or coherence-determined restrictions.

Additionally, interfaces for mobile are mainly divided into web access and native applications. We do not restrict our study to a specific kind of interface. Again, the goal is to elaborate a guideline which is independent of specific technologies. The interaction between users and mobile interfaces is similar regardless of the piece of software they are using.

3. RESULTS AND DISCUSSION

3.1. Problem Scope Definition

Users are increasingly adopting mobile devices. According to statistics of Pew Internet & American Life Project [18], only in the USA 35% of adults own smartphones and 83% of adults own a cell phone of some kind. Additionally, 87% of smartphones owners access the Internet or email on their handheld—68% on a typical day. A further 25% say that they mostly go online using their smartphone, rather than a computer. This survey shows that phones operating on the Android platform are currently the most prevalent type, followed by iPhones and Blackberry devices.

Mobile usability involves different kind of devices, contexts, tasks, and users. The compilation of a new heuristic guideline needs a restriction and definition of the scope of the user-interface interaction.

Devices can be divided in three types [19]:

I.   feature phones: they are basic handsets with tiny screens and very limited keypads that are suitable mainly for dialing phone numbers;

II. Smartphones: phones with midsized screens and full A–Z keypads;

III. touch phones/touch tablets: devices with touch-sensitive screens that cover almost the entire front of the phone.

In our study, we have ruled out feature phones because the interaction and interface design are deeply restricted and they are gradually being abandoned by a wide range of users. We have also ruled out smartphones because interaction is dramatically different due to the keyboard and they are commonly constrained to enterprise use. This study focuses on the ubiquitous touch phones and touch tablets. In this work, we use the term "touch phones" to refer to both phones and tablets because they share a similar interaction paradigm and the constraints we describe in Figure 2.

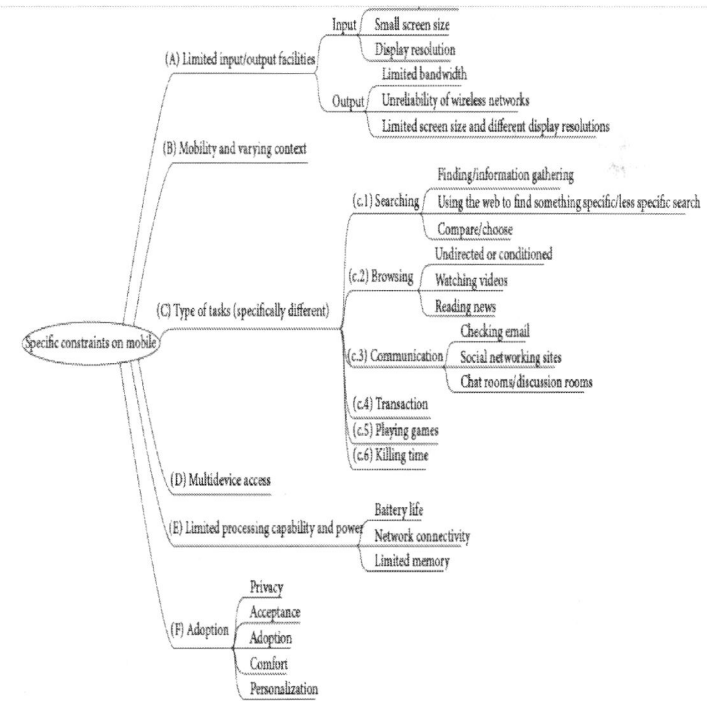

Figure 2: Specific constraints on mobile.

Mobile interactions define a new paradigm characterized by a wide range of specific constraints: hardware limitations, context of use, and so forth. All these restrictions have been studied in the bibliography in order to define the issues that must be overcome to improve usability. According to the literature, the main constraints when designing for mobile devices are (Figure 2):

A. *limited input/output facilities* [20–24]: these limitations are imposed by data entry methods, small screen size, display resolution, and available bandwidth, as well as unreliability of wireless networks;

B. *mobility and varying context* [20–23]: traditional usability evaluation techniques have often relied on measures of task performance and task efficiency. Such evaluation approaches may not be directly applicable to the often unpredictable, rather opportunistic and relatively unstable mobile settings. Mobile devices use is on-the-run and interactions may take from a few seconds to minutes, being highly context-dependent. Environmental distractions have a significant effect on mobile interfaces usability and hence they need to be taken into account [25]. Context of use involves background noise, ongoing conversations, people passing by, and so on. Distractions can be auditory, visual, social, or caused by mobility. The context of use is so influent in the interaction that many authors propose testing in the field as indispensable to study interaction with mobile devices [26]. Laboratory testing seems incapable of completely assuring usability in this mobile paradigm. Some attempts to cover this contextual information have been documented in the literature: Po et al. [27] proposed inclusion of contextual information into the heuristic evaluation proposed by Nielsen and Molich [9]; Bertini et al. [28] discussed the capacity of expert-based techniques to capture contextual factors in mobile computing. Indeed, it is not trivial to integrate real-world setting/context into inspection methods which are conceived as laboratory testing techniques. In any case, laboratory testing and expert-based techniques are complementary. Both approaches can be used in preliminary analysis and design of prototypes but, even more in mobile than when dealing with old desktop interaction paradigms, they need to be complemented with users-based testing.

C. Type of Tasks: in mobile environments, typical tasks are relatively different from traditional desktop devices. From the origins of mobile devices, concepts such as "personal space extension" [29] previewed new uses of mobile terminals. The literature has tended to classify mobile tasks on the basis of searching/browsing categories and also according to management of known information or new information. It is important to note that pre-2007 literature does not widely consider touch terminals which incorporate new tasks. Having taken all this into account, we can classify tasks as follows:

    i.   search [29–35]:

        a.   information gathering [33];

b. using the web to more or less specific search;

c. compare/choose [32];

ii. browsing [30, 31, 33–35]:

    a. undirected or conditioned browsing [31];

    b. watching videos [14];

    c. reading news [14];

iii. communication [14, 33, 35]:

    a. checking email [14];

    b. social networking sites [14];

    c. chat rooms/discussion rooms [33];

iv. transaction [29, 33, 34]: although it is more common to use a mobile device to browse the web or to perform some shopping-related search than shopping in earnest [14].

v. playing games [14];

vi. killing time [14]. Some literature includes other kinds of task like "maintenance" [35] or "housekeeping" [33] that have not been included in our classification because the frequency of realization is too low and these kinds of tasks do not define new kinds of interactions.

D. Multidevice access: user's familiarity with a web page [34] helps them to construct a mental model based on the structural organization of the information, such as visual cues, layout, and semantics. When a site is being designed for multidevice access, a major concern is to minimize user effort to reestablish the existing mental model. This new way of working around structured information that must be delivered through so many different interface restrictions has been studied as a new paradigm known as Responsive Design [36].

E. Limited processing capability and power [21–24]: these limitations include battery life, network connectivity, download delays, and limited memory.

F Adoption [22]: adoption of mobile technology by users is based on perceived privacy, acceptance of technology, comfort, and capacity of personalization. Different levels of adoption determine different group of users interacting in a very different way with the interface. This may not seem to be a mobile-specific restriction but the wide variety of mobile devices, touchable or keyboard-based, with different sizes and presentation models, makes the range of users requiring different approaches much broader.

3.2. Rearrangement of Traditional Heuristics

The first rearrangement of traditional heuristics in this step is mainly based on the review of the literature by Torrente [7] where the author selected the most influent heuristics guidelines [9, 37–44]. This compilation gives a total of 9 heuristics guidelines consisting globally of 83 heuristics and 361 subheuristics. We need to rearrange this list of items into a new classification which is coherent to our purpose. In this step, we only take into account "heuristics" and no "subheuristics" and this gives us the heuristic list shown in Figure 3.

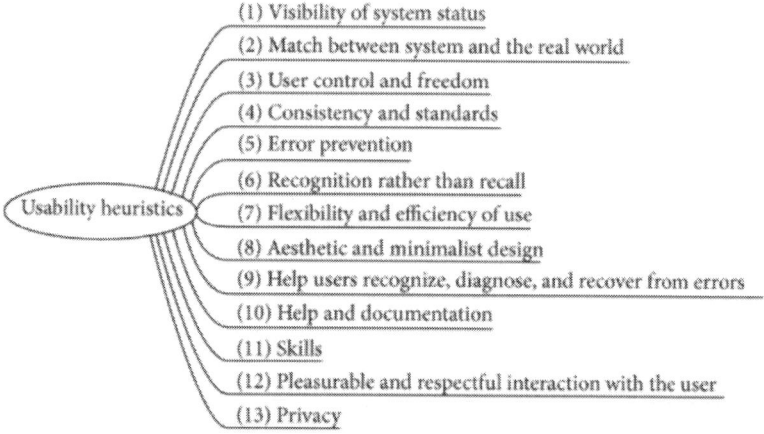

(1) Visibility of system status

(2) Match between system and the real world

(3) User control and freedom

(4) Consistency and standards

(5) Error prevention

(6) Recognition rather than recall

Usability heuristics

(7) Flexibility and efficiency of use

(8) Aesthetic and minimalist design

(9) Help users recognize, diagnose, and recover from errors

(10) Help and documentation

(11) Skills

(12) Pleasurable and respectful interaction with the user

(13) Privacy

Figure 3: Proposed heuristic list.

Our rearrangement focuses on literature coincidences (i.e., when the same concept or category is included in different works in the literature, perhaps under different names) and tries to propose a coherent, exhaustive, and complete framework of heuristics that could be used to arrange further identified subheuristics. Literature coincidences for each heuristics are as follows:

1) visibility of system status [9, 38, 44]: other bibliography references include this concept as "Track State" [41], "Give feed-back" [37], or "Feedback" [40];

2) match between system and the real world [9, 38];

3) user control and freedom [9, 38, 42, 43]: other bibliography references include this concept as "Support the user control" [37], "The feedback principle" [39], "Autonomy" [41], or "Visible Navigation" [41];

4) consistency and standards [9, 38, 41, 44] also cited as "Maintain Consistency" [37], "Structure Principle" [39], "Reuse Principle" [39], "Consistency" [40], or "Learnability" [41];

5) error prevention [9, 38, 40] also cited as "Tolerance Principle" [39];

6) recognition rather than recall [9, 38] also cited as "Reduce recent memory load for users" [37], "Structure principle" [39], "Reuse principle" [39], "Minimize the users' memory load" [40], or "Anticipation" [41];

7) flexibility and efficiency of use [9, 38] also cited as "Simplicity principle" [39] or "Look at the user's productivity not the computer's" [41];

8) aesthetic and minimalist design [9, 38];
9) help users recognize, diagnose, and recover from errors [9, 37, 38] also cited as "Good error messages" [40] or "In case of error, is the user clearly informed and not over-alarmed about what happened and how to solve the problem?" [42];(10)help and documentation [9, 38, 40, 42–44];
10) skills [38] also cited as "Prepare workarounds for frequent users" [37], "Shortcuts" [40], or "Readability" [41];
11) pleasurable and respectful interaction with the user [38] also cited as "Simplicity principle" [39], "Simple and natural dialog," or "Speak the user's language" [40]: this point also includes any accessibility questions that could enrich usability allowing a more universal access, such as "Color blindness" [41];
12) privacy [38].

3.3. Compilation of Subheuristics from Traditional General Heuristic Checklists

As defined before, "heuristic" in this paper refers to a global usability issue which must be evaluated or taken into account when designing. In contrast, the term "subheuristic" refers to specific guidelines items. In this third step, we focus on locating subheuristics from the literature.

The first group of potential heuristics is the 361 subheuristics proposed in the 9 references selected by Torrente [7]. Among these sub-heuristics we exclude those that do not fit well with the previously described mobile constraints. For example, subheuristics referred to desktop data entry methods is obviously discarded. In contrast, this referring to screen use optimization is particularly relevant. Other discarded amounts of subheuristics include some proposed [38] with specific response times which do not apply in a mobile and varying context. We also discard coincidences between different authors proposals.

Thus, from a total of 361 amounts of subheuristics proposed by the 9 references [9, 37–44] selected by Torrente [7], in this study, we obtain a first selection of 158 subheuristics.

In order to maintain consistency in our classification, some subheuristics has been moved from their original heuristic parents, and new subcategories have been added so that semantically related amounts of subheuristics are grouped together. The final framework, shown in Figure 4, builds on that presented in the previous section.

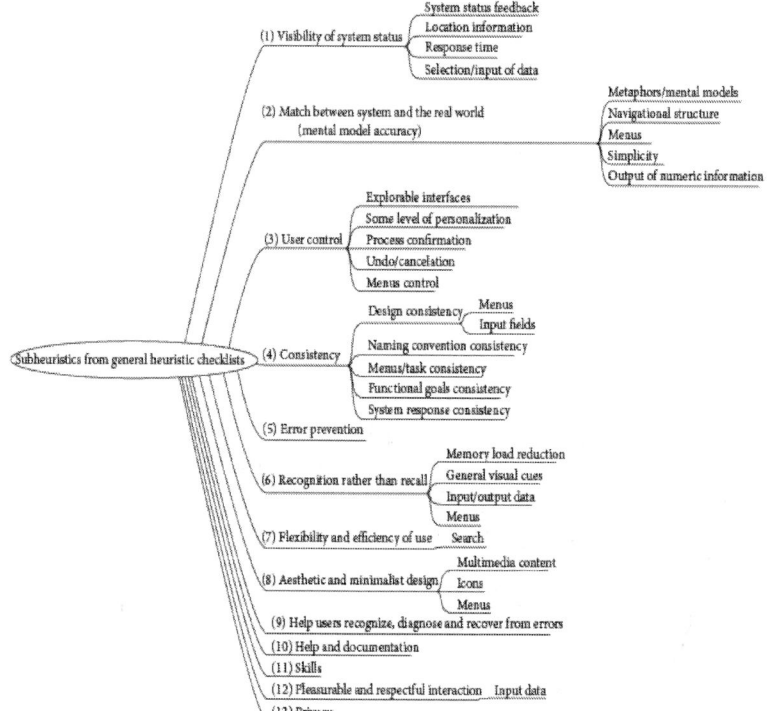

Figure 4: First framework for classification of detected subheuristics.

It is also important to recall that at this stage, subheuristics redactions have been kept unchanged from their corresponding references. In the final compilation, these redactions will be modified in order to homogenize the whole guideline as we planned for step 4 in our methodology.

The final list of subheuristics is as follows:

(1) visibility of system status: system status feedback:

1. is there some form of system feedback for every operator action? [38]
2. if pop-up windows are used to display error messages, do they allow the user to see the field in error? [38]
3. in multipage data entry screens, is each page labeled to show its relation to others? [38]
4. are high informative contents placed in high hierarchy areas? [42] location information:
5. is the logo meaningful, identifiable, and sufficiently visible? [42]

6. is there any link to detailed information about the enterprise, website, webmaster ... ? [42]
7. are there ways of contacting with the enterprise? [42]
8. in articles, news, reports ... are the author, sources, dates, and review information shown clearly? [42] response times:
9. are response times appropriate for the users cognitive processing? [38]
10. are response times appropriate for the task? [38]
11. if there are observable delays (greater than fifteen seconds) in the system's response time, is the user kept informed of the system progress? [38]
12. latency reduction [41]; Selection/input of data:
13. is there visual feedback in menus or dialog boxes about which choices are selectable? [38]. We will merge this statement with the following: "Do GUI menus make obvious which item has been selected?" [38], "Do GUI menus make obvious whether deselection is possible?" [38], "Is there visual feedback in menus or dialog boxes about which choice the cursor is on now?" [38], and "If multiple options can be selected in a menu or dialog box, is there visual feedback about which options are already selected?" [38]
14. is the current status of an icon clearly indicated? [38]
15. is there visual feedback when objects are selected or moved? [38]
16. are links recognizable? Is there any characterization according to the state (visited, active,)? [42]

(2) match between system and the real world (Mental model accuracy): metaphors/mental models:

17. use of metaphors [41];

18. are icons concrete and familiar? [38]

19. if shape is used as a visual cue, does it match cultural conventions? [38]

20. do the selected colours correspond to common expectations about color codes? [38] navigational structure:

21. if the site uses hierarchical structure, are depth and height balanced? [42]

22.navigation map [44], also known as site map or table of contents; menus:

23. are menu choices ordered in the most logical way, given the user, the item names, and the task variables? [38]

24. do menu choices fit logically into categories that have readily understood meanings? [38]

25. are menu titles parallel grammatically? [38]

26. in navigation menus, are the number of items and terms by item controlled to avoid memory overload? [42] simplicity:

27. do related and interdependent fields appear on the same screen? [38]

28. for question and answer interfaces, are questions stated in clear, simple language? [38]

29 is the language used the same target users speak? [42]. We will merge this statement with the following: "Is the menu-naming terminology consistent with the user's task domain?" [38]

30. is the language clear and concise? [42]. We will merge this statement with the following: "Does the command language employ user jargon and avoid computer jargon?" [38]

31.does the site follow the rule "1 paragraph = 1 idea"? [42] output of numeric information:

32.does the system automatically enter leading or trailing spaces to align decimal points? [38]

33.does the system automatically enter a dollar sign and decimal for monetary entries? [38]

34.does the system automatically enter commas in numeric values greater than 9999? [38]

35.are integers right-justified and real numbers decimal-aligned? [38]

(3) user control: explorable interfaces:

36.can users move forward and backward between fields or dialog box options? [38]

37.if the system has multipage data entry screens, can users move backward and forward among all the pages in the set? [38]

38.if the system uses a question and answer interface, can users go back to previous questions or skip forward to later questions? [38]

39.clearly marked exits [40];

40.is the general website structure user-oriented? [42]

41. is there any way to inform user about where they are and how to undo their navigation? [42] some level of personalization:

42. can users set their own system, session, file, and screen defaults? [38] process confirmation:

43. when a user's task is complete, does the system wait for a signal from the user before processing? [38]

44. are users prompted to confirm commands that have drastic, destructive consequences? [38] undo/cancelation:

45. can users easily reverse their actions? [38] Also found as "Do function keys that can cause serious consequences have an undo feature?" [38] and "Is there an "undo" function at the level of a single action, a data entry, and a complete group of actions?" [38]

46. can users cancel out of operations in progress? [38] menus control:

47. if the system has multiple menu levels, is there a mechanism that allows users to go back to previous menus? [38]

48. are menus broad (many items on a menu) rather than deep (many menu levels)? [38]

49. if users can go back to a previous menu, can they change their earlier menu choice? [38]

(4) consistency: designing consistency:

50. are attention-getting techniques used with care? [38]

51. intensity: two levels only [38];

52. color: up to four (additional colors for occasional use only) [38];

53. are there no more than four to seven colors, and are they far apart along the visible spectrum? [38]

54. sound: soft tones for regular positive feedback, harsh for rare critical conditions [38];

55. if the system has multipage data entry screens, do all pages have the same title? [38]

56. do online instructions appear in a consistent location across screens? [38] 57. have industry or company standards been established for menu design, and are they applied consistently on all menu screens in the system? [38]

58. are there no more than twelve to twenty icon types? [38]

59. has a heavy use of all uppercase letters on a screen been avoided? [38]

60. is there a consistent icon design scheme and stylistic treatment across the system? [38] menus:

61. are menu choice lists presented vertically? [38]

62. if "exit" is a menu choice, does it always appear at the bottom of the list? [38]

63. are menu titles either centered or left-justified? [38] input fields:

64. are field labels consistent from one data entry screen to another? [38]

65. do field labels appear to the left of single fields and above list fields?

66. are field labels and fields distinguished typographically? [38] naming convention consistency:(67)is the structure of a data entry value consistent from screen to screen? [38]

68. are system objects named consistently across all prompts in the system? [38]

69. are user actions named consistently across all prompts in the system? [38] menu/task consistency:

70. are menu choice names consistent, both within each menu and across the system, in grammatical style and terminology? [38]

71. does the structure of menu choice names match their corresponding menu titles? [38]

72. does the menu structure match the task structure? [38]

73. when prompts imply a necessary action, are the words in the message consistent with that action? [38] functional goals consistency:

74. where are the website goals? Are they well defined? Do content and services delivered match these goals? [42]

75. does the look & feel correspond with goals, characteristics, contents and services of the website? [42]

76. is the website being updated frequently? [42] system response consistency:

77. is system response after clicking links predictable? [42]

78. are nowhere links avoided? [42]

79. are orphan pages avoided? [42]

(5) error prevention: (80)are menu choices logical, distinctive, and mutually exclusive? [38](81)are data inputs case-blind whenever possible? [38](82)does the system warn users if they are about to make a potentially serious error? [38](83)do data entry screens and dialog boxes indicate the number of character spaces available in a field? [38](84)do fields in data

entry screens and dialog boxes contain default values when appropriate? [38]

(6) recognition rather than recall: memory load reduction:

85. high levels of concentration are not necessary and remembering information is not required: two to fifteen seconds [38];

86. are all data a user needs on display at each step in a transaction sequence? [38]

87. if users have to navigate between multiple screens, does the system use context labels, menu maps, and place markers as navigational aids? [38]

88. after the user completes an action (or group of actions), does the feedback indicate that the next group of actions can be started? [38]

89. are optional data entry fields clearly marked? [38]

90. do data entry screens and dialog boxes indicate when fields are optional? [38]

91. is page length controlled? [42] general visual cues:

92. for question and answer interfaces, are visual cues and white space used to distinguish questions, prompts, instructions, and user input? [38]

93. does the data display start in the upper-left corner of the screen? [42]

94..have prompts been formatted using white space, justification, and visual cues for easy scanning? [38]

95. do text areas have "breathing space" around them? [42]

96. are there "white" areas between informational objects for visual relaxation? [42]

97. does the system provide visibility; that is, by looking, can the user tell the state of the system and the alternatives for action? [38]

98. is size, boldface, underlining, colour, shading, or typography used to show relative quantity or importance of different screen items? [38]

99. is colour used in conjunction with some other redundant cue? [38]

100. is there good colour and brightness contrast between image and background colours? [38]

101. have light, bright, saturated colours been used to emphasize data and have darker, duller, and desaturated colours been used to deemphasize data? [38]

102. is the visual page space well used? [42] input/output data:

103. on data entry screens and dialog boxes, are dependent fields displayed only when necessary? [38]

104. are field labels close to fields, but separated by at least one space? [38] Menus

105. is the first word of each menu choice the most important? [38]

106. are inactive menu items grayed out or omitted? [38]

107. are there menu selection defaults? [38]

108. is there an obvious visual distinction made between "choose one" menu and "choose many" menus? [38]

(7) flexibility and efficiency of use: search:

109. is the searching box easily accessible? [42]

110. is the searching box easily recognizable? [42]

111. is there any advanced search option? [42]

112. are search results shown in a comprehensive manner to the user? [42]

113. is the box width appropriated? [42]

114. is the user assisted if the search results are impossible to calculate? [42]

(8) aesthetic and minimalist design:

115. sFitt's Law [41]: the time to acquire a target is a function of the distance to and size of the target;

116. is only (and all) information essential to decision making displayed on the screen? [38]

117. are field labels brief, familiar, and descriptive? [38]

118. are prompts expressed in the affirmative, and do they use the active voice? [38]

119. is layout clearly designed avoiding visual noise? [42] multimedia content:

120. does the use of images and multimedia content add value? [42]

121. are images well sized? Are they understandable? Is the resolution appropriate? [42]

122. are cyclical animations avoided? [42] icons:

123. has excessive detail in icon design been avoided? [38]

124. is each individual icon a harmonious member of a family of icons? [38]

125. does each icon stand out from its background? [38]

126. are all icons in a set visually and conceptually distinct? [38] menus:

127. is each lower-level menu choice associated with only one higher level menu? [38]

128. are menu titles brief, yet long enough to communicate? [38]

(9) help users recognize, diagnose and recover from errors;

(10) help and documentation:

129. are online instructions visually distinct? [38]

130. do the instructions follow the sequence of user actions? [38]

131. if menu choices are ambiguous, does the system provide additional explanatory information when an item is selected? [38]

132. if menu items are ambiguous, does the system provide additional explanatory information when an item is selected? [38]

133. is the help function visible, for example, a key labeled HELP or a special menu? [38, 42]

134. is the help system interface (navigation, presentation, and conversation) consistent with the navigation, presentation, and conversation interfaces of the application it supports? [38]

135. navigation: is information easy to find? [38]

136. presentation: is the visual layout well designed? [38]

137. conversation: is the information accurate, complete, and understandable? [38]

138. is the information relevant? ([38], Help and documentation) [42] It should be relevant in the following aspects [38]: goal-oriented (what can I do with this program?), descriptive (what is this thing for?), procedural (how do I do this task?), interpretive (why did that happen?), and navigational (where am I?);

139. is there context-sensitive help? [38, 42](140)can the user change the level of detail available? [38](141)can users easily

switch between help and their work? [38][142)is it easy to access and return from the help system? [38]

143. can users resume work where they left off after accessing help? [38]

144. if a FAQs section exists, are the selection and redaction of questions and answers correct? [42]

(11) skills:

145. do not use the word "default" in an application or service; replace it with "Standard," "Use Customary Settings," "Restore Initial Settings," or some other more specific terms describing what will actually happen [41];

146. if the system supports both novice and expert users, are multiple levels of error message detail available? [38]

147. if the system supports both novice and expert users, are multiple levels of detail available? [38]

148. are users the initiators of actions rather than the responders? [38]

149. do the selected input device(s) match user capabilities? [38]

150. are important keys (e.g., ENTER, TAB) larger than other keys? [38]

151. does the system correctly anticipate and prompt for the user's probable next activity? [38]

(12) pleasurable and respectful interaction:

152. protect users' work [41], also as "For data entry screens with many fields or in which source documents may be incomplete, can users save a partially filled screen?" [38]

153. do the selected input device(s) match environmental constraints? [38]

154. are typing requirements minimal for question and answer interfaces? [38]

155. does the system complete unambiguous partial input on a data entry field? [38]

(13) privacy:

156. are protected areas completely inaccessible? [38]

157. can protected or confidential areas be accessed with certain passwords [38]

158. is there information about how personal data is protected and about contents copyright? [38]

3.4. Compilation of Mobile-Specific Subheuristics

The fourth step in this work is to enrich the list with mobile-specific subheuristics. The subheuristic list obtained in the previous section does not include many mobile specific questions because, as mentioned before, traditional heuristics does not usually cover these issues. New mobile-specific questions have been added into this list, taken from mobile usability studies and best practices that actually do not provide HE. Our approach allows us to include these new items into their corresponding categories, enriching the heuristic with mobile-specific issues. Some new categories had to be added to the original heuristic framework to include new mobile-specific subheuristics. The final framework is shown in Figure 5.

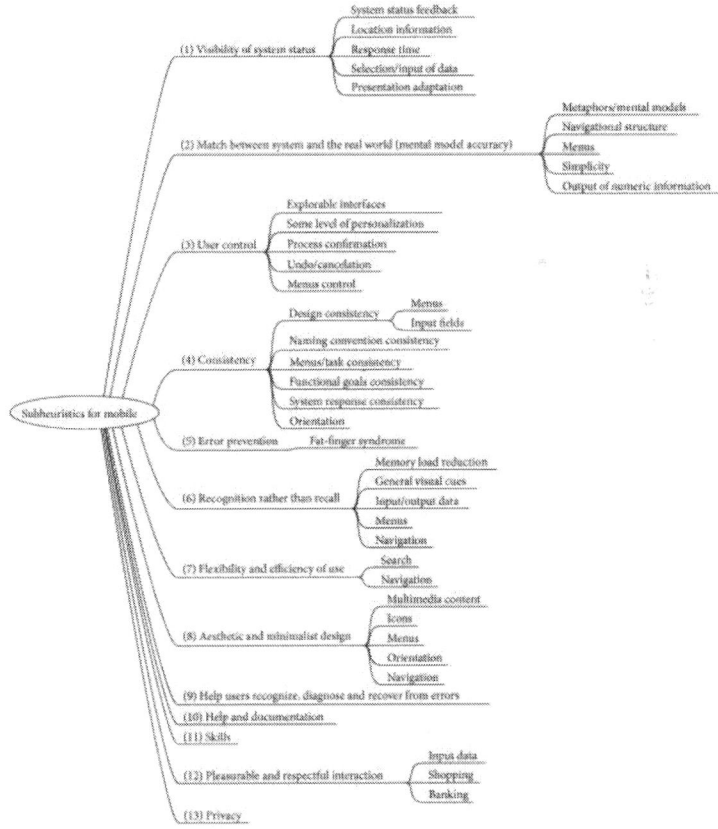

Figure 5: Second framework for classification of detected subheuristics.

As we mentioned earlier, not all mobile devices have been considered; we discarded featured phones because they are rarely used for tasks other than phone calls and short message services (SMS) and they are gradually being abandoned apart from specific groups of users such as elderly or cognitively impaired people. We also discarded smartphones (phones with midsized screens and full A–Z keypads) because the interactivity with these devices is dramatically different from that of touch phones and they are commonly constrained to enterprise use. This study is centred in touch phones and tablets which are very popular nowadays and similar from a usability point of view.

This fourth step adds 72 new subheuristics to the compilation:

(1) visibility of system status: System status feedback:

1. All the items on a list should go on the same page: if the items are text-only and if they are sorted in an order that matches the needs of the task [24];
2. if a list of items can be sorted according to different criteria, provide the option to sort that list according to all those criteria [24];
3. if a list contains items that belong to different categories, provide filters for users to narrow down the number of elements that they need to inspect [24];
4. if the list contains only one item, take the user directly to that item [24];
5. if the list contains items that download slowly (e.g., images), split the list into multiple pages and show just one page at a time [24];
6. if an article spans several pages, use pagination at the bottom. Have a link to each individual page, rather than just to the previous and the next ones [24]; location information:
7. whenever you have physical location information on your website, link it to a map and include a way of getting directions [24]; response time:
8. splash screens too long [14];
9. download time [14]: "Progress bar is preferable" and "Alternative entertainment if download time is greater than 20 seconds"; selection/input of data:
10. low discoverability (active areas that do not look touchable): users do not know that something is touchable unless it looks as if it is [14];
11. swiping [14]: swiping is still less discoverable than most other ways of manipulating mobile content, so we recommended including a visible cue when people can swipe. And swipe ambiguity should be avoided: the same swipe gesture should not be used to mean different things on different areas of the same screen:

12. expandable menus should be used sparingly. Menu labels should clearly indicate that they expand to a set of options [14]; presentation adaptation:
13. detect if users are coming to your site on a mobile phone and direct them to your mobile site [24];
14. include a link to your mobile site on your full site. It can direct mobile users who were not re-directed to your mobile site [24];
15. include a link to the full site on the mobile page [24];

(2) match between system and the real world: navigational structure:

16. too much navigation (TMN) [14];

(3) user control and freedom: explorable interfaces:

17. accidental activation (lack of back button) [24];

18. include navigation on the homepage of your mobile website [14];

(4) consistency and standards: orientation:

19. about constraining orientation: users tend to switch orientation when an impasse occurs and, if the application does not support them, their flow is going to be disrupted, and they are going to wonder why it is not working [14];

20. navigation (horizontal and vertical) must be consistent across orientations. Some applications use a different navigation direction in the two orientations; for instance, they use horizontal navigation in landscape and use vertical navigation in portrait [14];

21. inconsistent content across orientations [14]: "Same content," "Keep location," and "If a feature is only available in one orientation, inform users";

(5) error prevention

22. accidental activation (lack of back button) [14]; fat-finger syndrome:

23. touchable areas are too small [14]. Research has shown that the best target size for widgets is 1 cm × 1 cm for touch devices [14];

24. crowding targets: another fat-finger issue that we encountered frequently is placing targets too close to each other. When targets

are placed too close to each other, users can easily hit the wrong one [14];

25. padding: although the visible part of the target may be small, there is some invisible target space that if a user hits that space, their tap will still count [14];

26. when several items are listed in columns, one on top of another (see the time example below), users expect to be able to hit anywhere in the row to select the target corresponding to that row. Whenever a design does not fulfil that expectation, it is disconcerting for users [14];

27. do not make users download software that is inappropriate for their phone [24];

28. JavaScript and Flash do not work on many phones; do not use them [24];

(6) recognition rather than recall: Memory load reduction:

29. the task flow should start with actions that are essential to the main task. Users should be able to start the task as soon as possible [14];

30. the controls that are related to a task should be grouped together and reflect the sequence of actions in the task [14]; navigation:

31. use breadcrumbs on sites with a deep navigation structure (many navigation branches). Do not use breadcrumbs on sites with shallow navigation structures [24];

(7) Flexibility and efficiency of use: search:

32. a search box and navigation should be present on the homepage if your website is designed for smartphones and touch phones [24];

33. the length of the search box should be at least the size of the average search string. We recommend going for the largest possible size that will fit on the screen [24];

34. preserve search strings between searches. Use autocompletion and suggestions [24];

35. do not use several search boxes with different functionalities on the same page [24];

36. if the search returns zero results, offer some alternative searches or a link to the search results on the full page [24]; navigation:

37. use links with good information scent (i.e., links which clearly indicate where they take the users) on your mobile pages [24];

38. use links to related content to help the user navigate more quickly between similar topics [24];

(8) aesthetic and minimalist design:

39. recognizable application icons to be found in the crowded list of applications [14]; multimedia content:

40 getting rid of Flash content [14];

41. carousels [24]: avoid using animated carousels, but if they must be used, users should be able to control them;

42. do not use image sizes that are bigger than the screen. The entire image should be viewable with no scrolling [24];

43. for cases where customers are likely to need access to a higher resolution picture, initially display a screen-size picture and add a separate link to a higher resolution variant [24];

44. when you use thumbnails, make sure the user can distinguish what the picture is about [24];(45)use captions for images that are part of an article if their meaning is not clear from the context of the article [24];(46)do not use moving animation [24];

47. if you have videos on your site, offer a textual description of what the video is about. [24];

48. clicking on the thumbnail and clicking on the video title should both play the video [24];

49. indicate video length [24];

50. specify if the video cannot be played on the user's device [24];

51. use the whole screen surface to place information efficiently [14]: "Popovers for displaying information restricts size of frame where information will be shown" and "Small modal views present the same size constraints"; orientation:

52. desktop websites have a strong guideline to avoid horizontal scrolling. But for touch screens, horizontal swipes are often fine [19]; navigation:

53. do not replicate a large number of persistent navigation options across all pages of a mobile site [24];

(9) Help users recognize, diagnose, and recover from errors:

> 54. To signal an input error in a form, mark the textbox that needs to be changed [24];

(10) help and documentation:

> 55. focus on one single feature at a time. Present only those instructions that are necessary for the user to get started [14];

(11) skills:

> (12) pleasurable and respectful interaction: input data:
>
> 56. users dislike typing. Compute information for the users. For instance, ask only for the zip code and calculate state and town; possibly offer a list of towns if there are more under the same zip code [14];
>
> 57. be tolerant of typos and offer corrections. Do not make users type in complete information. For example, accept "123 Main" instead of "123 Main St." [14];
>
> 58. save history and allow users to select previously typed information [14];
>
> 59. use defaults that make sense to the user [14];
>
> 60. If the application does not store any information that is sensitive (e.g., credit card), then the user should definitely be kept logged in (log out clearly presented) [14];
>
> 61. minimize the number of submissions (and clicks) that the user needs to go through in order to input information on your site [24];
>
> 62. When logging in must be done, use graphical passwords at least some of the time, to get around typing [24];
>
> 63. Do not ask people to register on a mobile phone; skipping registration should be the default option [24];
>
> 64. When logging in must be done, have an option that allows the user to see the password clearly [24]; shopping:
>
> 65. when you present a list of products, use image thumbnails that are big enough for the user to get some information out of them [24];

66. on a product page, use an image size that fits the screen. Add a Link to a higher resolution image when the product requires closer inspection [24];

67. offer the option to email a product to a friend [24];

68. offer the option to save the product in a wish list [24];

69. on an e-commerce site, include salient links on the homepage to the following information: locations and opening hours (if applicable), shipping cost, phone number, order status, and occasion-based promotions or products [24]; banking and transactions:

70. whenever users conduct transactions on the phone, allow them to save confirmation numbers for that transaction by emailing themselves. If the phone has an embedded screen-capture feature, show them how to take a picture of their screen [24];

(13) privacy:

71. for multiuser devices, avoid being permanently signed in on an application [14];

72. If the application does store credit card information, it should allow users to decide if they want to remain logged in [24]. Ideally, when the user opts to be kept logged in, he/she should get a message informing of the possible risks

3.5. Final New Mobile-Specific Heuristics

The final compilation of heuristics and subheuristics, which is shown in Appendix A (Supplementary Material available online at http://dx.doi.org/10.1155/2014/434326), gives a total of 13 heuristics and 230 subheuristics (158 + 72). In this final compilation, we have omitted intermediate classifications introduced during the discussion. Also, semantically related items have been merged into a single item following the most common presentation of heuristics guidelines in literature. Wording has been corrected to offer a homogeneous collection of heuristics questions.

This final mobile heuristics can be used as a tool to evaluate usability of mobile interfaces. In its current version, possible answers for the proposed questions are "yes/no/NA." The number of "yes" answers provides a measure of the usability of the interface. Other approaches in the literature include more elaborates ratings that have to be agreed between evaluators [45].

3.6. Empirical Test of the New Mobile-Specific Heuristics

The goal of our test was to perform an evaluation of the usefulness of the proposed heuristics as a tool for designers and software engineers with no specific knowledge and experience of usability.

The use case design was as follows: two software engineers without any specific knowledge about usability were asked to design an interface for a tablet application having a functional description in a low-fidelity prototype designed for a desktop version of the application. Over their proposed interface design they used our heuristics as an evaluation and reflexion tool. In view of the results of the evaluation, they were asked to develop a new prototype. Finally, both interfaces were tested with a small group of users to compare their usability.

This empirical test of usefulness of the proposed usability list was divided into the following phases:(1)prototype 1: developing an interface prototype oriented to tablet access from a given PC-desktop low-fidelity functional design (prototype 1, P1);(2)HE of P1 using the proposed heuristics as the basis for an oriented discussion between designers;(3)prototype 2: evolution of P1 fixing usability gaps detected in phase 2 (prototype 2, P2);(4)Empirical comparison of prototypes: users' testing of P1 and P2.

3.6.1. Prototype 1 Developing

The functional description of the desktop version used to build the prototypes evaluated in this testing was provided by Project PROCUR@ [46], an e-care and e-rehabilitation platform focused on neurodegenerative diseases patients, their carers, and health professionals. The project is based on the deployment of three social spaces for research and innovation (SSRI) [47] in the three validation scenarios: Parkinson's disease SSRI, acquired brain damage (ABD) SSRI, and Alzheimer's disease SSRI. The functional description corresponds to this latter SSRI and provides five low-fidelity interface descriptions from the point of view of five profiles: patients, relatives, doctors, caregivers, and sanitary personnel, respectively.

The subjects of these experiments were two software engineering students preparing their end of degree project. They had never been trained in usability but had knowledge about software life-cycles and design techniques. P1 was the result of a first tablet-interface adaptation without usability training. The tablet format was imposed because a bigger screen size is specially convenient for the target users (i.e., elderly people with low vision capability and motor control).

This first adaptation included two main groups of changes: functional refinement and new interface adaptation. Functional refinement required changes that were not particularly relevant to this work. However, adaptations to the new interface involved decisions adopted by designers without knowledge of usability, guided only by their common sense. At a later stage, some of these decisions were confronted with the HE new tool and not all of them were maintained. These decisions are described in Figure 6.

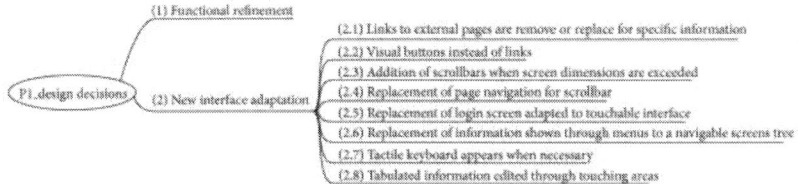

Figure 6: Design decisions in prototype 1.

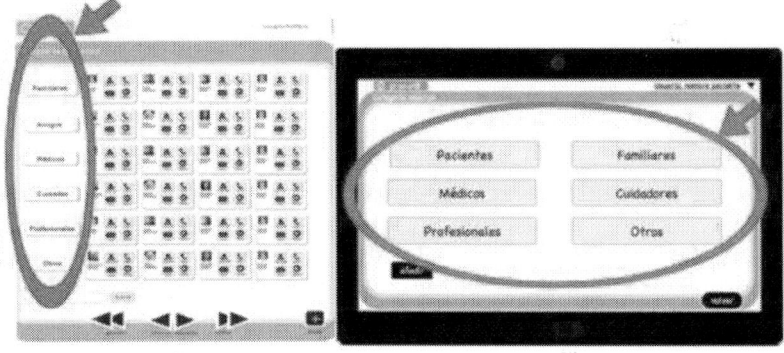

Figure 7: Shows an example of the interface change.

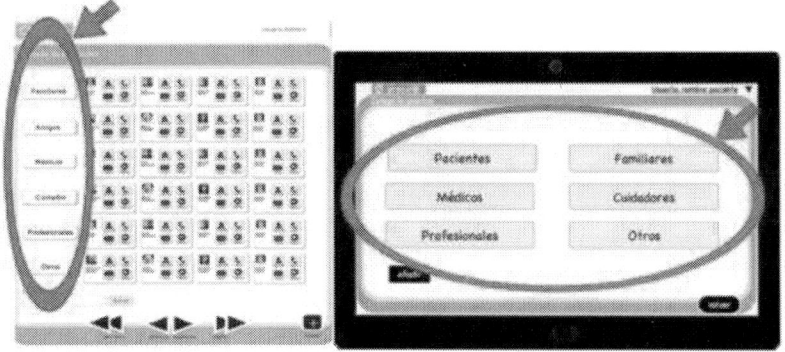

Figure 7: Prototype 1 from desktop version description.

3.6.2. Prototype 1: Heuristic Evaluation

Once Prototype 1 was designed, the next step was to evaluate its usability. The objective was not the evaluation itself but how the designers reflected on its usability.

When performing a HE using such tool, one has to make certain decisions about the scoring of each subheuristics. In this case, the experts were asked to use a ponderation which would allow the prioritization of heuristic item relevance for the specific evaluated interface. Experts are marked with values from 1 to 4: 1 for accomplished heuristic items, 2 for those corresponding to usability gaps, 3 for heuristic items which were not evaluable in the actual software life-cycle phase, and 4 for questions not applicable to the interface.

Applying a Delphi-based [48] approximation, both experts were asked to independently evaluate the interface using the list. Afterwards, the results of the evaluations were confronted and the experts had to agree in the case of items with different scorings.

In the independent evaluation, the level of coincidence of the experts was moderate and in the final HE scoring, where both experts agreed, the results were as follows: 68 items scored 1, 33 items scored 2, 41 scored 3, and 98 scored 4. This final result established a huge number of items as "not applicable." This may have been because the heuristics was intended to be as general as possible, not focusing on any specific kind of application, and it therefore included an exhaustive list of checks.

The most important result from this evaluation was that experts were forced to reflect on each item in the heuristic guideline. For each not accomplished question, they learnt which usability gaps had to be avoided in the interface design. This learning provided a wider knowledge background when it came to designing next prototype.

3.6.3. Prototype 2 Building

Prototype 2 was not only a series of modifications to Prototype 1 but also a complete revision of the whole interface concept. This global reflexion was guided by the expert discussion from the previous section.

The most specific changes which fix detected usability gaps are shown in Figure 8 but, as mentioned, the overall appearance and design have changed dramatically (Figure 9).

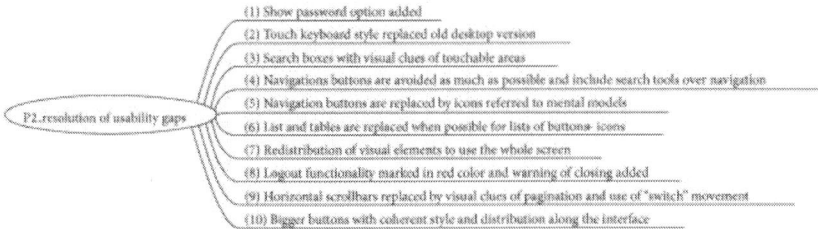

Figure 8: Prototype 2 main changes.

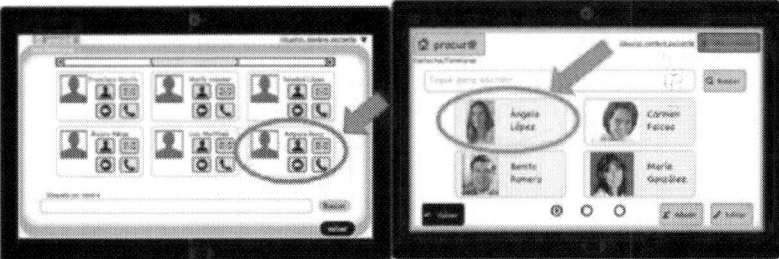

Figure 9: Prototype 2 global concept changes from the first version.

3.7. Empirical Comparison of Prototypes

The empirical comparison of the two prototypes was intended to evaluate whether P2 designed using the proposed HE tool was better in any way than P1.

This empirical study involved users so the experiment had to be designed carefully to obtain valid results. The approach included a test design, a pilot phase to check the test design, the execution of the test itself, and a phase to analyze the collected data.

Several decisions were taken in the design phase.(1)Wizard of Oz [49] (WO) was chosen as the evaluation technique because the prototypes are developed on paper and are well suited to presentation through human intervention to the users.(2)To develop WO technique, users were asked to perform a task-guided interaction. The experts selected three functional tasks that users had to carry out interacting with the interface. The tasks were representative enough to be useful in this test. They were briefly described to the users so that they were able to accomplish them by exploring the interface without step-by-step guides.(3)Ten users were selected with the characteristics shown in Table 1.(4)The experimentation adopted an inner group [50] design: half of the users interacted with P1 first and the other half with P2 first. This was to avoid as much as possible correlations due to learning of the interfaces.(5)Lastly, users were asked to give feedback about their overall feelings about each interface to provide us with some conclusions related to user experience beyond usability.

Table 1: Users of the experiment.

	Gender	Age	Kind of mobile devices they are used to	Adoption of technology
USER 1	M	40–50	Touch phone	Basic
USER 2	F	40–50	Smartphone	None
USER 3	M	40–50	Smartphone	None
USER 4	F	40–50	Touch phone	Basic
USER 5	F	40–50	Touch phone	Basic
USER 6	F	40–50	Touch phone	Basic
USER 7	M	40–50	Touch phone	Basic
USER 8	F	40–50	Touch phone	Basic
USER 9	M	50–60	None	None
USER 10	F	50–60	Touch phone	Basic

The pilot phase consisted of a simulation of the final experiment using two dummy users. This phase was very useful for consolidating of task description and helping to improve Wizard's skills managing prototypes and for the whole experiment.

Final test execution detected 6 serious usability gaps in P1 and 3 serious gaps in P2 (which are also included in the first 6) as can be seen in Table 2. When asked about general satisfaction, 100% of users stated they were more satisfied with P2 prototype interaction.

Table 2: Results of empirical user-based evaluation of prototypes.

	Prototype 1. Usability gaps	Prototype 2. Usability gaps	Description
1	Authentication method inappropriate for the targeted users	Authentication method inappropriate for the targeted users	The boxes "user" and "password" should appear independently
2	Information screen confusing	Information screen confusing	It was maintained because the functional description included it
3	Chatting function not localizable	Returning to main menu not localizable	Even after changing the graphical clue
4	Personal profile function not localizable		
5	Returning to main menu not localizable		
6	Close session function not localizable		

4. CONCLUSIONS AND FUTURE WORK

In this paper, we have presented a compilation of heuristic evaluation checklists readapted to mobile interfaces. We started our work by reusing heuristics from desktop heuristics evaluation checklists, which is allowed because "heuristic checklists change very slowly because they derive from human behaviour, not technology" [14]. In fact, in the final proposal of this work, the amount of reused heuristics from the literature is 69% of the total proposed subheuristics. The rest are best-practices and recommendations for mobile interfaces not initially conceived as part of a usability tool.

In the final collection of 13 heuristics, the most influent author is Nielsen [9, 37–44]. While it is not a long list of heuristics, it is exhaustive enough to be a useful categorization for further research. However, in our work, Nielsen's heuristics has been rearranged taking into account other proposals in the literature which emphasize concepts such as skills adaptation and pleasurable and respectful interaction with the user and privacy, elevating them to the category of heuristic item.

The added mobile-specific subheuristics in this proposal focus specifically on overcoming specific constraints on mobile such as limitations in input/output, limited processing capabilities, and power. Additionally, it focuses on favouring usual tasks in mobile and issues related to the adoption of this kind of devices (privacy, acceptance, comfort, personalization).

The main original contributions of our work include (a) rearrangement of existing desktop heuristics into a new compilation, including detailed subheuristics, adapted to the new mobile paradigm; (b) enriching the list with mobile-specific subheuristics, mainly taken from mobile usability studies and best-practices proposed in the literature; (c) homogenization of the redaction and format of subheuristics in order to make it a useful and comprehensive tool for nonexperts; and (d) user-evaluation of the usefulness of the tool as an aid in designing for mobile.

Future work includes mobility and varying context and multidevice access, constraints that are not considered with enough detail in this work. Indeed, these two questions constitute specific areas of study. The typical mobility and varying context of this kind of devices highlight the limitations of laboratory testing: to fully test mobile interfaces, some field-testing is required. Multidevice access questions deal with Responsive Design [36], a discipline that manages access to a given source of information from different devices in a coherent and comprehensive manner.

Regarding rating, in this study, no weighting for categories was established. We mentioned the nonnegligible amount of items scored as nonapplicable in our experiment. Weighting specific categories or subsets of subheuristics according to the kind of application being evaluated represents a highly interesting area for future work and one which is closely related to certain advances in the work of Torrente [7].

The heuristic checklist we have proposed needs to be thoroughly validated in future research in relation to different aspects. The preliminary test and results obtained in this work appear to indicate that the proposed HE guideline is a useful tool for engineers, designers, and technicians with no specific knowledge in usability. A first hypothesis to explain this result is that more specific heuristic guidelines named subheuristics in this work are easier to manage for nonexpert evaluators. The specificity of the items collected in the tool means that it can be used as a reference guide to help conceive more usable interfaces and not just as a reactive evaluation tool for existing prototypes. Future work should look into this to confirm this partial result.

Furthermore, other aspects related to the suitability of this guideline need to be validated. For instance, an experts-guided review could evaluate the completion, coherence, and adequacy of the heuristic checklist. This review could be carried out through questionnaires, experts panels, or some kind of Delphi-based surveys [48]. Another highly interesting question is the empirical comparison of general heuristics and this mobile-specific heuristics when analysing mobile interfaces.

REFERENCES

1. International Organization for Standardization (ISO), "ISO 9241-11:1998 Ergonomic requirements for office work with visual display terminals (VDTs)—part 11: guidance on usability," http://www.iso.org/iso/iso_catalogue/catalogue_tc/catalogue_detail.ht m?csnumber=16883.

2. R. Inostroza, C. Rusu, S. Roncagliolo, C. Jiménez, and V. Rusu, "Usability heuristics for touchscreen-based mobile devices," in Proceedings of the 9th International Conference on Information Technology (ITNG '12), pp. 662–667, Las Vegas, Nev, USA, April 2012.

3. J. Wang and S. Senecal, "Measuring perceived website usability," Journal of Internet Commerce, vol. 6, no. 4, pp. 97–112, 2007.

4. J. Karat, "User-centered software evaluation methodologies," in Handbook of Human-Computer Interaction, M. Helander, T. K. Landauer, and P. Prabhu, Eds., pp. 689–704, Elsevier, New York, NY, USA, 1997.

5. Zhang, "Overview of usability evaluation methods," http://www.usabilityhome.com.

6. J. Heo, D. Ham, S. Park, C. Song, and W. C. Yoon, "A framework for evaluating the usability of mobile phones based on multi-level,

hierarchical model of usability factors," Interacting with Computers, vol. 21, no. 4, pp. 263–275, 2009.

7. M. C. S. Torrente, Sirius: Sistema de evaluación de la usabilidad web orientado al usuario y basado en la determinación de tareas críticas [Ph.D. thesis], Universidad de Oviedo, Oviedo, España, 2011.

8. M. Y. Ivory and M. A. Hearst, "The state of the art in automating usability evaluation of user interfaces," ACM Computing Surveys, vol. 33, no. 4, pp. 470–516, 2001.

9. J. Nielsen and R. Molich, "Heuristic evaluation of user interfaces," in Proceedings of the SIGCHI Conference on Human Factors in Computing Systems (CHI '90), J. C. Chew and J. Whiteside, Eds., pp. 249–256, ACM, New York, NY, USA, 1990.

10. J. Nielsen, "Guerrilla HCI: Using Discount Usability Engineering to Penetrate the Intimidation Barrier," Nielsen's Alertbox, 1994, http://www.nngroup.com/articles/guerrilla-hci/.

11. R. Agarwal and V. Venkatesh, "Assessing a firm's Web presence: a heuristic evaluation procedure for the measurement of usability," Information Systems Research, vol. 13, no. 2, pp. 168–186, 2002.

12. C.-M. Karat, R. Campbell, and T. Fiegel, "Comparison of empirical testing and walkthrough methods in user interface evaluation," in Proceedings of the SIGCHI Conference on Human Factors in Computing Systems (CHI '92), pp. 397–404, ACM, New York, NY, USA, May 1992.

13. J. Nielsen, "Heuristic evaluation," in Usability Inspection Methods, J. Nielsen and R. L. Mack, Eds., John Wiley & Sons, New York, NY, USA, 1994.

14. R. Budiu and N. Nielsen, Usability of iPad Apps and Websites, Nielsen Norman Group, 2nd edition, 2011.

15. C. Rusu, S. Roncagliolo, V. Rusu, and C. Collazos, "A methodology to establish usability heuristics," in Proceedings of the 4th International Conferences on Advances in Computer-Human Interactions (ACHI '11), pp. 59–62, IARIA, 2011.

16. Android Design Guidelines, https://developer.android.com/design/index.html.

17. iOS Human Interface Guidelines. Designing for iOS7, 2014, https://developer.apple.com/library/ios/documentation/userexperience/conceptual/MobileHIG/index.html.

18. Smartphone Adoption and Usage, http://www.pewinternet.org/2011/07/11/smartphone-adoption-and-usage/.

19. J. Nielsen, Mobile Usability Update, Nielsen's Alertbox, 2011, http://www.nngroup.com/articles/mobile-usability-update/.

20. M. Dunlop and S. Brewster, "The challenge of mobile devices for human computer interaction," Personal and Ubiquitous Computing, vol. 6, no. 4, pp. 235–236, 2002.

21. Zhang and B. Adipat, "Challenges, methodologies, and issues in the usability testing of mobile applications," International Journal of Human-Computer Interaction, vol. 18, no. 3, pp. 293–308, 2005.

22. R. Looije, G. M. te Brake, and M. A. Neerincx, "Usability engineering for mobile maps," in Proceedings of the 4th International Conference on Mobile Technology, Applications, and Systems and the 1st International Symposium on Computer Human Interaction in Mobile Technology (Mobility '07), pp. 532–539, ACM, New York, NY, USA, 2007.

23. D. Zhang, "Web content adaptation for mobile handheld devices," Communications of the ACM, vol. 50, no. 2, pp. 75–79, 2007.

24. R. Budiu and J. Nielsen, Usability of Mobile Websites: 85 Design Guidelines for Improving Access to Web-Based Content and Services Through Mobile Devices, Nielsen Norman Group, 2008.

25. S. Tsiaousis and G. M. Giaglis, "An empirical assessment of environmental factors that influence the usability of a mobile website," in Proceedings of the 9th International Conference on Mobile Business and 9th Global Mobility Roundtable (ICMB-GMR '10), pp. 161–167, June 2010.

26. H. B. Duh, G. C. B. Tan, and V. H. Chen, "Usability evaluation for mobile device: a comparison of laboratory and field tests," in Proceedings of the 8th International Conference on Human-Computer Interaction with Mobile Devices and Services (MobileHCI '06), pp. 181–186, ACM, Helsinki, Finland, September 2006.

27. S. Po, S. Howard, F. Vetere, and M. Skov, "Heuristic evaluation and mobile usability: bridging the realism gap," in Mobile Human-Computer Interaction—MobileHCI 2004, vol. 3160 of Lecture Notes in Computer Science, pp. 49–60, Springer, Berlin, Germany, 2004.

28. Bertini, S. Gabrielli, and S. Kimani, "Appropriating and assessing heuristics for mobile computing," in Proceedings of the Working Conference on Advanced Visual Interfaces (AVI '06), pp. 119–126, ACM, New York, NY, USA, May 2006.

29. Y. Cui and V. Roto, "How people use the web on mobile devices," in Proceeding of the17th International Conference on World Wide Web 2008 (WWW '08), pp. 905–914, New York, NY, USA, April 2008.

30. L. D. Catledge and J. E. Pitkow, "Characterizing browsing strategies in the World-Wide web," Computer Networks and ISDN Systems, vol. 27, no. 6, pp. 1065–1073, 1995. View at Publisher · View at Google Scholar · View at Scopus

31. C. W. Choo, B. Detlor, and D. Turnbull, "A behavioral model of information seeking on the web," in Proceedings of the ASIS Annual Meeting Contributed Paper, 1998.

32. J. B. Morrison, P. Pirolli, and S. K. Card, "A Taxonomic analysis of what world wide web activities significantly impact people's decisions and actions," in Proceeedings of the Conference on Human Factors in Computing Systems (CHI EA '01), pp. 163–164, ACM, New York, NY, USA, April 2001.

33. J. Sellen, R. Murphy, and K. L. Shaw, "How knowledge workers use the web," in Proceedings of the SIGCHI Conference on Human Factors in Computing Systems (CHI '02), pp. 227–234, ACM, New York, NY, USA, 2002.

34. MacKay, C. Watters, and J. Duffy, "Web page transformation when switching devices," in Mobile Human-Computer Interaction—MobileHCI 2004: Proceedings of 6th International Symposium, MobileHCI, Glasgow, UK, September 13–16, 2004., vol. 3160, pp. 228–239, 2004.

35. M. Kellar, C. Watters, and M. Shepherd, "A goal-based classification of web information tasks," Proceedings of the American Society for Information Science and Technology, vol. 43, pp. 1–22, 2006.

36. E. Marcotte, "A list apart," Responsive Web Design, 2010, http://alistapart.com/article/responsive-web-design/.

37. Shneiderman and C. Plaisant, Designing the User Interface: Strategies for Effective Human-Computer Interaction, Addison-Wesley, Reading, Mass, USA, 1987.

38. Pierotti, "Heuristic evaluation—a system checklist," Tech. Rep., Xerox Corporation, Society for Technical Communication, 2005.

39. L. Constantine, Devilish Details: Best Practices in Web Design, Constantine and Lockwood, 2002.

40. K. Instone, "Usability engineering for the web," W3C Journal, 1997.

41. B. Tognazzini, "First principles of interaction design," Interaction Design Solutions for the Real World, 2003, http://www.asktog.com/.

42. Y. Hassan Montero and F. J. Martín Fernández, "Guía de Evaluación Heurística de sitios web," 2003, http://www.nosolousabilidad.com/articulos/heuristica.htm.

43. Y. Hassan, M. Fernández, J. Francisco, and G. Iazza, Diseño Web Centrado en el Usuario: Usabilidad y Arquitectura de la Información, vol. 2, 2004, http://www.upf.edu/hipertextnet/.

44. L. Olsina, G. Lafuente, and G. Rossi, "Specifying quality characteristics and attributes for websites," in Web Engineering, Software Engineering and Web Application Development, S. Murugesan and Y. Deshpande, Eds., pp. 266–278, Springer, London, UK, 2001.

45. J. Nielsen, "Severity ratings for usability problems," Nielsen's Alertbox, 1995, http://www.nngroup.com/articles/how-to-rate-the-severity-of-usability-problems.

46. November 2013, http://innovation.logica.com.es/web/procura.

47. 2013, http://www.researchspaces.eu.

48. A. Linstone and M. Turoff, Delphi Method: Techniques and Applications, Addison-Wesley, Reading, Mass, USA, 1975.

49. N. O. Bernsen, H. Dybkjær, and L. Dybkjær, "Wizard of Oz prototyping: when and how?" Working Papers in Cognitive Science and HCI, 1993.

50. J. Lazar, J. H. Feng, and H. Hochneiser, Research Methods in Human-Computer Interaction, John Wiley & Sons, New York, NY, USA, 2010.

CHAPTER 9

Heuristics and Pattern Recognition in Complex Geo-Referenced Systems

Gilbert Ahamer[1], Adrijana Car, Robert Marschallinger, Gudrun Wallentin and Fritz Zobl

[1] *Institute for Geographic Information Science at the Austrian Academy of Sciences, ÖAW/GIScience, Schillerstraße 30, A-5020 Salzburg, Austria*

1. LET'S START TO THINK

1.1 Our world is the entirety of perceptions. (Our world is not the entirety of facts.)

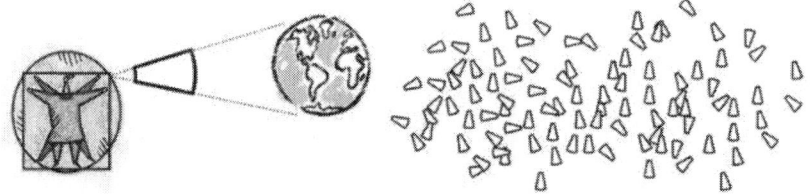

Figure 1. At left: The human being perceives the world. At right: The "primordial soup" of living, before the advent of (social) organisms: uncoordinated perspectives, uncoordinated world views.

Hence, every individual lives in a different world (Fig. 1 at left).

1.2 The "indivisible unit", the atom (ατομος[1] -) of reality, is equal to one (human) perspective. Our world is made up of a multitude of perceptions, not of a multitude of realities and not of a multitude of atoms (Fig. 1 at right).

1.3 In order to share one's own conception with others, "writing" was invented. Similarly, complex structures, such as landscapes, are "mapped". To map means to write structures.

1.4 Writing helps to become aware. We ask: Is it possible to map = write

1. the distribution of material facts and elements in geometric space? (physics)
2. the distribution of factual events in global time? (history)
3. the distribution of real-world objects across the Earth? (geography)
4. the distribution of elements along material properties? (chemistry)
5. the distribution of growth within surrounding living conditions[2] - ? (biology)
6. the distribution of persons acting in relationships? (sociology)
7. the distribution of individuals between advantage and disadvantage of trade? (economics)
8. the distribution of perspectives within feasible mindsets? (psychology)
9. the distribution of living constructs along selectable senses? (theology)

We see: awareness results from reflection (Fig. 2).

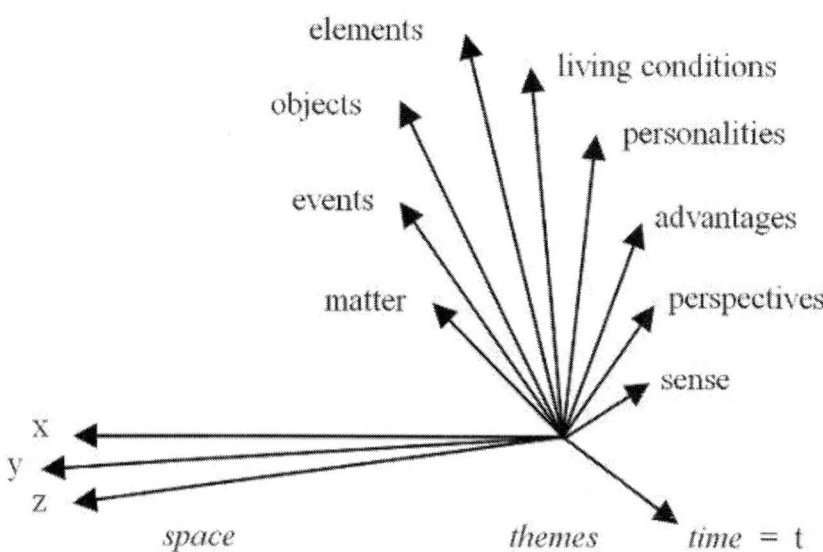

Figure 2. Fundamental dimensions, along which to coordinate individual world views when reflecting.

2. TIME CAN BE

1. an attribute of space (a very simple historic GISystem)
2. an independent entity (Einstein's physics)
3. the source of space (cosmology).

In terms of GIS item 2.1 is expressed as "t is one of the components of geo-data"[3] - (Fig. 3).

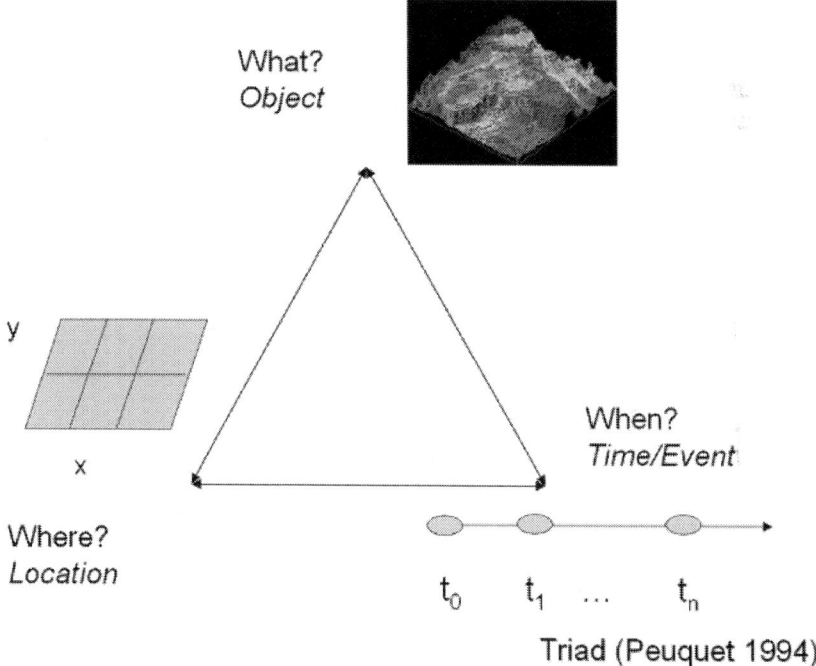

Triad (Peuquet 1994)

Figure 3. The where-what-when components of geo-data, also known as triad (Peuquet 2002: 203).

Time can be understood as

- establishing an ordinal scale for events
- driving changes ($=\Delta$) of realities
- something that unfortunately does not appear on paper.

A proposed solution is to map changing realities (Δ) instead of mapping time.

Time is replaced by what it produces. This is indicated in Fig. 4.

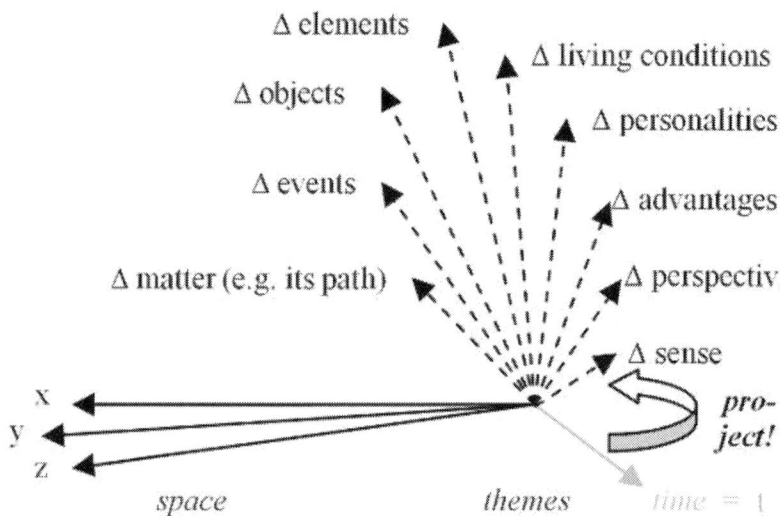

Figure 4. The projection of time (t) onto the effects of time (the changes) can apply to any science.

This idea flips = projects the t axis onto one of the vertical axes. Time means then: how maps are changed by the envisaged procedures.

Such procedures modify the variables along the axes, be they of physical (gravity force) or of social nature (war).

A classical example is Minard's map of Napoleon's 1812 campaign into Russia[4] - (Fig. 5a, b).

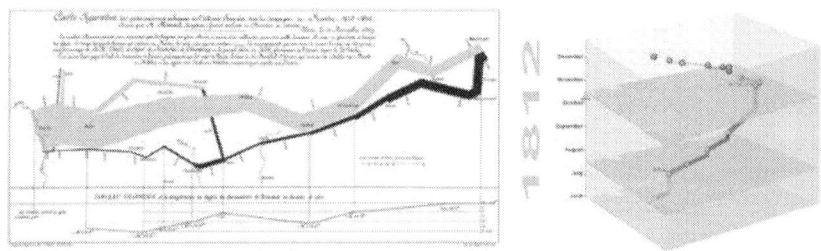

Figure 5. Notions of path in a geo-space: (a) Minard's map of human losses during Napoleon's 1812 campaign into Russia; and (b) its geovisualisation in a time cube (Kraak, 2009).

Further examples such as landslides in geology, growth of plants, energy economics, economics will be shown in chapter 7.

For implementing the idea to project the t axis onto the axis we need to have clear insight how time quantitatively changes reality.

In other words: we need a model, which (explaining how processes occur) determines the representation of time (Fig. 6). Examples are sliding geology, GDP/cap, plant growth.

One cannot perceive time (never!), only its effects: what was perceived in this time span (duration)[5] - ? This is why the t axis is projected onto another axis denoting the effect of elapsed time; what this means to the individual sciences is shown in Fig. 4.

Very similarly, in physics nobody can feel force, only its effect (deformation, acceleration), and still forces have been undisputedly a key concept for centuries.

What is time? Just a substrate for procedures.

What is space? Just hooks into perceived reality.

We retain from this chapter 2 that we need a clear model of how elapsing time changes reality. Then we can map time as suggested: by its effects.

3. HOW TO WRITE TIME?

The big picture shows us various examples:

1. as a wheel (see the Indian flag): revolving zodiacs, rounds in stadiums, economic cycles, Kondratieff's waves
2. as an arrow (see Cartesian coordinates): directed processes, causal determinism, d/dt, d^2/dt^2
3. as the engine for further improvement (evolutionary economics): decrease vs. increase in global income gaps, autopoietic systems, self-organisation
4. as the generator of new structures (institution building, political integration, progressive didactics): new global collaborative institutions, peer-review, culture of understanding, self-responsible learning, interculturality
5. as evolving construct (music).

From this chapter 3 we only keep in mind that the concepts to understand and represent time are fundamentally and culturally different.

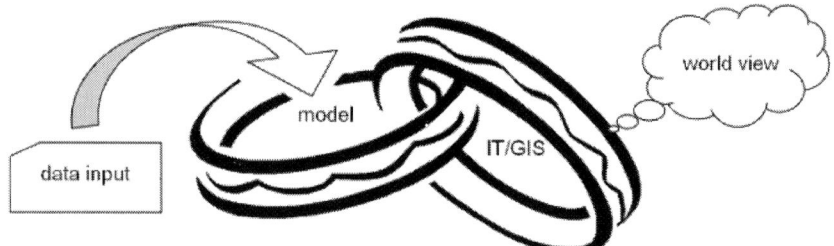

Figure 6. All data[6] - representations require models.

4. HOW TO WRITE SPACE?

The big picture shows us various examples:

1. as a container of any fact and any process (geography and GIS)
2. as result of human action (landscape planning)
3. as evolving construct (architecture).

Examples span space as

- received and prefabricated versus

- final product of one's actions, namely:

  o spaces as the key notion for one's own science: everything that can be geo-referenced means GIS

  o space as the product of human activity

  o expanding space into state space: the entirety of possible situations is represented by the space of all "state vectors" which is suitable only if procedures are smooth.

The main thesis here is: the "effects of time" are structurally similar in many scientific disciplines, and they often imply "changes in structures" too. Information Technology (IT) is already providing scientific tools to visualise such structures.

5. HOW TO MAP SPACE AND TIME?

The detailed picture: it is obvious that a choice must be made for one mode of representation and for one view of one scientific discipline:

- 1. $(x, y; t)$: cartography, GIS (Fig. 7)

- 2. $(x, y, z; t)$: geology

- 3. $(x, y, z; v_x, v_y, v_z; t)$: landslides

- 4. $(x, y, z;$ biospheric attributes; $t)$: ecology, tree-line modelling

- 5. (countries; economic attributes; GDP/cap) or (social attributes; structural shifts; elapsing evolutionary time): economic and social facts in the "Global Change Data Base"[7] - (Fig. 8)

- 6. perceiving rhythms and structures: (only) these are "worth recognising": music, architecture, fine arts.

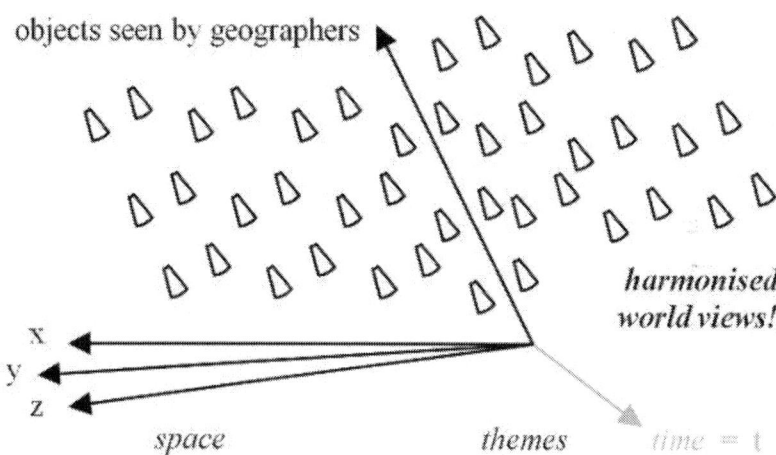

Figure 7. Harmonising world views: GIS reunites world views by relating everything to its location.

Different sciences may have considerably different outlooks on reality (Fig. 8). A humble attitude of recognising facts[6] instead of believing in the theories one's own discipline offers can empower people to survive even in the midst of other scientific specialties: Galileo's (1632) spirit: give priority to observation, not to theories!

This is the essential advantage of geography as a science: geographers describe realities, just as they appear. Such a model-free concept of science has promoted the usefulness of GIS tools to people independent of personal convictions, scientific models or theories.

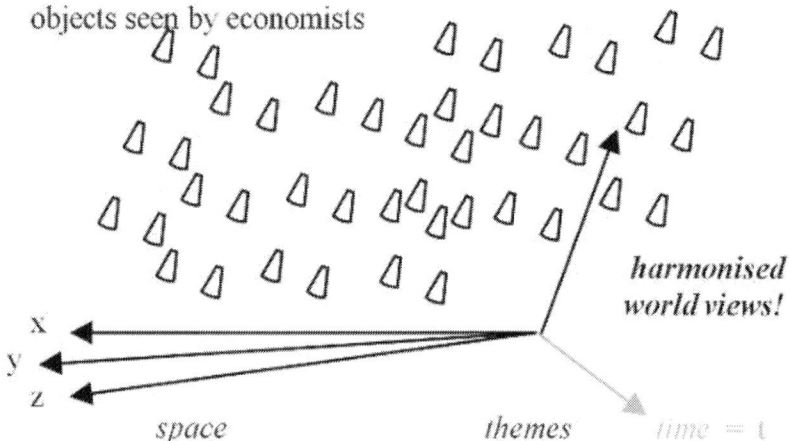

Figure 8. Different but again internally harmonised world views: explain facts from another angle.

6. WHAT IT DOES, DID, AND COULD DO

6.1 IT Helps To Organise the Multitude of Views (= Perceptions) Onto Data that are Generated by Humans:

- IT constructs world views, such as: GIS, history, economics, geology, ecology etc.
- IT has already largely contributed to demolishing traditional limitations of space and time:
  - Space: tele(-phone, -fax, -vision), virtual globes (Longley et al., 2001)
  - Time: e-learning, asynchronous web-based communication, online film storage (Andrienko & Andrienko 2006).

6.2 This Paper Investigates Non-Classical Modes of Geo-Representation.

We would like to point out that there are two already well-established fields that offer solutions to mapping (space and time, Fig. 9) views: Scientific and information visualisation are branches of computer graphics and user interface design which focus on presenting data to users, by means of interactive or animated digital images. The goal of this field[8] - is usually to improve the understanding of the data presented. If the data presented refers to human and physical environments, at geographic scales of measurement, then we talk about Geovisualisation, e.g. (MacEachren, Gahegan et al. 2004; Dykes, MacEachren et al. 2005, Dodge et al., 2008).

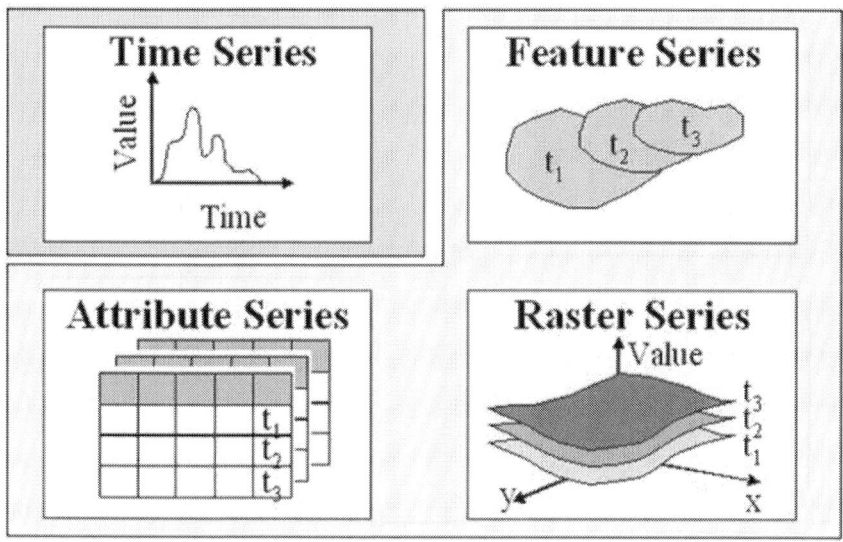

Figure 9. Time series and three spatio-temporal data types (http://www.crwr.utexas.edu/gis/gishydro05/).

6.3 IT could develop tools that are then interchangeable across scientific disciplines, e.g. landslides that may structurally resemble institutional and economic shifts (see 7.1).

IT could prompt scientists to also look at data structures from other disciplines.

Whatever the disciplines may be, the issues are structures and structural change!

7. EXAMPLES

The authors are members of the "Time and Space" project at their institution named "Geographic Information Science"[9] - , a part of which explores the cognitive, social, and operational aspects of space & time in GIScience.

This includes models of both social and physical space and consequences thereof for e.g. spatial analysis and spatial data infrastructures. We investigate how space and time are considered in these application areas, and how well existing models of space and time meet their specified needs (see e.g. Fig. 9 left). This investigation is expected to identify gaps. Analysis of these gaps will result in improved or new spatio-temporal concepts particularly in support of the above mentioned application areas.

7.1. Sliding Realities: Geology

The notion of the path in geography (x, y, t) is extended by the z axis (see item 5.2) which produces a map of "time": Fig. 9 left (Zobl, 2009).

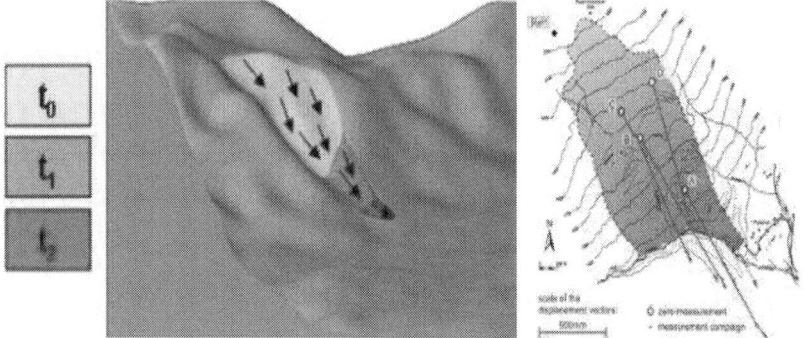

Figure 10. Left: Geology takes the (x, y, z; t) world view. Right: These effects of time occur in space, most helpfully. Source: Brunner et al. (2003).

The "effect of time" is sliding (luckily in the same spatial dimensions x, y, z): we take the red axis in Fig. 10 right. Space itself is sufficiently characteristic for denoting the effects of time.

Figure 11. This series of understandings of bedrock shows how data are stepwise combined with model results in order to reach approximation of understandings (Klima & Zobl, 2009).

7.2. Slices of Realities: Geology

Despite the lucky coincidence that the effect of time (x, y, z) occurs in the same space (x, y, z) we try to produce slides carrying more information (item 5.3) and hence recur to the so-called attributes mentioned in Fig. 10 such as grey shades or colours.

The speed of sliding (d/dt x, d/dt y, d/dt z) is denoted both by horizontal offsets and whitish colours in the spaghettis (Marschallinger, 2009) of Fig. 12.

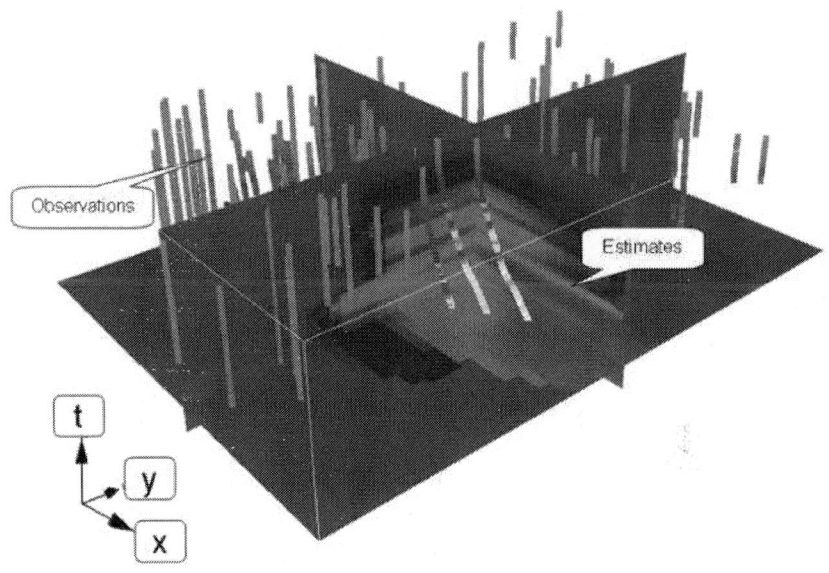

Figure 12. The (x, y, z; v_x, v_y, v_z; t) view of a landslide process (shades of grey mean speed v).

7.3. Slide Shows

How to map spatial realities that are not any longer isotropic displacement vectors of space itself? For the example of changing tree lines in the Alps (Wallentin, 2009) a slide show is used to present the change of growth patterns made up of the multitude of individual agents (= trees = dots in Fig. 13). Moving spatial structures are depicted as a film of structures (item 5.4).

Figure 13. The (x, y, z; biospheric attributes; t) view of the Alpine tree line (above) and its shift induced by climate change as a slide show (below), as computed by the TREELIM model.

In such processes which involve independent behaviour of autonomous agents (here: trees) it becomes seemingly difficult to apply a transformation of space itself, e.g. d/dt(x, y, z).

The model used and its background is briefly described: alpine tree line responses dynamically to changes in the environment. Currently a substantial upwards shift of the alpine tree line can be observed due to land use and climate change. The spatio-temporal tree patterns are modelled in an individual-based approach, where system-level attributes emerge from ecological processes of single trees, their mutual interactions and reactions to environmental factors such as climate or the elevation gradient.

TREELIM is an individual based model that was developed to get a better understanding of the alpine tree line dynamics in respect to land-use change. The model was validated for a case study in Längenfeld, Ötztal (Tyrol, Austria) over a period of 52 years (Wallentin et al. 2008). Individual based models that are structurally realistic model the real-world processes that drive landscape patterns (Grimm & Railsback 2005). Thus TREELIM is designed as a generic model that can be extrapolated in space and time.

Traditionally, individual based models are validated through a model-reality comparison of spatial patterns at a certain point in time. However, a dynamic system does not merely have characteristic spatial patterns, but rather can be described through spatio-temporal patterns. Whereas in non-spatial process models of ecosystems, the temporal aspect is commonly considered as a crucial point in the model validation, this is not the case for spatial models. For spatial models, i.e. models that result in maps as the central model output, the model validation focuses on spatial aspects at a certain point in time (a snapshot situation). Although the temporal aspect may be included to some extent by considering time steps, (i.e. distribution of a spatial feature at several points in time) there is no consideration of spatio-temporal patterns and temporal model uncertainties in the model output variables, as e.g. the average tree line elevation.

7.4 Global Deforestation

One key driver for global change is deforestation; easy to map as change of land use category of a given area (Fig. 14).

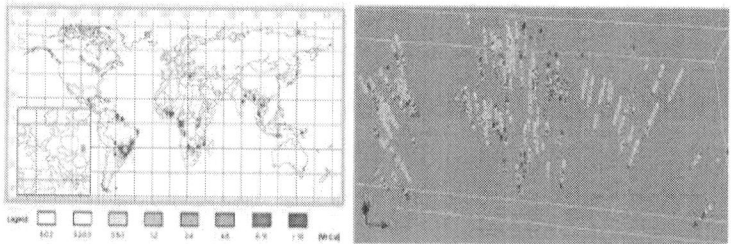

Figure 14. The (x, y, z; biospheric attributes; t): view of the global deforestation process in megatons carbon. Above: map of carbon flow, below: time series of GCDB data per nation symbolically geo-referenced by the location of their capitals.

This representation is analogous to Fig. 12. In both, the focus shifts from maps(t) maps(t, t). Interest includes temporal dynamics:

t = colour (above); t = height+colour (below), enriching the purely spatial interest.

Even if to the aim is to enlarge the scope of the information delivered from the static map (Fig. 14 above) to the "dynamic map" (Fig. 14 below), readers will remain unsatisfied because no insight into the dynamic properties of deforestation is provided (Fig. 16).

Increasingly, the viewer's focus turns further from "facts" to "changes of facts", to "relationships with driving parameters[10] - " and to (complex social and political) "patterns[11] - ".

7.5. Space as Text

Studying geo-referenced data sets (GIS) can help to facilitate bridging interperceptional gaps. Critical GIS literature (e.g. Sarah Elwood) suggests that GIS tools are combined with gaining power over space (Fischer, 2008a). Theories going back to Ferdinand de Saussure (Fischer, 2008b) understand space as "text". Societies write space and then read it.

GIS systems are the expression of the ability to "write space", they express the "space views of those being able to write space" (Fischer, 2008c).

7.6. Spaces Constituted By Social Media

The concept of Manuel Castell's "space of flows" (Castells, 2004, Ahamer & Strobl, 2009) sees space as constitutes by communication. Urban planning (Crang, 2000) sees that "electronic media has raised issues of political action, community formation and changing identities". "The metaphorical adoption of urban models is co-determined by electronic sociality and suggests four principle approaches: cities set in or against world flows, suburbanised telecities, communitarian visions and accounts that appeal to a renewed public sphere – all of them shape an electronic architecture. Spatial metaphors and electronic practices are seen as entangled and shaping each other." "In this sense, the 'real' city is the indefinable complexity and folding of spaces—lying outside the visualisations offered of cyberspace."

Urban social relations are co-determined by (electronic) communication (Purcell, 2001; Kirby, 2008).

Users and Non-Users of Social Network Sites exhibit distinct social patterns (Hargittai, 2008) and construct their spaces (and times) in distinct manner (Zheng & Niu, 2007, Ahamer, 2010).

7.7. Design of Social Processes

Design can be perfomed in and of these following substrates :

- Time = theatre
- Space = architecture
- Geometry = graphics design
- Functionality = design in the narrower sense of industrial design)
- Interests = technology assessment (and administration)
- Structures = arts, science
- Love = new life.

In any of these cases it is necessary to en-act reality in and along time: time is the sequencer for all structures. Finally, en-acting means also the act of love: to "enact" life. Create life. New life.

Design is creative *generating* of structures. Hence it means the human share in continuous creation. Structures are created in an evolutionary way (Ahamer & Wahliss, 2008). On the other hand, science is (only) the *cognition* of structures.

Restrepo & Christiaans (2004: 3) find after several so-called ethnographic studies where they watch and analyse designers at work that "design is a *discursive* activity" i.e. a self-referenced process. "Designers propose design issues, reflect upon and discuss them and for each issue propose answers (also called positions). A discussion about which position to accept (design moves)." "However, problem structuring is not a clearly distinguishable phase of the design process but instead an activity that reoccurs regularly (...) and can contribute to either further structure the problem or to solve it."

Thomas & Carroll (1979) discovered that designers tend to treat all problems as though they were ill-defined. They do so by changing the problems constraints and goals – even if the product was well-defined. Designers will be designers even if they can be problem solvers.

Dorst (2004) sees as the main "design problem" that "this process of reasoning is non-deductive". There are "two ways in which a design problem is underdetermined:

- a description in terms of needs, requirements and intentions can never be complete
- 'needs, requirements and intentions' and 'structure' belong to different conceptual worlds.

He cites Dorst & Cross (2001) viewing creativity in the design process as a co-evolution of problem and solution spaces.

As a result, we suggest:

- Rhythmisation of time (examples: theatre, evolution)
- Rhythmisation in space and in opinion (examples: court, perspectives, politics, evolution of societal institutions, institution building).

Both these suggested strategies are implemented in the negotiation game "Surfing Global Change" (Ahamer, 2004).

Rhythmisation represents the theatrical strategy and *multi-perspectivism* represents the geo-locating strategy in the "space of perspectives".

7.8. Space and Time for Consensus Building

A suitable case study for a heuristic pattern for consensus building evolving in space and time is the five level web based negotiation game (Figure 15), its rules were published as (Ahamer, 2004).

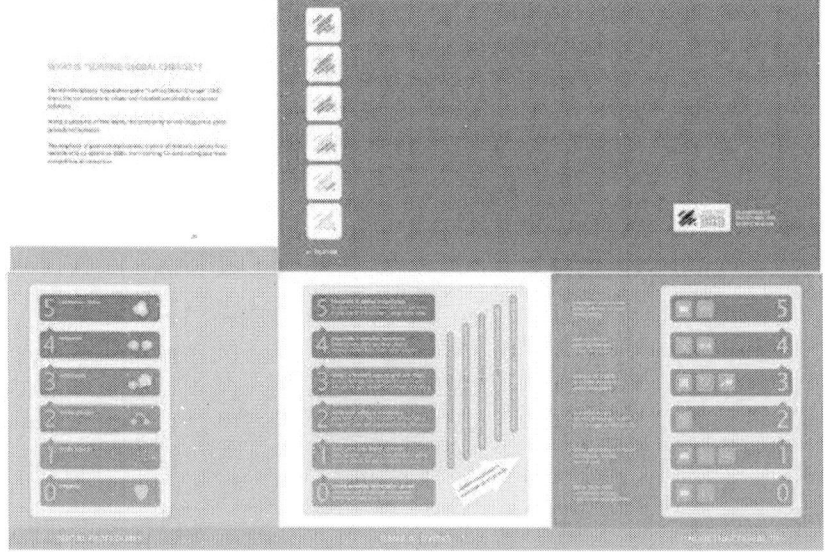

Figure 15. The „unfolder" gives way to successively deeper levels of detail after embarking on reading (© G. Ahamer).

8. TRANSFORMATION OF COORDINATES

8.1 All the above examples have shown that

- various "spaces" can be thought of
- it would be suitable to enlarge the notion of "time".

8.2 Suitably, a transformation of coordinates from time to "functional time" may be thought of.

8.3 In chapter 2, we suggested already to regard time as the substrate for procedures. Consequently, different "times" can be applied to different procedures. As an example, in theoretical physics, the notion of "Eigentime[12] - " is common and means the system's own time.

8.4 Similar to the fall line in the example of landslides in chapter 7.1 (red in Fig. 10) the direction of the functional time is the highest gradient of the envisaged process. This (any!) time axis is just a mental, cultural construction.

8.5 According to chapter 2 (Fig. 6) a clear understanding (mental model) is necessary to identify the main "effect of time". We see that such an understanding can be culturally most diverse. Just consider the example of economic change:

- optimists think that the global income gap decreases with development

- pessimists believe that it increases, hampering global equity.

8.6 Therefore, any transformation of coordinates bears in itself the imponderability of complex social assumptions about future global development and includes a hypothesis on the global future.

8.7 Still, a very suitable transformation is

t GDP/capita

(Fig. 16) both because of good data availability and increased visibility of paths of development. GDP/cap resembles evolutionary time.

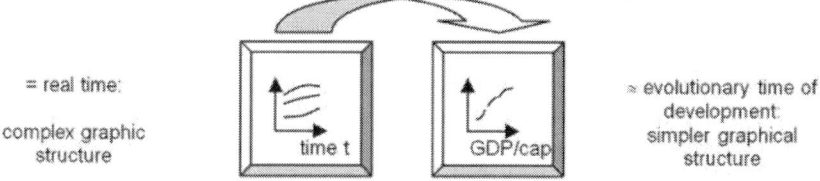

Figure 16. A suitable transformation of time uses the economic level, measured as GDP per capita.

8.8 The strategic interest of such a transformation is "pattern recognition", namely to perceive more easily structures in data of development processes. Examples for such "paths of development" are shown in Fig. 17 for the example of fuel shares in energy economics.

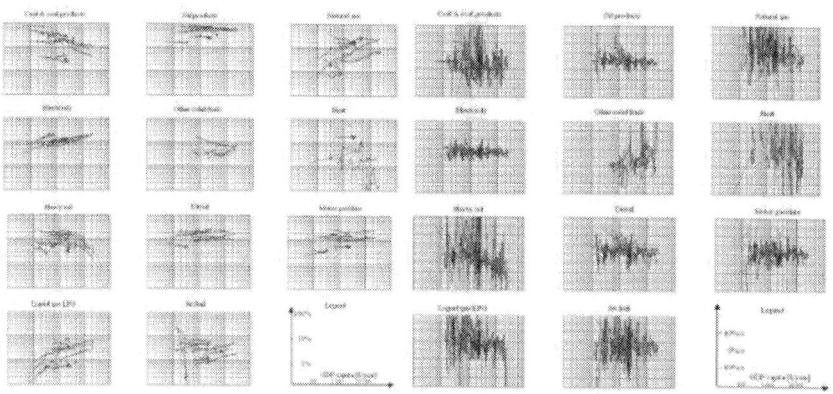

Figure 17. Structural shift of percentages (left) and change rates of percentages (right) of various fuels in all nations' energy demand 1961-91. Data source: GCDB (Ahamer, 2001).

8.9 It is suggested here that implicitly during many mapping endeavours such transformation occurs. This is legitimate, but care must be taken to take into account the (silently) underlying model of human development.

8.10 Suitable transformation of coordinates can facilitate to see and communicate evolutionary structures, as it enables common views of humans and is therefore helpful for global consensus building.

8.11 Also the "effects of time" are projected into a common system of understanding which might give hope to facilitate common thinking independently of pre-conceived ideologies.

This plan creates the "common reference system of objects".

8.12 This paper suggests enlarging the concept of

- "globally universal geo-referencing" (one of the legacies of IT)

to

- "globally universal view-referencing"
- or "globally universal referencing of perspectives"[13] - .

Fig. 20 illustrates this step symbolically.

9. A FUTURISTIC VISION

9.1 Building on the vision of "Digital Earth" (Gore, 1998), the deliberations in this paper might eventually lead to the vision of "Digital Awareness": the common perspective on realities valid for the global population, aided by (geo)graphic means.

9.2 The primordial element of (human and societal) evolution is consensus building. Without ongoing creation of consensus global "evolutionary time" is likely to fall back.

The futuristic vision is to map global awareness.

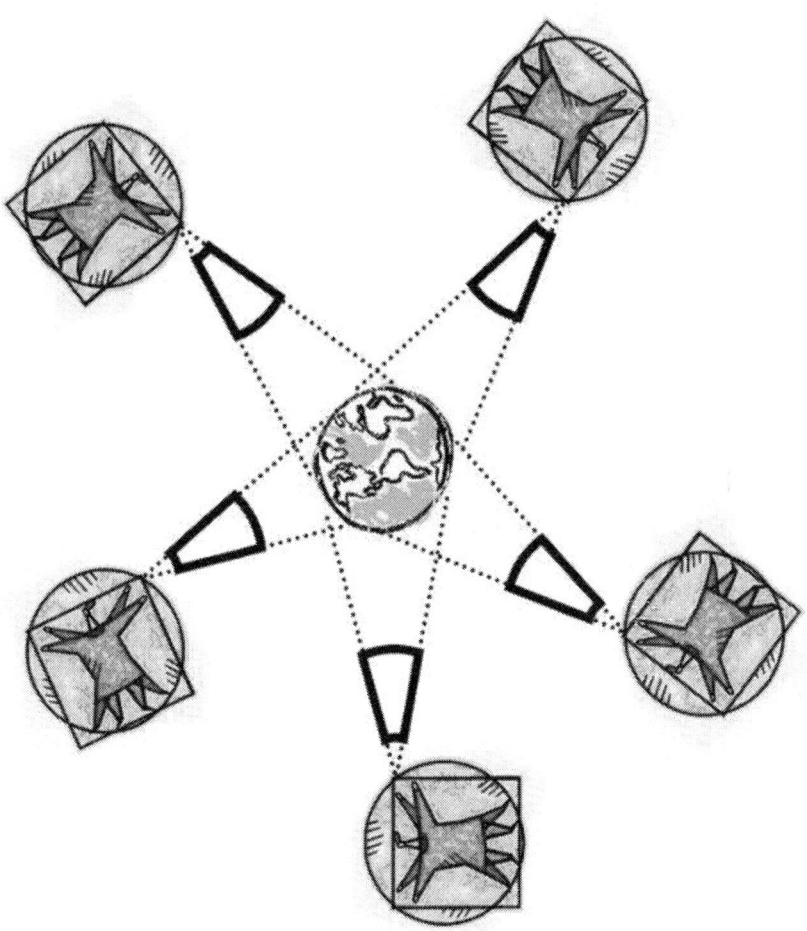

Figure 18. The global society perceives the world.

9.3 Much like the georeferenced satellites which circulate around the world produce a "Google, Virtual [or similar] Earth", the individual spectators in Fig. 18 circle around the facts – and they create a "common virtual perception": an

IIS = Interperspective Information System.

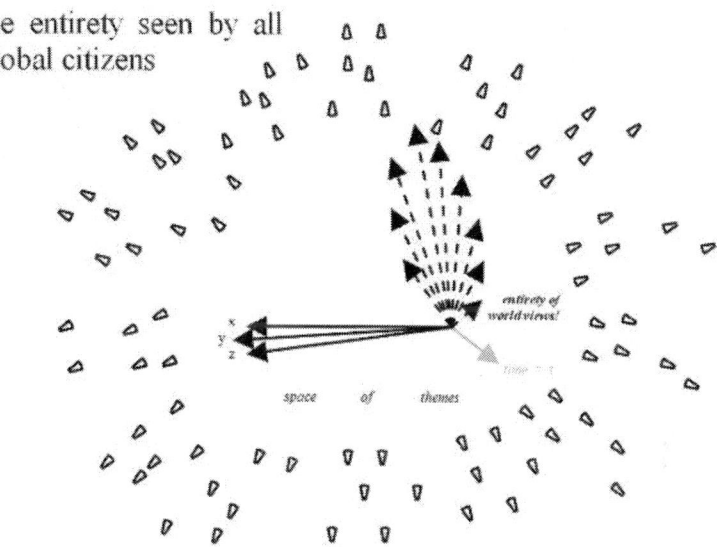

the entirety seen by all global citizens

Figure 19. Divergent perceptions circulate around earthen realities. The entirety of world views creates the IIS (Interperspective Information System).

9.4 Do we just mean interdisciplinarity? No. Nor do we simply refer to people looking into any direction. Fig. 20 shows the difference to IIS.

9.5 The science of the third millennium will allow dealing with a multitude of world views and world perspectives (see Tab. 1) with an emphasis on consensus building.

When learning, the emphasis lies on social learning and may also make use of game-based learning (such as the web-based negotiation game "Surfing Global Change") which allows to experimentally experiment with world views without any risk involved.

9.6 A suitable peaceful "common effort[14] - " for a peaceful future of humankind would involve developing tools and visual aids in order to understand the opinions of other citizens of the globe.

The future is dialogue.

Or else there will be no future.

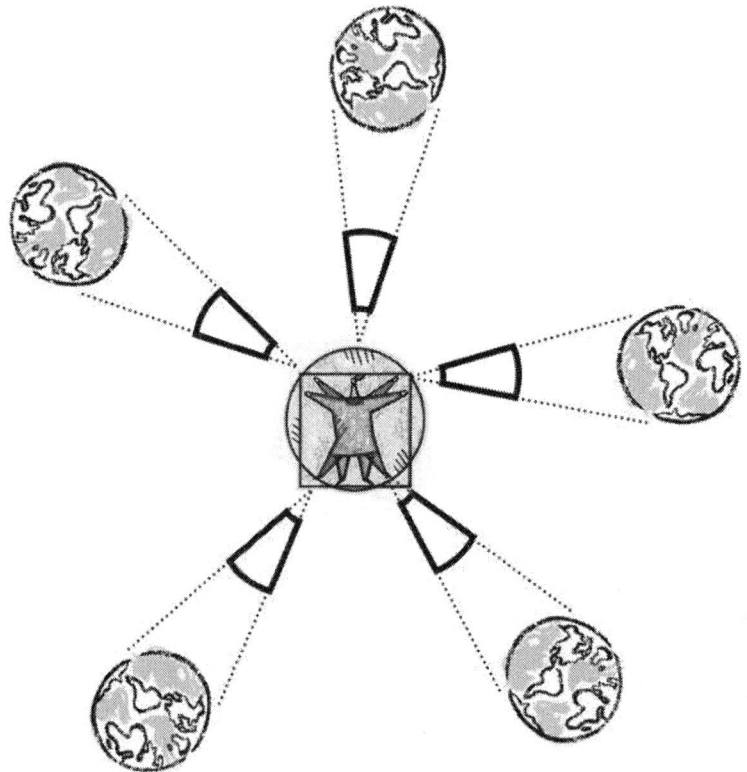

Figure 20. This is not IIS.

Table 1. The science of the third millennium encompasses multiple perspectives.

	element • •	interaction ↙•	perspective ◁ ↙•
single ones	Mechanics	Logics	Teaching
manifold	Thermo-dynamics (19th cent.)	Systems analysis (20th cent.)	Social learning gaming, IIS (21st cent.)

10. CONCLUSION

Sciences are similar to "languages" spoken by people, they differ globally. Understanding for others' languages is essential for global sustainable peace.

Human perceptions are also strongly influenced by underlying models, assumptions and preconceived understandings.

Studying geo-referenced data sets (GIS) can help to facilitate bridging interperceptional gaps.

For the transformation of world views – to make them understandable – it is necessary to know about

- the "effect of time", namely the "path along the continuum of time" which a variable is expected to take

- the speakers' underlying model of a complex techno-socio-economic nature

- the resulting perception of other humans.

A future task and purpose of IT could be to combine the multitude of (e.g. geo-referenced) data and to rearrange it in an easily understandable manner for the viewpoints and perspectives of another scientific discipline or just another human being. Such a system is called Interperspective Information System IIS.

Merging a multitude of perspectives to form a common view of the entire global population is the target of an IIS.

Symbolically, a "Google Earth"-like tool would eventually develop into a "Google World Perspective"-like tool, or a "Virtual Earth"-like tool would become a "Virtual Perspective" tool encompassing all (scientific, social, personal, political, etc.) views in an easily and graphically understandable manner.

In the above futuristic vision, IT can/should(!) become a tool to facilitate consensus finding. It can rearrange the same data for a new view.

Symbolically speaking: similar to Google Earth which allows one to view the same landscape from different angles, a future tool would help to navigate the world concepts, the world views and the world perspectives of the global population.

IT can reorganise extremely large data volumes (if technological growth rates continue) and could eventually share these according to the viewpoint of the viewer.

Such a step of generalisation would lead from "Geographic Information Science" to "Interperspective Information Science", implying the change of angles of perception according to one's own discipline.

Dialogue represents the ultimate heuristics for complex interdisciplinary and intercultural problems. Science does not offer more than such dialogue.

NOTES

[1] - what cannot be split any further (Greek)

[2] - životné prostredie (Slovak): living environment

[3] - GIScience goes way beyond this view of time and space (considering time as function) because it allows for much more complex queries and analyses.

[4] - Patriotic War (in Russian): Отечественная война

[5] - T. de Chardin's (1950) concept of durée (French).

[6] - datum (Latin): what is given (unquestionable)

[7] - This GCDB is described in Ahamer (2001)

[8] - http://en.wikipedia.org/wiki/Scientific_Visualization

[9] - The overarching aim of the GIScience Research Unit is to integrate the "G" into Information Sciences (GIScience, 2009)

[10] - see the suggested scenarios for water demand, water supply and water quality (Ahamer, 2008)

[11] - Patterns: name of the journal of the American Society for Cybernetics ASC

[12] - literally (German): the own time (of the system)

[13] - The facts themselves may well be delivered by endeavours such as Wikipedia but here it refers to the perspective on facts! A huge voluntarily generated database on people's perceptions, views and opinions would be needed.

[14] - (jihad in Arabic) also means: common effort of a society

REFERENCES

1. G. Ahamer, 2001 A Structured Basket of Models for Global Change. In: Environmental Information Systems in Industry and Public Administration (EnvIS). ed. by C. Rautenstrauch and S. Patig, Idea Group Publishing, Hershey, 101 136 , http://www.oeaw-giscience.org/ProjectFactSheets/ProjectFactSheet_GlobalChange.pdf.

2. G. Ahamer, W. Wahliss, 2008 Baseline Scenarios for the Water Framework Directive. Ljubljana, WFD Twinning Project in Slovenia,

http://www.oeaw-giscience.org/ProjectFactSheets/ ProjectFactSheet_ EU_SDI.pdf.

3. G. Ahamer, J. Strobl, 2009 Learning across social spaces. In: Cases on Technological Adaptability and Transnational Learning, IGI Publishing, Hershey, USA, 1 24 .

4. G. Ahamer, 2010 Heuristics of social process design, 265 298 .

5. N. Andrienko, G. Andrienko, 2006 Exploratory Spatial Analysis, Springer

6. F. K. Brunner, F. Zobl, G. Gassner, 2003 On the Capability of GPS for Landslide Monitoring. Felsbau 2/2003 51 54 .

7. M. Castells, 1998 The Information Age: Economy, Society and Culture. Trilogy containing three volumes. Cambridge, MA; Oxford, UK: Blackwell.

8. M. Crang, 2000 Public Space, Urban Space and Electronic Space: Would the Real City Please Stand Up? Urban Studies, 37(2), 301 317.

9. T. de Chardin, 1950 La condition humaine [Der Mensch im Kosmos]. Beck, Stuttgart.

10. M. Dodge, M. McDerby, M. Turner, (Eds.) 2008 Geographic Visualisation, Wiley

11. C. H. Dorst, N. G. Cross, 2001 Creativity in the design process: co-evolution of problem-solution, Design Studies, 22 425 437 .

12. K. Dorst, 2004 On the Problem of Design Problems: problem solving and design expertise. Journal of Design Research, 4(2).

13. J. Dykes, A. MacEachren, et al. 2005 Exploring Geovisualization. Oxford, Elsevier.

14. F. Fischer, 2008a Location Based Social Media- Considering the Impact of Sharing Geographic Information on Individual Spatial Experience. In: Car A., Griesebner G. a. J.Strobl (Eds.): Geospatial Crossroads @ GI_Forum'08. Proceedings of the Geoinformatics Forum Salzburg, 90 96.

15. F. Fischer, 2008b Implications of the usage of mobile collaborative mapping systems for the sense of place. In: M. Schrenk. et al. (Eds.): REAL CORP 008: Mobility Nodes as Innovation Hubs. Proceedings of 13th International Conference on Urban Planning, Regional Development and Information Society, 583 587 .

16. F. Fischer, 2008c Microsoft Virtual Earth. Integrating Geospatial Technology in Everyday Life. Geoinformatics, 4(11), 6-9.

17. G. Galileo, 1632 Dialogo sopra i due massimi sistemi del mondo, tolemaico, e copernicano. Fiorenza.

18. GIScience, 2008 Connecting Real and Virtual Worlds. Poster at AGIT'08, http://www.oeaw-giscience.org/index.php?option=com_content&task=blogcategory&id=43&Itemid=29.

19. A. Gore, 1998 Vision of Digital Earth, http://www.isde5.org/al_gore_speech.htm.

20. V. Grimm, S. F. Railsback, Eds. 2005 Individual-based Modeling and Ecology. Princton Series in Theoretical and Computional Biology. Princton University Press, Princton and Oxford.

21. E. Hargittai, 2008 Whose Space: Differences Among Users and Non-Users of Social Network Sites. Journal of Computer-Mediated Communication, 13(1), 276 297 .

22. A. Kirby, 2008 The production of private space and its implications for urban social relations. Political Geography, 27(1), 74 95.

23. K. Klima, F. Zobl, 2009 in press). Herausforderung der geologischen Erkundung und der Untergrundmodellierung für die geotechnische Analyse Herausforderung der geologischen Erkundung und der Untergrundmodellierung für die geotechnische Analyse. Key note paper published in: Tagungsband Computerorientierte Geologie 2009 in Salzburg, Wichmann.

24. Kraak 2009 Minard's map. www.itc.nl/personal/kraak/1812/3dnap.swf

25. P. A. Longley, et al. 2001 Geographic Information. Science and Systems. Wiley

26. A. M. MacEachren, M. Gahegan, et al. 2004 Geovisualization for Knowledge Construction and Decision Support. IEEE Computer Graphics & Applications 2004 (1/2): 13-17.

27. R. Marschallinger, 2009 Analysis and Integration of Geo-Data. http://www.oeaw-giscience.org/.

28. D. J. Peuquet, 2002 Representations of Space and Time. New York, The Guilford Press.

29. J. Restrepo, H. Christiaans, 2004 Problem Structuring and Information Access in Design. Journal of Design Research, 4(2).

30. J. Thomas, J. Carroll, 1979 The psychological study of design. Design Studies, 1(1), 5-11.

31. G. Wallentin, U. Tappeiner, J. Strobl, E. Tasser, 2008 Alpine tree line dynamics: an individual based model. Ecological Modelling, 218(3-4). 235 246 , doi:10.1016/j.ecolmodel.2008.07.005.

32. G. Wallentin, 2009 Ecology & GIS. Spatiotemporal modelling of reforestation processes. See http://www.oeaw-giscience.org/images/stories/Downloads/pecha%20kucha%20technoz%20day.pdf

33. J. Zheng, J. Niu, 2007 Unified Mapping of Social Networks into 3D Space. IMSCCS 2007, Second International Multi-Symposiums on Computer and Computational Sciences. See http://ieeexplore .ieee.org/xpl/conferences.jsp.

34. F. Zobl, 2009 Mapping, Modelling and Visualisation of georelevant processes. http://www.oeaw-giscience.org/.

Dependability Evaluation Based on System Monitoring

Janusz Sosnowski[1] and Marcin Król

[1] Institute of Computer Science, Warsaw University of Technology, Poland

1. INTRODUCTION

Recently dependability, maintainability and performance are becoming challenging issues for system designers and users. This results from the increasing complexity of hardware and software. In consequence these issues triggered various measurement-based studies in the literature, in particular they relate to the detection or prediction of critical situations. Most published results are focused on restricted and fragmented problems encountered in the systems or applications considered by the authors e.g. (Daniel et al., 2008; Ganapathi & Patterson, 2005: Hoffmann et al., 2007; Li et al., 2006; Makanju et al. 2008). In contemporary computer systems various monitoring mechanisms are provided, usually they relate to event and performance monitoring (John & Eckhout, 2006; Simache & Kaaniche 2001; Stearley, 2004; Zhang 2008; Ye, 2008). Such monitoring can generate a huge quantity of various data. An important issue is the selection and exploration of this data, to characterise the system operation. This is a non-trivial task even for experienced system administrators and analysts. Hence, it needs further investigations in the following aspects: system observation techniques, selecting the most sensitive observation parameters, creating the model of normal and abnormal (dangerous) behaviour of the system to facilitate problem identification and applying appropriate reactions.

In the literature most papers concentrate on finding some characteristic patterns in log files related to well defined critical problems encountered in the considered systems e.g. leading to system crash (Kalyanakrishman et al., 1999; Sahoo et al., 2004; Xu et al., 1999). Various performance parameters have been monitored to predict specified network or processor bottlenecks

(Cherkasova et al., 2008; Li et al., 2006; Reinders, 2007; Simache & Kaaniche, 2001), to detect attacks (Ye, 2008) or asses system dependability (Heath et al., 2002; Malek, 2008). Some statistical or data mining models have been developed for specific problems, however they are hardly applicable or irrelevant to other systems (Bertino et al., 1998; Hoffman et al., 2007, Lim et al., 2008). Hence, further studies covering various systems are still required to get better knowledge of monitoring capabilities and limitations.

We have faced many dependability and maintainability problems in computer systems used by students and scientists within the university for didactic and research purposes. Moreover, the load of the systems changes in time or place, hardware and software are updated or tuned, various maintenance and administrative activities occur sporadically, etc. Hence, some operational problems or configuration inconsistencies arise. Such systems create a good basis for studying monitoring techniques. We have also some experience with other commercial systems handling many customers with fluctuating usage profiles. Long-term observations of these systems allowed us to improve monitoring techniques and dependability. For this purpose we have developed some special software modules, collected a lot of data and performed various analyses.

The paper outlines the main features of possible measurements (related to system operation) and the scope of collected data. On the basis of this survey we formulate problems of selecting and processing the collected data in relevance to dependability issues. We concentrate on software implemented monitoring systems, which provide combined exploration of event logs and performance counters. As opposed to other approaches the developed monitoring systems are interactive and adjusted to appearing problems. Moreover, we deal with a wider scope of observations, so we rely on many data sources simultaneously (e.g. event logs, performance logs, and exceptions). We have combined two approaches: identifying normal operation features and exploring long term trends (neglected in the literature); detecting various abnormalities. In both approaches we take into account correlation with environment and configuration changes. The paper describes this in relevance to two monitoring techniques based on various event logs and performance data. The presented considerations are illustrated with practical monitoring results. They relate to long term observations of many workstations and servers.

In section 2 we give an outline of event logs collected in computer systems. They are illustrated with some statistical data derived from long term observations of computers in didactic laboratories, some comments on data exploration are also included. Section 3 describes performance objects and related performance variables, which are usually monitored. The capabilities and problems with performance monitoring are presented in relevance to results from real systems. Final conclusions are given in section 4.

2. EVENT LOGS IN COMPUTER SYSTEMS

2.1. Event Specifications and Statistics

Computer systems are instrumented to provide various logs on their operation. These logs comprise huge amounts of data describing the status of system components, operational changes related to initiation or termination of services, configuration modifications, execution errors, etc. In Windows various events are stored in one of the three log files:

- *security log* comprises events related to system security and auditing processes,

- *system log* is used primarily to store diagnostic messages, abnormal conditions, events generated by system components (e.g. services, drivers),

- *application event log* reports errors that occur during the application execution (e.g. failing to allocate memory, aborting the transfer of a file, etc.).

Each event log record comprises the following fields:

- *event specification* - specifies 5 event types related to event severity level: error, warning, information, success audit, failure audit; this is supplemented with the event category, ID, date and time,

- *event source* – name of the user and the computer that generated the event,

- *description* – event details.

The list of possible events in Windows systems exceeds 10000 (Sosnowsk & Poleszak, 2006). In Unix and Linux systems over 10 sources of events and more priority levels are distinguished (*Syslog*). During normal operation of workstations or servers a large amount of events is registered in the logs. Hence, we have developed a special software system *LogMon* which collects data logs from specified computers within LAN and performs predefined processing to identify critical, abnormal and other situations (e.g. unavailability, warnings). *LogMon* co-operates with standard services (e.g. *Eventlog*) and provides some statistical and data exploration techniques. The performed analysis can be targeted at individual computers or specified computer subsets to find various correlations, etc.

The event files can be filtered preliminary according to specified rules related to event identifier, source, type, system user, computer identifier, date and time (specified intervals by two points in time, specified month, week day, etc.). Complex multi step filtering is also possible, we can combine filtered files in one file, etc. Typical statistics relate to:

- 1) *Event counting* – the distribution (e.g. in decreasing order) of the number of registered events;

- 2) *Time between events* - time distribution between events of the same or different types;

- 3) *Event occurrence distribution* – statistics of the number of the selected event type in relevance to months, weeks, days or hours of the day;

- 4) *Event frequency profile* – the frequency of a selected event in the considered time period.

The calculated statistics are visualised in graphical forms, including a scatter plot where x axis is the time and y axis represents different event categories, or system components, etc. Such visualisations are useful to interpret the collected data, e.g. identification of significant patterns. The collected events can also be presented according to some ranking features e.g. frequency of appearance, entropy, etc.

The developed tool *LogMon* collects data from logs of many computers via internet. It provides many possibilities of filtering, searching specified event sequences, and visualising results. The log analysis can be targeted at different problems e.g. identifying critical situations, evaluating system availability, activity, system load, power problems. This process needs some knowledge of log specificity and experience with the used tool. We illustrate this in the subsequent section.

2.2. Illustrative Results

While analysing the registered events we should identify the system start up and shutdown. When the Windows system is booted, event 6009 is logged and then it is followed by event 6005, which corresponds to *EventLog* service start-up. The termination of *EventLog* service registers event 6006. Event *6006* should be the last one registered in the system log after shutting down the system (clean shutdown). Nevertheless we observed some unexpected events registered after 6006 (anomalous situation). The event 6008 is recorded when a dirty shutdown ("blue screen") occurs. The description part of this event comprises system time stamp (date and time). It may happen that the system cannot record 6006 or 6008 event, however 6009 and 6005 events are recorded. This complicates identification of system restarts, etc. After the system restart (e.g. in consequence of power supply outage) caused by event 6008 other registered events may give more details.

Within the events, which are correlated with system restarts, we can distinguish 4 groups: system and application updates, errors in applications and system services, hardware errors, unidentified restarts. Update events relate to restarts forced by installing new programs, system updating or recovery of the previous version (with deletion of the updated ones).

Typically they are initiated by: *Automatic Updates, NtService Pack, MsiInstaler, WindowsMedia, Print*. The events specify types of updates, information if it has been successfully accomplished or not, etc. For example for one of the computers the distribution of antivirus data base updates was as follows: for 270 registered events of this type (4570; *McUpdate*) 62 appeared in time period less than 1 day, 34 in the period from 1 to 2 days, etc. The analysis covered the log of 668 days. For some computers this frequency was sporadically disturbed – due to some configuration problems. Updates of different programs were performed successfully in over 80% cases, however for some computers non-successful updates were reported. The deeper analysis proved configuration inconsistency and network problems. Not successful clock synchronisation appeared on average in 10% cases.

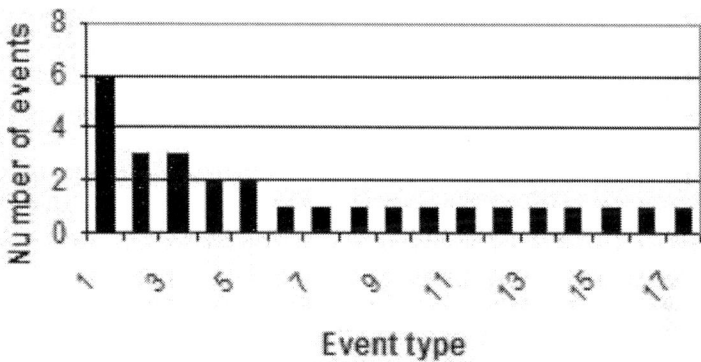

Figure 1. Distribution of event types before restarts.

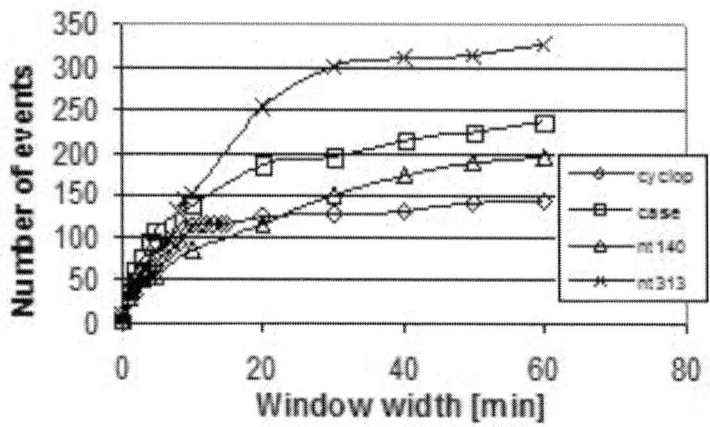

Figure 2. The number of events within the specified time window.

Looking for the sources of restarts we can analyse the distribution of events registered in the specified window before the event sequence related to reboot (6006, 6009, 6005) This is illustrated in Figure 1. The x-axis specifies different events In particular: x=1 relates to event *2013Srv* (the disk is almost full, you may need to delete some files); x=2 - event *54w32time* (the Windows Time Service was not able to find a Domain Controller, a time and date update was not possible); x=13 – event *21automatic program updates*; x=15 – event 26 *application error,* etc. Such graph facilitates to identify the most frequent sources of restarts. Complete distribution of all registered events in a decreasing order of occurrence is also useful to identify other problems. Typically 90% of registered events related to only 36 different event types (from the total list of over 10000 possible events).

To find sources of some event A it is useful to check events before this event within a specified time window . For this purpose *LogMon* provides the capability of finding such statistics for a specified window .Figure 2 shows the number of registered events for 4 servers in function of the time window width before restarts. We can observe some kind of saturation for $10<<30$ minutes, depending upon the system. Basing on the registered events we can identify restarts. For the considered servers (Figure 2) we have identified 24-60 restarts (on average 5 events per restart in the window).

The developed system *LogMon* provides various data exploration capabilities. In particular it can identify reasons of restarts and dirty restarts. In the case of restarts we have defined some regular expressions describing events most probably related to specified situations e.g. program update restarts. Table 1 shows restart statistics for 4 laboratories (L1-L4) each comprising 17 workstations and 3 servers (S1-S3). It gives the percent of restarts caused by program updates, application errors and hardware errors. In some cases the restart source is ambiguous - related to more than one source (mostly a program update and some other source). Quite significant percentage of detected restarts (unknown cause) did not comprise additional events facilitating their identification. They relate to power downs and restarts initiated by the user in response to some messages appearing on the screen, some of them can be identified from the application log. The table comprises the restart frequency (RF) expressed in the number of restarts per day (per single computer). Relatively low values of RF for two servers (S2 and S3) result from the stable profile of their usage.

Table 1. Restart statistics for workstations and servers.

	L1	L2	L3	L4	S1	S2	S3
RF	0.20	0.21	0.19	0.18	0.19	0.04	0.05
updates	20.2%	16.5%	22.6%	24.3%	32.5%	53.8%	3.9%
applic.	19.2%	29.7%	12.0%	29.6%	10.4%	15.4%	11.8%
hardware	1.2%	0.5%	2.0%	0.4%	9.1%	0%	0%
ambig.	1.4%	5.6%	5.0%	5.4%	6.5%	26.9%	1.3%
unknown	59.4%	53.4%	63.4%	45.7%	48.1%	30.8%	84.2%

Special attention is needed to dirty restarts. Dirty restarts mostly relate to such events as 6008, 1000, 1001 *save dump*. Pressing RESET button also causes dirty restart. At the system level power down is treated in the same way as fast switching off the computer. In logs with power down closing event 6006 was missing. Analysing events in the window before the dirty restart we observed small number of events. This relates mostly to the situation with no possibility of recording the event due to the restart problem. In a period of 100 days we have identified typically 2-5 dirty restarts per computer. However, for a few computers this was in the range of 20-50 (old computers). In the case of servers practically the dirty restarts were caused by hardware errors. In the case of workstations dirty restarts were caused mostly by application errors, which caused system hang-ups or operational instability (due to developing and testing various pogrom modules). Moreover, many student projects comprised critical errors leading to restarts.

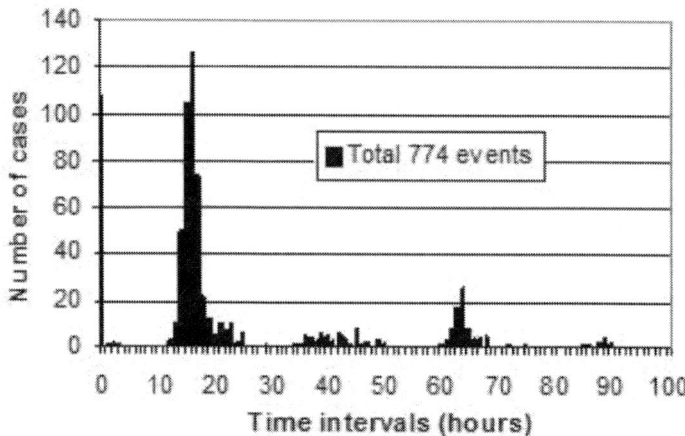

Figure 3. Distribution of non activity periods in a workstation.

Figure 3 shows distribution of time periods between event A and B, which correspond to closing and starting the system. The x-axis of the plot has the granularity of single hours. The first bar (108 cases) relates to short time intervals (less than 1 hour) and it corresponds to short operation breaks related to restarts. The next group of higher bars relates to the periods of 15-16 hours, corresponding to switching off the computer for the evening and night periods. Subsequent groups of bars relate to longer non activity periods e.g. weekends, holidays, etc. Power supply problems can be directly identified by checking the time between events 3230:UPS (notification of power down and supply delivered from the batteries) and 3234:UPS (power recovery) or 3231:UPS (power switched off in the system) generated by UPS power supply. For an illustration we give some power statistics related to 2 servers SA and SB. In particular we specify the number of power down events per month for 9 subsequent months (October- June):

SA: 1; 1; 3; 1; 2; 2; 0; 0; 1 (total 9 events) and SB: 1; 9; 8; 7; 12; 10; 12; 1; 0 (total 60 events)

For all power down events the servers were supplied from UPS batteries, due to short power outages (power outages were tolerated by UPS). Power outages longer than 17 minutes and 150 sec result in SA and SB server switching off, respectively. The distribution of the duration of power outages was as follows: for server SA - (6 events) < 1 minute, 2 minutes (2 events) < 7 minutes and 1 event with 17 minutes duration; for server SB - (13 events) < 1s, 1s (33 events) < 2s, 2s (5 events) <3 s, 3s (4 events) < 4s and 5s (4 events) 12s. Computers without UPS crashed and needed restarting. Some crashes appeared simultaneously in several computers (common power failure). In 3 cases the power downs signalled by UPSs of SA and SB servers were correlated (caused by the same power network outage), however the attributed timestamps differed by about 20 minutes, due to the lack of the clock synchronisation in both servers (such anomalies need identification).

An important issue is to evaluate the behaviour of various used programs within longer periods and correlate it with system upgrades, reconfiguration, load (number of users, processor stress). For an illustration Table 2 shows the frequency distribution of registered program errors per day for two workstation (WS1 and WS2) within the one-year period. WS1 is less reliable due to higher number of used programs.

Table 2. Distribution of program errors for workstation WS1 and WS2.

Computer	Number of errors per day							
	0	1	2	3	4	5	6	6
WS1	68%	10%	6%	3%	3%	2%	0%	1%
WS2	82%	5%	1%	2%	1%	2%	1%	0%

Events related directly to hardware faults appear before restarts. The most frequent relate to memory media, network cards, printers and other I/O devices. For example event 26 with description that the system could not write or read data from a specified file, device, etc. Other examples are: faulty block of CD ROM, timeout situation, IP address conflict, failure to load specified drivers, application errors, etc. It is worth noting that many events do not comprise descriptions, on the other hand some descriptions are ambiguous. Many faults can be identified from sequences of events. For this purpose some knowledge database can be systematically developed taking into account the gained experience from the system exploitation and maintenance.

2.3. Exploring Data in Event Logs

Analysing logs is the basis for automatic system management and helpful in assuring high dependability. The registered reports may be related to different formats, the text messages are usually relatively short, contain a free format description of events (using a large size vocabulary), and quite often are ambiguous. Hence, data mining is not trivial and needs many preliminary studies to identify specificity and abnormal behaviour of the monitored systems. In this process some categorisation of text messages into a set of common classes over various system components is required. Moreover, an important issue is to visualise various statistics, temporal dependencies, etc.

Simple data mining can be targeted at discovering frequent and some specific well defined patterns (Lim et al., 2008; Makanju et al., 2008; Peng et al., 2005; Razavi & Kontogiannis, 2008; Vaarandi, 2008). In more advanced analysis of log reports we should take into account not only the individual messages but also their temporal dependencies, which can provide supplementary context information. For example a massage on starting a program update may be followed by some errors due to inconsistency in the system configuration. Having transformed messages in some concise categorised form simplifies further data processing and finding characteristic patterns. Unfortunately, different systems use different log formats, etc. Hence, data collection and analysis has to be tuned to these systems. This is sufficient for individual system monitoring. In practice we are also interested in general properties of many systems, so some specification of similarities has to be defined to identify common characteristics, etc.

The log pre-processing may involve visualisation of event types or categories in relation to their appearance (time stamps). From such plot it is easy to identify some general properties e.g. the fact that event A usually happens after event B, the time distance between such events (it can be deterministic or random). An event may appear with some periodicity (e.g. antivirus updates, system heartbeat) or randomly. Some events may form a loop in a circular pattern or an event chain (e.g. related to a problem progress in predictable way). An event may appear simultaneously with other events. Various temporal relationships can be represented by appropriate graphs (Peng et al., 2007). Looking for temporal dependencies we analyse the distribution of time distance between events or compare the unconditional probability of the waiting time for some event with a conditional probability in relevence to some other event. Various event patterns may signal system problems or confirm its health (e.g. heartbeats, successful program updates). Their interpretation can be simplified by correlating them with performance properties (section 3).

An important issue in tracing event logs is appropriate system configuration If we are interested in monitoring system activity it is important to activate writing into logs supplementary information on user logins, file openings, processor usage, etc. It is also important to assure completeness of collected data. Some danger arises if logs are read periodically, so overwriting may happen. Hence, it is reasonable to collect this data systematically and storing it in a separate server. Complete information on all computers simplifies finding various correlations.

3. PERFORMANCE MONITORING

3.1. Performance Objects and Variables

In most computer systems various data on performance can be collected in appropriate counters (e.g. provided by Windows, Linux) and according to some sampling policy (e.g. in 1-minute periods) (John & Eckhout, 2006; Reinders, 2007). These counters are correlated with performance objects such as processor, physical memory, cache, physical or logical discs, network interfaces, server of service programs (e.g. web services), I/O devices, etc. For each object many counters (variables) are defined characterising its operational state, usage, activities, abnormal behaviour, performance properties, etc. Special counters related to developed applications can also be added. These counters provide data useful for evaluating system dependability, predicting threats to undertake appropriate corrective actions, etc. The list of counters, which can be configured, is very long. For an illustration we describe some representative counters related to Windows systems.

Processor Time is the percentage of elapsed time that the processor spends to execute a non-idle thread. This counter is the primary indicator of processor activity, and displays the average percentage of busy time observed during the sample interval. *User Time* and *Privileged Time* relate to the percentage of elapsed time the processor spends in the user and in privileged mode, respectively. *Processor Queue Length* is the number of ready threads in the processor queue. *Processes* is the number of processes at the time of data collection. Similarly are counted threads, events, semaphores, etc. *Context Switches/sec* is the combined rate at which all processors on the computer are switched from one thread to another e.g. when a running thread voluntarily relinquishes the processor (is pre-empted by a higher priority ready thread), or switches between user-and privileged (kernel) mode to use an executive or subsystem service.

Interrupts/sec is the average rate, in incidents per second, at which the processor received and serviced hardware interrupts. It does not include deferred procedure calls (DPCs), which are counted separately. This value is an indirect indicator of the activity of devices that generate interrupts, such as the system clock, the mouse, disk drivers, network interface cards, and other peripheral devices. These devices normally interrupt the processor when they have completed a task or require attention. The system clock typically interrupts the processor every 10 milliseconds, creating a background of interrupt activity. This counter displays the difference between the values observed in the last two samples. *Interrupt Time* is the time the processor spends receiving and servicing hardware interrupts during sample intervals.

System Up Time is the elapsed time (in seconds) that the computer has been running since it was last started till the current time. *C1 Time* is the percentage of time the processor spends in the C1 low-power idle state (enables the processor to maintain its entire context and quickly return to the running state), similar times are measured for C2 (a lower power and higher exit latency state than C1, it maintains the context of system cache) and C3 (a lower power and higher exit latency state than C2, is unable to maintain the coherency of its caches) states. There are also counters related to transitions to these states.

Available Bytes is the amount of physical memory, in bytes, available to processes running on the computer. It is calculated by adding the amount of space on the *Zeroed, Free,* and *Standby* memory lists. *Free* memory is ready for use; *Zeroed* memory consists of pages of memory filled with zeros to prevent subsequent processes from seeing data used by a previous process; *Standby* memory is memory that has been removed from a process working set (its physical memory) on route to disk, but is still available to be recalled. This counter displays the last observed value.

Free Space is the percentage of total usable space on the selected logical disk drive that was free. *Avg. Disk Bytes/Read* is the average number of bytes transferred from the disk during read operations, similar counter on write operations is available also.

Page Faults/sec is the average number of pages faulted per second (a referenced page in virtual memory is not available in the working area). Hard faults require disk access and soft faults cover faulted pages found elsewhere in physical memory. Most processors can handle large numbers of soft faults without significant consequences. However, hard faults, which require disk access, can cause significant delays. Similarly *Cache Faults/sec* is the rate at which faults occur when a page sought in the file system cache is not found and must be retrieved from elsewhere in memory or disk.

Page Reads/sec is the rate at which the disk was read (the number of read operations, without regard to the number of pages retrieved in each operation) to resolve hard page faults. *Pages Output/sec* is the rate at which pages are written to disk to free up space in physical memory. A high rate of *Pages Output* might indicate a memory shortage. *Pool Paged Failures* is the number of times allocations from paged pool have failed. It indicates that the computer's physical memory or paging file are too small.

File Read Operations/sec is the combined rate of file system read requests to all devices on the computer, including requests to read from the file system cache. This counter displays the difference between the values observed in the last two samples, divided by the duration of the sample interval. *File Control Operations/sec* is the combined rate of file system operations that are neither reads nor writes, such as file system control requests and requests for information about device characteristics or status. *Split IO/Sec* reports the rate at which I/Os to the disk were split into multiple I/Os. It may result from requesting data of a size that is too large to fit into a single I/O or that the disk is fragmented.

Server performance counters give: the number of bytes the server has received (or sent) from the network (indicates the server load); the number of sessions that have been closed due to unexpected error conditions or sessions that have reached the autodisconnect timeout and have been disconnected normally; failed logon attempts to the server (password guessing programs are being used to crack the security); the number of sessions that have been forced to logoff (due to logon time constraints); the number of sessions that have terminated normally (this allows to find percentage of the sessions time outs or errors). Other counters provide some statistics on file operations such as: the number of times accesses to files opened successfully were denied (improper access authorisation, etc.), the number of failed file opens (attempting to access files not properly protected), the number of files currently opened in the server, the number of searches for files currently active, the number of sessions currently active in the server (indicates current server activity).

There are many counters characterising network traffic or TCP/IP protocol activity. Here are given some examples. *BytesReceived/sec* is the rate at which bytes are received over each network adapter, including framing characters. *Current Bandwidth* is an estimate of the current bandwidth of the network interface in bits per second. *Packets Received Errors* is the number of inbound packets that contained errors preventing delivery to a higher-layer protocol. *Packets Received Discarded* is the number of inbound packets that were discarded even though no errors had been detected (e.g. to free up buffer space). *Packets Received Unknown* is the number of packets received through the interface that were discarded because of an unknown or unsupported protocol. *Output Queue Length* is the length of the output packet queue, if this is longer than two, there are delays and the bottleneck should be found and eliminated. *Connection Failures* is the number of times TCP connections have made a direct transition to the CLOSED state from the SYN-SENT or SYN-RCVD state, and to the LISTEN state from the SYN-RCVD state.

There are also counters related to I/O devices e.g. for printers they count current number of jobs in a print queue, number of references (open handles) to this printer, peak number of references, current or maximal number of spooling jobs in a print queue. Accumulated statistics comprise data since the last restart e.g. number of out of paper errors, not ready errors and job errors in a print queue.

Resuming we can state that the number of possible performance variables is quite big and monitoring all of them is too expensive due to the additional load to the system processors and memory. Hence, an important issue is to select those variables which can provide the most useful information. This depends upon the goals of monitoring, the sensitivity of variables to the monitored properties of the system, the system usage profile, etc. To deal with this problem some preliminary studies of the system behaviour are needed. They facilitate tuning the monitoring tasks to the current needs and system specificity. We outline this problem in the next section.

3.2. Performance Monitoring Goals

Depending upon the goal of monitoring we have to select and configure appropriate counters within the objects of interest, to evaluate how well they are performing. Too large number of counters results in some additional load to the system and more complex data analysis. Hence, an important issue is to check which counters are most sensitive to the monitored problems. We have performed such studies in relevance to hardware and software failures as well as configuration or maintenance inconstancies, effectiveness of some services, etc. Moreover, the applications can also use counter data to determine how much system resources to consume. For example, to determine how many data to transfer without competing for network bandwidth with other network traffic. The application could adjust its transfer rate as the bandwidth usage from other network traffic increases or decreases. Having specified performance counter thresholds we can generate alert notifications, query performance data, create event tracing sessions, capture a computer's configuration, and trace the API calls in some of the Win32 system DLLs.

Most authors concentrate on well-defined critical problems e.g. cyberattacks or system availability. We have extended the scope of analysis to checking the normality of system operations e.g. periodicity of backups, program updates, acceptable level of signalled errors (e.g. rejected packets) and to detect abnormalities which may result in future problems, this relates mostly to long term observations and detecting dangerous trends e.g. decreasing of free memory. We correlate performance counters with event logs as well as with changes in configurations, system load, temporal disturbances in the operational environment (system maintenance and updates).

Some performance measures are directly used to balance system loads, etc. To assure this we have to analyse short term and long-term trends, correlate them with working hours, weekends, summer months (seasonal system behaviour), user activity profiles, etc. The considered systems were specific due to frequent configuration changes, many users with different profiles (students and different courses, projects, used programming environment, etc.) or servicing thousands of customers with random activities, influenced by various events (e.g. dynamic changes of the stock market).

Some problems are relatively easy to identify e.g. decreasing free memory in relevance to systematically increasing number of users and their higher engagement in more complex calculation problems, bigger data bases, etc. However, new not known problems are not evident and need deeper data multidimensional exploration. For example a higher rate of application warnings in the log was correlated with an installation of a new version of the operating system and the increased number of users. This related to configuration inconsistencies, which were alleviated later on.

Analysing the performance variables we can look at their instantaneous values, statistical properties, correlation with other variables or events. These statistics may relate to specified time periods. Moreover, we can target the analysis at averaged variable values (within specified periods, etc.) or analyse spikes, their frequencies, time distribution, periodicity, etc. All this depends upon the monitoring goal. For example in detection of cyber attacks we can try to find characteristic statistical deviations caused by the attack as compared with normal workload. Interesting studies have been presented in (Ye, 2008) for Windows systems. The authors give statistical properties of various performance variables related to different objects for 10 known cyber attacks. Analysing these results we have checked the observability properties of these attacks.

Table 3. The impact of attacks on performance parameters.

Property	Objects	Variables	Attacks	Sensitivity
M+	3-16 (7.7)	10-362 (105)	1-9 (7.7)	77/100
M-	3-11 (5.9)	17-182 (60)	1-10 (5.6)	59/180
DUL	1-4 (2.5)	1-33 (10.3)	1-8 (3.1)	25/80
DUR	3-9 (5.7)	9-52 (27.8)	1-10 (3.9)	58/110
DMM	3-9 (5.7)	24-52 (43.3)	2-9 (5.0)	57/120

Table 3 shows minimal - maximal (average) numbers of monitored performance objects and related variables (counters) which reveal characteristic statistical properties during attacks. They related to an increase (M+) or decrease (M-) in the mean value, unimodal left skewed (DUL), unimodal right skewed (DUR) and multimodal (DMM) distribution properties as compared with normal workload statistics. The 4-th table column gives the distribution (minimal, maximal and average) of the number of attacks affected by the considered objects (over each object class). The last column specifies the number of nonzero entries in the matrix correlating objects and attacks. Each entry in this matrix gives the number of object variables affecting (by specified statistical property) the appropriate attack. This gives some view on attack sensitivity (the number of all entries in the matrices is given after / character). For DUL, DUR and DMM statistics we observe lower number of affected objects and variables as compared with M+ and M-. For each distribution change we have found that *Process* object affects the most of attacks 7-10 (within all 10 attacks). Moreover, this object involves the biggest number of affected variables per single attack: up to 18, 21 and 23, for DUL, DUR and DMM distribution change, respectively. Some objects show maximal number of affected variables by specific attacks (dominating). Such sensitivity analysis allows the designer to minimise the number of monitored variables and assure good detection accuracy. For other problems we have to trace different properties, this is illustrated in the sequel.

3.3. Illustrative Results

To give a better view on performance monitoring we present some illustrative results related to monitoring selected system objects and performance variables. The dynamic properties of these variables depend upon active applications, system load, environment interactions, etc. Hence, their behaviour in time can be very diversified resulting in different shapes and characteristics of related time plots. This creates some challenges for data exploration.

In general we can be interested in short term or long term monitoring results. In the first case we collect many samples which assure high accuracy. The results can be presented graphically with specified average (horizontal line), minimal and maximal values (vertical lines) for each sample. This is illustrated in fig 4a and 4b, which give the number of disc writes per second (y-axis covers the range 0-100 operations/s partitioned in 10 segments) for the system with no active application and for an application displaying a film of about 30 minutes (from a file in HDTV standard - 1280x720 pixels). The samples were collected every second. It is worth noting low activity of disc writes, nevertheless even in no active system there is some background activity related to operational system and Internet tasks. The both plots differ in time and amplitude distribution of spikes. Bigger difference was observed for processor usage (0% vs 24%) and disc read operations/s (no disc reads with 8 short spikes of 30 operations/s vs continuous average activity of 15 operations/s with many additional smal spikes). The number of disc operations is much higher for disc defragmentation (on average 379 control operations per sec). Short-term observations are useful to find application properties, identify their disturbances etc. It is worth noting that the behaviour of performance variables may differ upon applications not only in the average values of analysed parameters (during the application run) but also in time (plot shapes). Quite often we observe some spikes in time plots of variables, their frequency and amplitudes may also characterise the applications (compare. Figure 4).

Long term observations give a view on general trends in the system. We illustrate this with some results in Figure 5-7. Figure 5 shows some increase (from 20000 to over 100000) of transmitted bytes on the network in 6-month period. This resulted from adding new users. Figure 6 presents the number of created connections with a www server, the middle pulse (100-350 connections) corresponds to day hours 8.00-17.00, the negative pulse at the end of the plot corresponds to the switch problem on the next day (time period 8.00-10.00). Figure 7 illustrates the number of connections related to 13 subsequent days. It is almost equal for all working days (a little bit over 200), much lower for Saturdays (below 100) and close to 0 for Sundays.

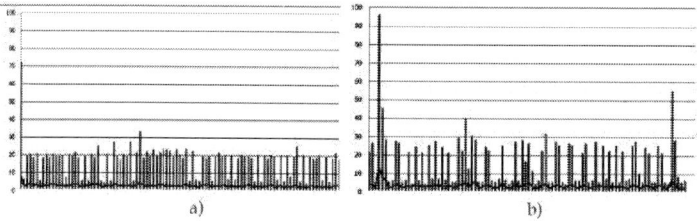

Figure 4. Disc writes operations: a) no active applications, b) playing a film.

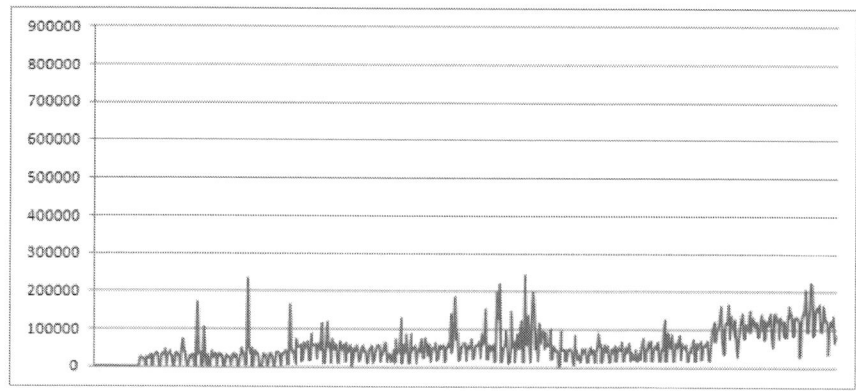

Figure 5. Sent out bytes per second (half year profile).

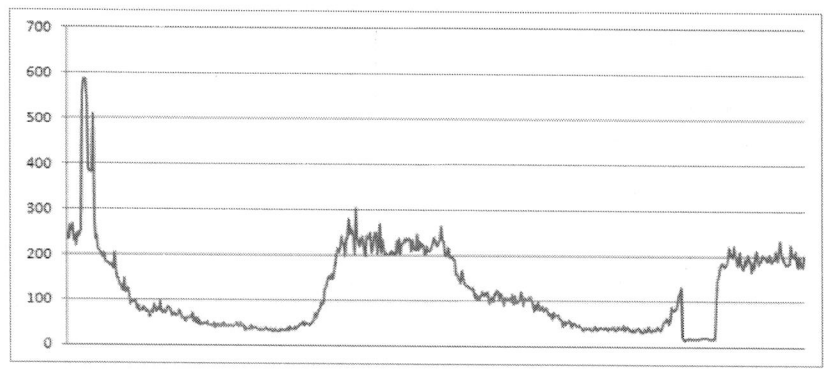

Figure 6. The number of established connections in TCP (3 day profile).

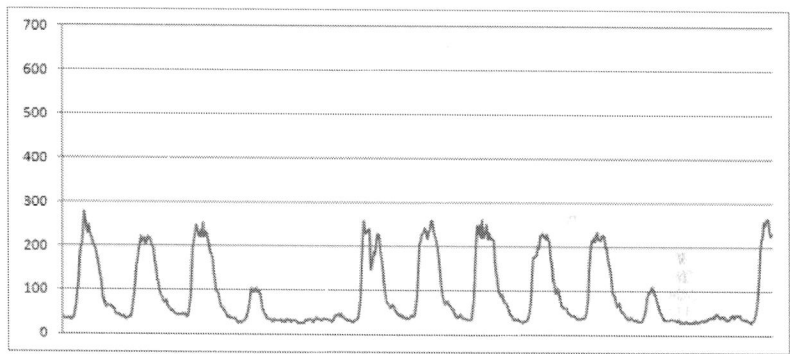

Figure 7. The number of established connections in TCP (2 week profile).

Figure 8 shows the number of active processes on Unix server *mion*, which handles Emails of students (about 3500 students within the Faculty of Electronics of our University) in the period 1st June till 31st December. The y-axis covers the range 0-25000 processes (partitioned into 5 equal segments – each 5500). The plot shows increasing trend of active processes. However on 12 September the system reconfiguration and restarting resulted in deleting many zombie and other not useful processes.

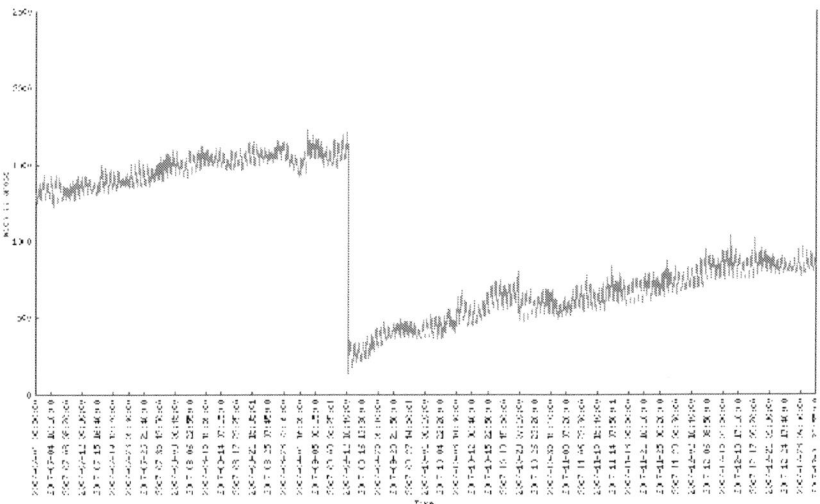

Figure 8. The number of processes in the communication server (*mion*).

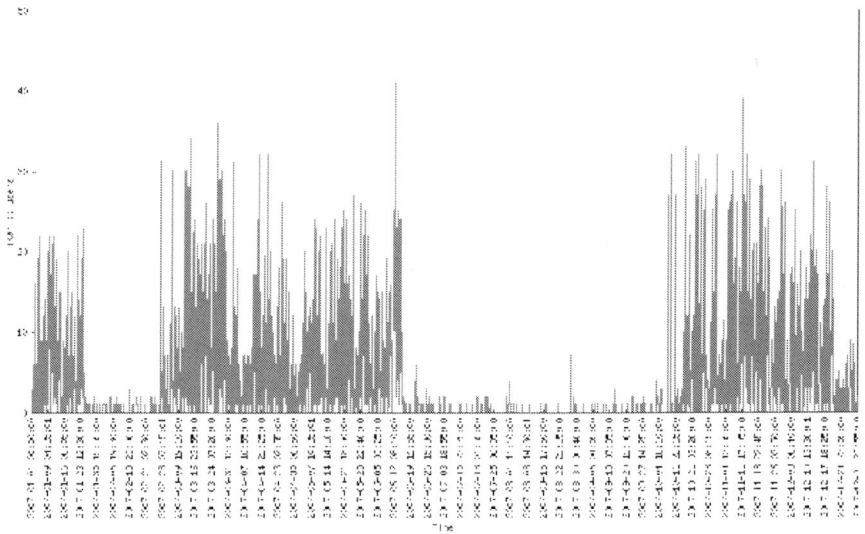

Figure 9. The number of processes in the data processing server (*ikar*).

Figure 9 shows the number of logged users to Unix server *ikar*, which handles programming laboratory with 16 workstations. The time scale on the x-axis covers the period of 12 months. The y-axis covers the range 0-50 users portioned in 5 equal segments (each 10 users). The first period of low activity relates to the winter vacation (February) and the second longer one corresponds to summer vacation (3.5 months).

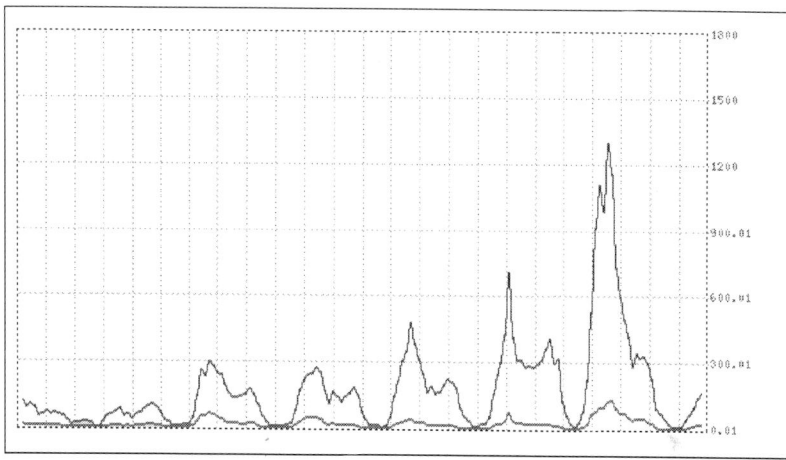

Figure 10. Distribution of incoming session requests.

Interesting observations were made for a farm of 16 servers providing some web services to many thousands of customers. The system uses quite sophisticated load balancing algorithm. Figure 10 shows one-week plot of the total number of sessions (lower line relates to the number of sessions created within the last minute) handled by the farm within the 14th server. The y-axis covers the range 0-1800 session requests partitioned into 6 segments (300 requests each). The subsequent peaks of the plot relate to 7 days. The highest peak corresponds to Friday, for the weekend low activity is visible. In Table 4 we give the server CPU usage (UP in percents) corresponding to the highest load on each day. Moreover we give also the outgoing traffic (OT in KB/sec) on the network port of the server. For each server the farm-managing program monitors its available resources (in particular processor and memory) and attributes user requests so as to achieve balanced load of all servers adapted to their functional capabilities and current activity. There are several classes of servers with different processors and architectural features. Hence, we take into account not only the current values of performance parameters but also architectural capabilities. In particular the same level of processor usage expressed in percents does not reflect the real available processing power, which also depends upon the processor speed, etc. Monitoring the operation of all servers confirmed the effectiveness of the used load-balancing algorithm. Moreover, it allows finding critical deviations in farm operation and identifying problems (e.g. to eliminate a faulty server and move the traffic to other servers).

Table 4. Outgoing traffic and processor usage in 14th server of a processing farm.

	Mon.	Tue.	Wed.	Thur.	Fri.	Sat.	Sun.
OT	700	600	500	550	620	150	30
UP	22%	23%	42%	38%	45%	8%	4%

The presented plots confirm a large diversity in possible shapes related both to normal and erroneous operation, hence their qualification needs advanced techniques, which take into account various correlation factors. In practice it is reasonable to correlate performance variables with event logs as well as environment changes, activities of administrators (maintenance, reconfigurations), software updates, user profiles, etc.

4. CONCLUSION

The available system logs and possible monitoring of various performance features provide enormous amount of data on system operation. This is a very useful source of information to evaluate and improve system dependability. However, selecting this information is not a trivial problem. It is possible to monitor and collect data on various aspects using pre-programmed counters, etc. Monitoring too many variables may result in system performance loss and high memory load with collected logs. So some optimisation is required here, in particular we can select the most sensitive variables related to various dependability issues. The next problem is interpretation of the collected data. This requires gaining some experience from long-term observations and correlating them with opinions of users and administrators. This simplifies creating procedures for automatic data exploration targeted at dependability issues. Hence, it is reasonable to enhance the available system mechanisms and software modules with an integrated database and advanced visualisation, statistical and data mining procedures (provided in the presented systems).

Further research is targeted at correlating various logs from many computers, identifying typical operational profiles, system loads, finding their changes in time and developing more efficient data exploration techniques to predict as soon as possible requirements of reconfigurations, detect inconstancies or usage anomalies, etc. Having itemised specific patterns we can formulate appropriate actions. The gained experience is useful in defining event reduction rules, correlation rules (identifying events which are symptoms of specific problems), and problem avoidance rules (for some problems several stages of progress can be distinguished, early detection can prevent critical situation). We also plan to enhance the collected data from the field with logs relevant to injected faults (Sosnowski & Gawkowski 2006).

REFERENCES

1. E. Bertino, E. Ferrari, G. Guerrini, 1998 An approach to model and query event based temporal data, Proc. of 5th Int. Workshop on Temporal Representation and Reasoning, 122 131 , IEEE Comp. Society
2. L. Cherkasova, K. Ozonar, N. Mi, J. Symons, E. Smirni, 2008 Anomaly? application change? or workload change?, Proc. of IEEE DSN Symposium, 452 461 , IEEE Comp. Society, 1-42442-398-9
3. E. Daniel, R. Lal, G. Choi, 1999 Warnings and errors: A measurement study of a UNIX server, FastAbstracts of IEEE Int. Symp. on Fault-Tolerant Computing, FTCS-29, www.crhc.uiuc.edu/FTCS-29/fastabs.html.
4. A. Ganapathi, D. Patterson, 2005 Crash data collection: A windows case study, Proc. of IEEE DSN Symposium, 772 184 , IEEE Comp. Society, 0-76952-282-3
5. T. Heath, R. P. Martin, T. D. Nguyen, 2002 Improving cluster availability using workstation validation. Proc. ACM SIG-METRICS Conf. Measurement and Modelling of Computer Systems, 217 227 .
6. G. A. Hoffmann, K. S. Trivedi, M. Malek, 2007 A best practice guide to resource forecasting for computing systems, IEEE Transactions on Reliability, 56 4 615 628 , ISNN 0018-9529
7. L. K. John, L. Eckhout, (editors) 2006 Performance evaluation and benchmarking, CRC Taylors & Francis, 100849336228
8. M. Kalyanakrishman, Z. Kalbarczyk, R. K. Iyer, 1999 Failure data analysis of a LAN of Windows NT based computers, Proc. of 18th IEEE Symposium on Reliable Distributed Systems, 178 188 , IEEE Comp. Society
9. M. Li, S. Wang, W. Zhao, 2006 A real-time and reliable approach to detecting traffic variations at abnormally high and low rates, In ATC 2006, LNCS 4158, 2006, L.T. Yang et al., (Ed.) 541 550 , Springer Verlag, 3-54069-294-0 New York.
10. Ch. Lim, N. Singh, S. Yainik, 2008 A log mining approach to failure analysis of enterprise telephony systems, Proc. of IEEE DSN Symposium, 388 403 , IEEE Comp. Society, 1-42442-398-9
11. A. Makanju, S. Brooks, A. N. Zincir-Heywood, E. E. Milin, 2008 LogView: visulizing event log clusters, Proc. of Annual Conf. on Privacy, Security and Trust, 99 108 , 978-0-76953-390-2
12. M. Malek, 2008 Online dependability assessment through runtime monitoring and prediction, Proc. of EDCC-7., 181 IEEE Comp. Society, 978-0-76953-138-0
13. M. Mansouri-Somani, H. Sloman, 1996 A configurable event service for distributed systems, Proc. of IEEE 3rd Int. Conf. on Configurable Distributed Systems, 210 217 , IEEE Comp. Society, 0-81867-395-8

14. W. Peng, Ch. Peng, T. Li, H. Wang, 2007 Event summarization for system management, Proc. of ACM KDD'07, 1028 1032 .
15. W. Peng, T. Li, S. Ma, 2005 Mining logs files for computing system management, SIGKDD Explorations, 7 1 44 51 .
16. A. Razavi, K. Kontogiannis, 2008 Pattern and policy driven log analysis for software monitoring, Proc. of IEEE Int. Computer Software and Applcations Conference, 108 111 , IEEE Comp. Society, ISNN 0730-3157
17. J. Reinders, 2007 VTune performance analyser essential, Intel Press, 0-97436-495-9
18. R. K. Sahoo, A. Sivasubramanian, M. Squillante, Y. Zhang, 2004 Failure data analysis of a large-scale heterogeneous server environment, Proc. of IEEE DSN Symposium, 283 285 , IEEE Comp. Society, 0-76952-052-9
19. C. Simache, M. Kaâniche, 2002 Event Log Based Dependability Analysis of Windows NT and 2K Systems, Proc. of IEEE Pacific Rim Int. Symposium on Dependable Computing, 311 315 , IEEE Comp. Society, 0-76951-852-4
20. C. Simache, M. Kaâniche, 2001 Measurement-based availability of Unix systems in a distributed environment, Proc. of 12th International Symposium on Software Reliability Engineering (ISSRE'01), 346 355 , IEEE Comp. Society
21. J. Sosnowski, P. Gawkowski, 2006 Enhancing fault injection test bench, Proc. of DepCos_RELCOMEX Conference, 76 83 , IEEE Comp. Society, 0-76952-565-2
22. J. Sosnowski, M. Poleszak, 2006 On-line monitoring of computer systems, Proc. of IEEE DELTA 2006 Workshop. 327 331 , IEEE Comp. Society, 0-76952-500-8 presentation in LATW 2006 and ICIT 2009)
23. J. Stearley, 2004 Towards informatic analysis of Syslogs, Proc. of IEEE Int. Conf. on Cluster Computing, 309 318 , IEEE Comp. Society, 00-803-8430X
24. R. Vaarandi, 2008 Mining event logs with SLCT and LogHound. Proceedings of the 2008 IEEE/IFIP Network Operations and Management Symposium, 1071 1074 .
25. J. Xu, Z. Kallbarczyk, R. K. Iyer, 1999 Networked Windows NT system field failure data analysis. Technical Report CRHC 9808, University of Illinois at Urbana- Champaign,
26. N. Ye, 2008 Secure Computer and Network Systems, John Wiley& Sons, Ltd. 978-0-47002-324-2 Chichester
27. Y. Zhang, A. Sivasubramanian, 2008 Failure prediction IBM BlueGene?L event logs, Proc. of Int. Symp. on Parallel and Distributed Processing, 1 5 , IEEE Comp. Society, 978-1-42442-030-8

Coupled Heuristic Prediction of Long Lead-Time Accumulated Total Inflow of a Reservoir during Typhoons Using Deterministic Recurrent and Fuzzy Inference-Based Neural Network

Chien-Lin Huang [1], Nien-Sheng Hsu [1,*] and Chih-Chiang Wei [2]

[1] *Department of Civil Engineering, National Taiwan University, No. 1, Sec. 4, Roosevelt Road, Taipei 10617, Taiwan*

[2] *Department of Marine Environmental Informatics, National Taiwan Ocean University, No.2, Beining Rd.; Jhongjheng District, Keelung City 20224, Taiwan*

ABSTRACT

This study applies Real-Time Recurrent Learning Neural Network (RTRLNN) and Adaptive Network-based Fuzzy Inference System (ANFIS) with novel heuristic techniques to develop an advanced prediction model of accumulated total inflow of a reservoir in order to solve the difficulties of future long lead-time highly varied uncertainty during typhoon attacks while using a real-time forecast. For promoting the temporal-spatial forecasted precision, the following original specialized heuristic inputs were coupled: observed-predicted inflow increase/decrease (OPIID) rate, total precipitation, and duration from current time to the time of maximum precipitation and direct runoff ending (DRE). This study also investigated the temporal-spatial forecasted error feature to assess the feasibility of the developed models, and analyzed the output sensitivity of both single and combined heuristic inputs to determine whether the heuristic model is susceptible to the impact of future forecasted uncertainty/errors. Validation results showed that the long lead-time–predicted accuracy and stability of the RTRLNN-based accumulated total inflow model are better than that of the ANFIS-based model because of the real-time recurrent deterministic routing mechanism of RTRLNN. Simulations show that the RTRLNN-based model with coupled heuristic inputs (RTRLNN-CHI, average error percentage (AEP)/average forecast lead-time (AFLT): 6.3%/49 h) can achieve better prediction than the model with non-heuristic inputs (AEP of RTRLNN-NHI and ANFIS-NHI: 15.2%/31.8%) because of the full consideration of real-time hydrological initial/boundary conditions. Besides, the RTRLNN-CHI model can promote the forecasted lead-time above 49 h with less than 10% of AEP which can overcome the previous forecasted limits of 6-h AFLT with above 20%–40% of AEP.

Keywords: accumulated total reservoir inflow; long lead-time hydrograph prediction; coupled heuristic inputs; real-time recurrent learning neural network; adaptive network-based fuzzy inference system

1. INTRODUCTION

Taiwan is located in the path of typhoons as they move in from the Western Pacific, and as a result, three to five typhoons hit Taiwan annually [1,2]. As the basins of the reservoir in Taiwan are mostly steep-sided, the concentration time is especially short and the reservoir inflow is extremely high under typhoon-induced precipitation [3]. The frequency of typhoons that bring heavy rain has been growing due to climate change [4,5,6], and inflows are more frequently surpassing original design and construction standards. Therefore, effective methods of ameliorating typhoon-related disasters need to include non-engineered disaster prevention programs, such as effective disaster forewarning and associated response mechanisms, which include the ability to identify the disaster before it occurs. The optimal releasing strategies for flood control are to minimize the maximum release and maximize the final storage under the principles of avoiding dam failure and overflow from the upstream riverbank, and keeping the water level lower than the dead storage level. Hence, we can expect that an accurate accumulated total reservoir inflow forecast model plays a most important role in determining whether the releasing decision can achieve optimization for flood control.

However, previous research into real-time long lead-time accumulated total reservoir inflow forecast during typhoons has been scarce, and it has proved difficult to achieve effective and accurate results because of future meteorological-hydrological uncertainty. The traditional method to derive real-time forecasted reservoir inflow hydrographs and the corresponding accumulated total inflow is firstly to forecast the typhoon precipitation hyetograph, and then the reservoir inflow hydrograph of the entire typhoon event is derived from the rainfall-runoff model. This type of rainfall-runoff modeling has been examined in the fields of the hydrological approach [7,8,9,10,11,12] and statistical approach [13,14]. However, studies like those above, regarding the real-time precipitation hyetograph forecast of an entire typhoon event, are scarce, so efficient and accurate long lead-time accumulated total inflow forecast is still in urgent need of development.

The other method regarding inflow forecast is to directly predict short lead-time reservoir inflow, because the model inputs only consider the real-time observed meteorological-hydrological information. These related works have been categorized under both the hydrological approach [15,16,17] and the statistical approach which mostly applied artificial neural networks (ANNs) such as the back-propagation neural network (BPNN) [18,19,20,21,22], the state space neural network [23], the adaptive network-based fuzzy inference system (ANFIS) [24], the recurrent neural network (RNN) [21], support vector machine [1], and the radial basis function [2] as construction tools. The advantage of the short lead-time forecast is that it is fairly accurate in medium-low reservoir inflow, whereas the disadvantages are that (1) the effective forecasted lead-time is only 6 h; (2) the forecasted error in the high flow periods is high, within the range of 10% to 40% [1,2]; and (3) the time-lag circumstances of the forecasted flow rate of a longer forecasted lead-time are significant. The main reason is that the previous models do not consider future reliable meteorological-hydrological factors as inputs. The feasible inputs include the delays from the current moment to the various key moments on the rainfall-runoff hydrograph, the accumulated total precipitation, and the observed-predicted inflow increase/decrease rate (OPIID rate), etc. Besides, the above studies concluded that the forecasted ability of RNN and ANFIS is better than the traditional ANNs like BPNN, and they have the potential to simulate longer lead-time inflow with larger tolerance ability for input errors. Moreover, among various ANN models, Chang et al. (2002) [25] indicated that real-time recurrent learning neural networks (RTRLNN) possesses dynamic real-time recurrent routing mechanisms that can simulate time-varying systems effectively.

In summary, the previous models seldom can achieve a reservoir inflow forecast with a long lead-time of up to 48–72 h considering the future highly varied meteorology-hydrology uncertainty of a typhoon. Because of the powerful capability of ANNs to model any kind of nonlinear relationship between inputs and output through a series of transfer functions without the need to make assumptions in advance, in recent years, ANNs have been used increasingly in applications for modeling hydrological processes. The advantages of using ANNs include the ability to derive accurate and fast real-time short-term forecasts with low building costs. However, the development and application of accurate and effective ANN-based models that have the most potential for long lead-time real-time inflow forecast (e.g., RTRLNN and ANFIS) with the other advanced novel heuristic techniques in accumulated total inflow forecasts during typhoons is a subject that urgently requires development and scientific breakthrough.

The purpose of this study is to apply RTRLNN and ANFIS with specially devised novel heuristic inputs such as observed-predicted inflow increase/decrease rate (OPIID rate), total precipitation (TP), duration from current time to the time of maximum precipitation, and direct runoff ending (DRE) to develop heuristic-type long lead-time accumulated total reservoir inflow forecast models. This study also utilized temporal-spatial forecasted error feature analysis to assess the feasibility of the developed long lead-time RTRLNN- and ANFIS-based models, and conducted output sensitivity analysis of single/combined heuristic inputs to determine whether the developed heuristic model is superior to the non-heuristic model and whether it is vulnerable to the impact of future forecasted uncertainty and error on inputs.

2. DEVELOPMENT OF METHODOLOGY

2.1. Procedures

The procedures used in this study are divided into three steps as shown in Figure 1. The detailed procedures are thoroughly described as follows:

Step 1: First, observed short lead-time meteorological-, precipitation-, and pattern-based reservoir inflow factors during previous typhoons were specified as non-heuristic candidate inputs, and future long lead-time total precipitation- and pattern-based duration factors were specified as heuristic inputs. The optimal inputs for the non-heuristic and heuristic typhoon total inflow forecast model were selected by using non-parametric statistical correlation analysis.

Step 2: The steepest gradient descent (SGD) and conjugate gradient algorithm (CG) were used to train the parameter of RTRLNN, and subtractive clustering (SC) with the least square estimator (LSE) were applied to train the parameter of ANFIS. On obtaining the best model by comparing the assessment index value of the individually developed model type, the forecasted outcome for the RTRLNN-CHI (Coupled Heuristic Inputs) model, RTRLNN-NHI (No Heuristic Inputs) model, ANFIS-CHI model, and ANFIS-NHI model were compared across long lead-times.

Step 3-1: The temporal and spatial forecasted error feature of the four best types of long lead-time models developed were respectively analyzed, and a superior model determined.

Step 3-2: The output sensitivity of single or combined heuristic inputs due to future forecast uncertainty of the selected candidate optimal model among the four model types was analyzed under the impact of input forecasted error. Following the assessment, the optimal total reservoir inflow forecast model during typhoons was determined.

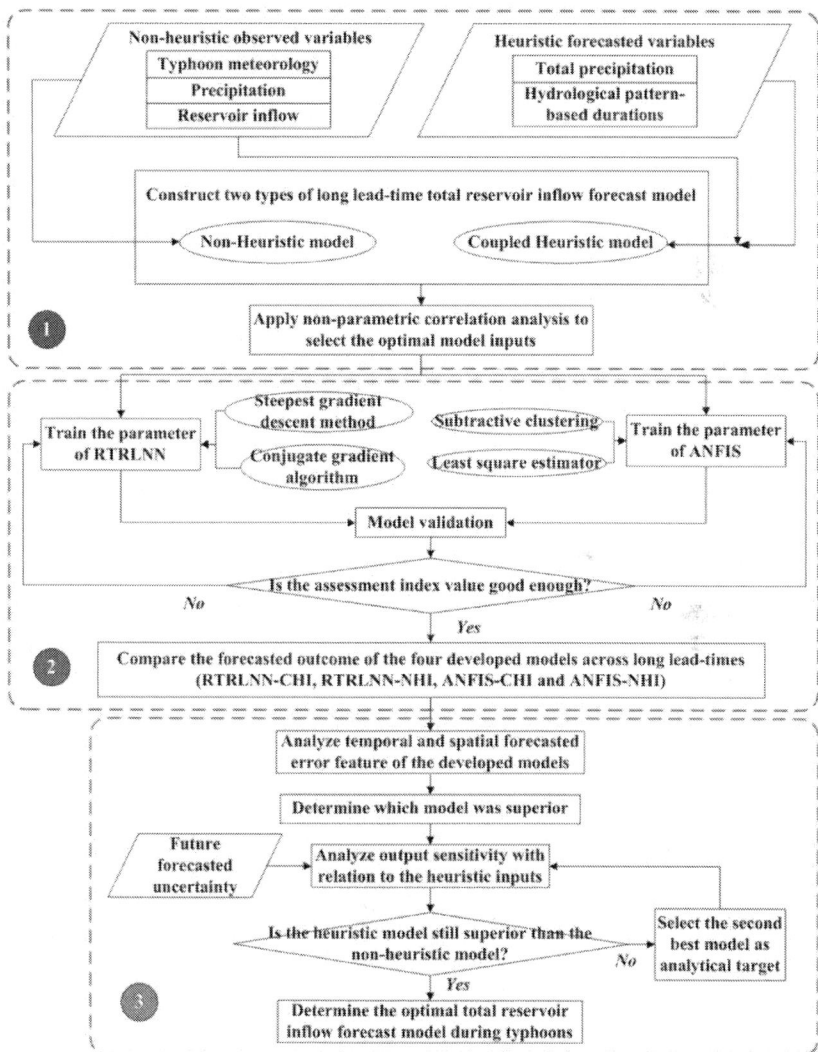

Figure 1. Flowchart of the methodology.

2.2. Developed Model Type of Accumulated Total Reservoir Inflow Forecast

This study designates the systematic operating mechanism of a reservoir of different stages as shown in Figure 2. To avoid dam failure and overflow from the upstream riverbank, the constraint can be expressed in Equation (1), and that to avoid a water level that is lower than dead storage is expressed in Equation (2).

$$A_2 - A_1 < S_{dam-safety}^{max} - x_0^S \tag{1}$$

$$A_1 < x_0^S - S^{dead} \tag{2}$$

where A_1, A_2, and A_3 are the increasing/reducing storage of Stage I, increasing storage of Stage II, and increasing/reducing storage of Stage III, respectively; $S_{dam-safety}^{max}$ is maximum safety storage for the dam; $xS0$ is the initial storage; and S^{dead} is the dead storage.

The releasing operating objectives of Stage I have to consider flood detention (expressed in Equation (3)) and final storage that at the same time (Equation (4)) are dominated by the future accumulated total inflow. Moreover, the constraint of Stage I involves avoiding the water level being lower than dead storage (Equation (2)). Hence, we can expect that the future accumulated total inflow is the key decision information of Stage I.

$$Max\{A_2\} \tag{3}$$

$$Min\{|A_1 + A_3 - A_2| - |S^{full} - x_0^S|\} \tag{4}$$

where $S_{dam-safety}^{max}$ is the maximum safety storage for the dam. In order to achieve optimal operation, the storage objective for the water supply is dominated primarily by Stage III and secondarily by Stage I, and the releasing operation of Stage II is used completely for flood detention (expressed in Equation (3)) that must subject to the safety constraint (Equation (1)). Hence, we can expect that the future total inflow is the key decision information of Stage II and Stage III.

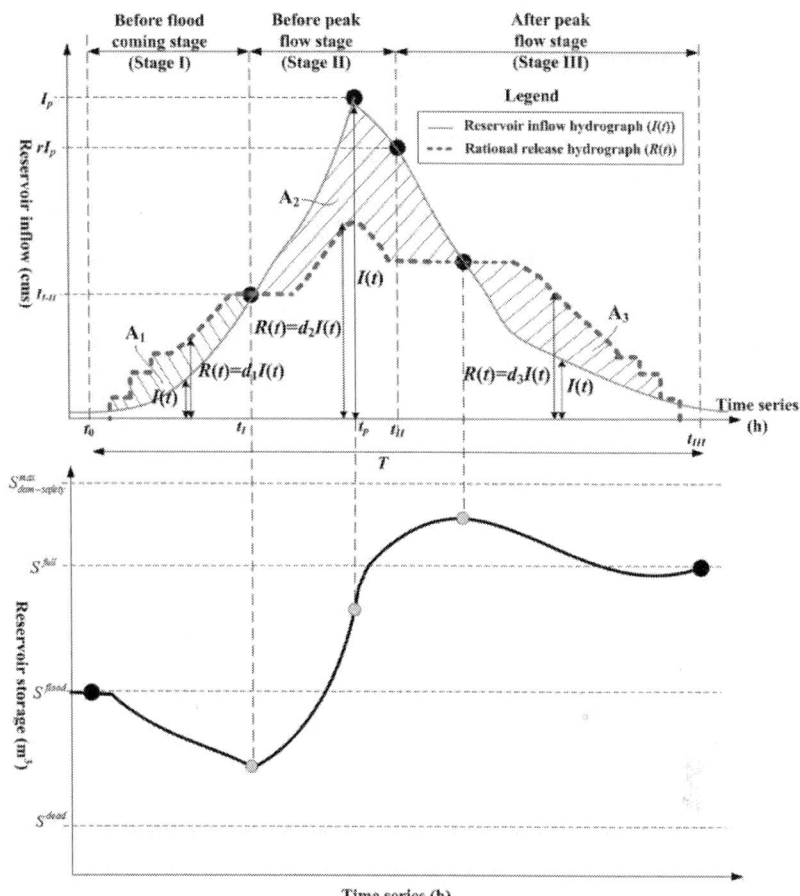

Figure 2. Schematic diagram of the flood operating mechanism of different stages in conjunction with the reservoir inflow.

The total reservoir inflow can be used as a criterion to determine the ideal amount of pre-discharge water and the benefit of flood detention under just filling the reservoir without overflowing the dam. Conventionally, the inflow can be calculated from the calculations of the rainfall-runoff simulation. The flow at the catchment outlet can be calculated using the unit hydrograph method, which is expressed as follows:

$$Q(t) = \int_A \int_0^t P(\tau) \cdot U(t - \tau) d\tau dA \tag{5}$$

where $Q(t)$ is the inflow at time t; $P(\tau)$ is the effective rainfall; and $U(t-\tau)$ is the flow path unit response function. Liu et al. (2003) [26] estimated the travel time at an arbitrary point in the catchment area by combining the diffusive wave model with the flow path unit response function. Molnar and Ramirez (1998) [27] used Manning's equation and energy dissipation theory to solve the approximate solutions to the diffusion waves, which can be expressed as follows:

$$U(t) = \frac{1}{\sigma \sqrt{2\pi \cdot \frac{t^3}{t_0^3}}} \exp\left[-\frac{(t-t_0)^2}{2\pi \cdot \frac{t}{t_0}} \right]$$

(6)

where t_0 is the average time of concentration for the water moving along the flow path from one point of the catchment area to the outlet; and σ is the standard deviation of the migration time. During the period of the typhoon, the effective rainfall in the future; $P(t+n)$, is related to the following atmospheric factors for the typhoon: distance between typhoon center and reservoir basin ($hc-w$), grade 7/10 typhoon radius (R), typhoon movement speed (vm), central wind speed ($Vmax$), and central pressure (pc). It can be expressed as the following:

$$P(t + \Delta t) = f(h_{c-w}, R, v_m, V_{max}, P_c)$$

(7)

where Δt is the forecasted lead-time. However, the uncertainty of the meteorology-hydrology relationship over a long lead-time is too high to make a determination as to the future typhoon atmospheric factors ahead of time. It is difficult to accurately forecast the rainfall hyetograph of the entire typhoon event in the future.

Hence, the rainfall-runoff model based on traditional hydrology was not used for real-time simulation and forecast of the reservoir inflow. A novel forecast method was developed and was found to be more reliable in forecasts. The new method adopted the total rainfall (P^{total}) method, and forecasted the various delays from the current moment to the key times along the rainfall-runoff hydrograph; for example, the delay from the current time to the maximum rainfall (T_{0-MP}), the delay to the end of the direct runoff (T_{0-DRE}), and the delay to the end of the water retreat (T_{0-EE}). The new method also used the observed-predicted inflow increase/decrease rate (OPIID rate) as the heuristic-type input. It is expected to be able to simulate the total reservoir inflow of the runoff hydrograph from the rainfall trend from a certain typhoon moving path in the future. A schematic diagram of hydrological key points within the rainfall-runoff hydrograph is shown in Figure 3. In this study, an original and innovative forecast model for the total reservoir inflow was developed with heuristic forecast inputs using ANFIS and RTRLNN. The model developed was analyzed and compared with the non-heuristic forecast model in which the input only included the real-time observed meteorology and hydrology information. The feasibility of the heuristic model for real-time forecast was also evaluated.

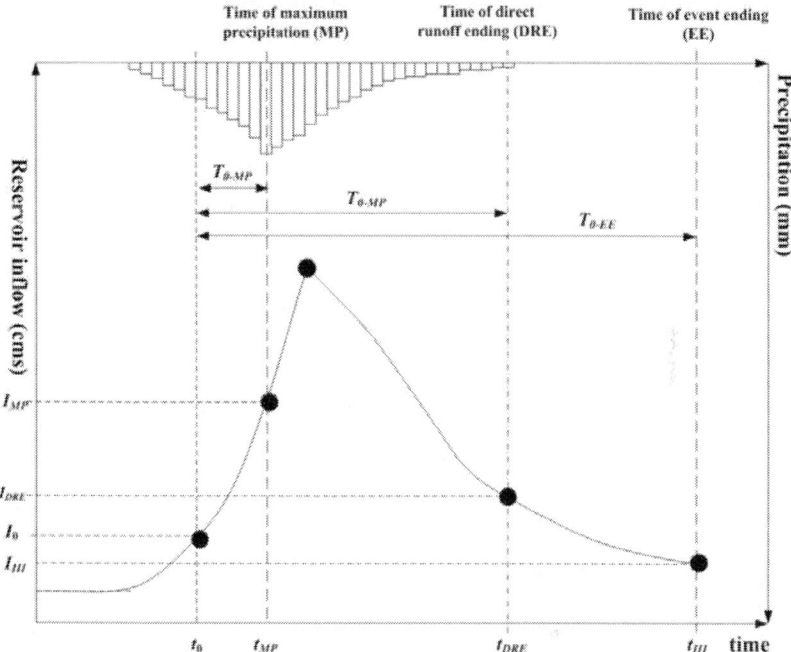

Figure 3. Schematic diagram of hydrological key points within the rainfall-runoff hydrograph.

2.2.1. Candidate Predictor

The choice of input candidates for the model is based on the theory of computation for the rainfall-runoff characteristics of the meteorology-hydrology relationship, such as the variables of reference Equations (5)–(7). In this study, four types of predictors which can be observed and predicted in real-time were used:

(1) Typhoon meteorological factor: longitude, latitude, central wind speed, central pressure, grade 7 typhoon radius, grade 10 typhoon radius, and typhoon movement speed, etc. The other feasible alternatives include relative humidity and temperature of the typhoon and basin, etc., but the other feasible alternatives are relatively not highly related to inflow and were not selected as candidate inputs.

(2) Rainfall station factor: observed hourly rainfall and total precipitation at the ground station. The other feasible alternatives include radar reflective information and satellite image from the typhoon, but the accuracy and correlation toward surface rainfall is not as high as that of the ground station. Hence, the alternatives are not adopted as candidate inputs.

(3) The distance between the typhoon center and the forecasting basin center (d(t)): this distance can be obtained using a conversion formula from longitude/latitude to distance:

$$y(t) = 111.1 \times (lat_c(t) - lat_{fos}(t)) \tag{8}$$

$$x(t) = 111.1 \times (lon_c(t) - lon_{fos}(t)) \times \cos(\frac{lat_c(t) + lat_{fos}(t)}{2}) \tag{9}$$

$$d(t) = \sqrt{(x(t))^2 + (y(t))^2} \tag{10}$$

where $lat_c(t)$ and $lat_{fos}(t)$ are the latitudes of the typhoon center and the forecasting basin center at time t, and $lat_c(t)$ and $lat_{fos}(t)$ are longitudes of the typhoon center and the forecasting basin center at time t.

(4) Runoff factor:

I. The delays from the current moment to the various key moments on the rainfall-runoff hydrograph in hydrology. For example, these include the delay from the current moment to the moment the maximum rainfall occurs ($T_{0\text{-}MP}$), the delay to the end of the direct runoff ($T_{0\text{-}DRE}$), and the delay to the end of the water retreat ($T_{0\text{-}EE}$). The feasible alternative includes the delay to the inflection point after peak flow, which is equal to the delay to rainfall excess ending plus the time of concentration. However, it is difficult to predict the delay to rainfall excess ending in real-time across a long lead-time, leading to this alternative not being adopted.

II. The real-time observed hourly reservoir inflow and the observed-predicted inflow increase/decrease rate (OPIID rate).

The total precipitation could be obtained by constructing a forecast database from the historical samples of the relationship between the center position of the typhoon and the rainfall in the catchment area using data mining techniques. Similarly, the delay of the future typhoon invasion could be obtained by constructing a forecast database from the historical samples of the distribution of the center position of the typhoon when the maximum rainfall occurred, the time when the direct runoff ended, and the time when the water retreated using data mining techniques. The above-mentioned heuristic inputs (total precipitation and delays) are estimated by the path and direction of the typhoon and the characteristic database. Besides, the output of the model was the total reservoir inflow during the period from the current moment to the end of the event. In this research, a heuristic forecast model was studied. The inputs to this heuristic model simultaneously comprised the real-time observed meteorology and hydrology information (typhoon characteristic factors, basin hourly precipitation, basin reservoir hourly inflow) and future forecasted heuristic meteorology and hydrology information (the total rainfall from the current moment to the end of the event, $T_{0\text{-MP}}$, $T_{0\text{-DRE}}$, $T_{0\text{-EE}}$, and OPIID rate). The input for the non-heuristic forecast model only included the real-time observed meteorology and hydrology information. The structure of the developed heuristic-type and non-heuristic accumulated total inflow forecast model is shown in Figure 4.

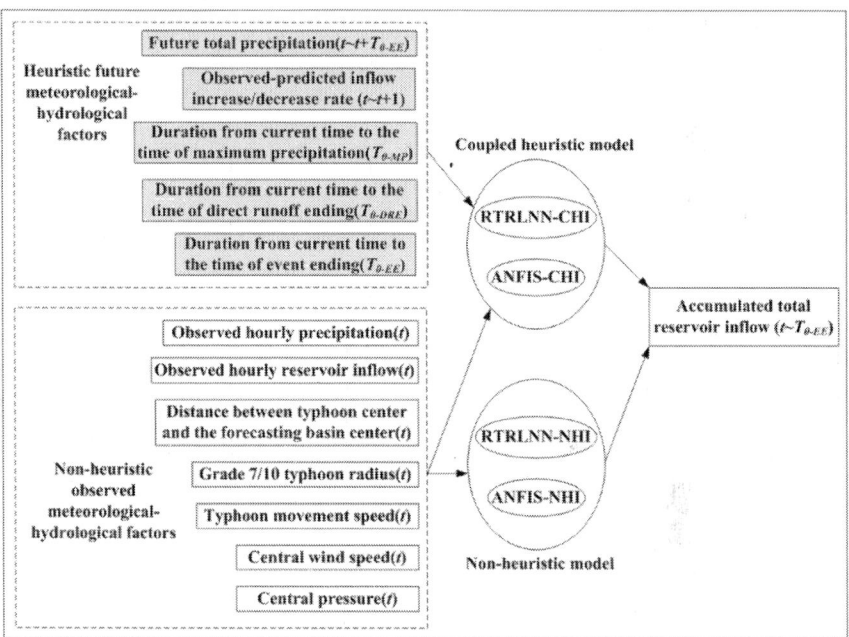

Figure 4. Structure of the developed coupled heuristic and non-heuristic accumulated total inflow forecast model.

2.2.2. Selection of Model Inputs

The feasible measures to select optimal model inputs include correlation analysis, principle component analysis, and the trial-and-error method. Among previous studies, the trial-and-error method is the most applied approach which is time-consuming. To effectively quantify the aptness for the large amount of candidate model inputs, this study uses correlation analysis for decision-making, and Spearman's rank correlation coefficient [28] is adopted as an analysis index. The analysis mechanism used for the correlation depends on the rank relationship of the time-series of two variables, and hence, this analysis can determine the correlation and suitability of input, regardless of the kind of relationship that exists between the candidate input and output, that is,

$$r_{rank} = 1 - \frac{6\sum_{i=1}^{n} D_i^2}{n(n^2 - 1)} \tag{11}$$

$$D_i^2 = (Rank_{x_i} - Rank_{y_i})^2 \tag{12}$$

where r_{rank} is Spearman's rank correlation coefficient, n is the number of data, x is the candidate input of the forecast model (predictor), y is the model output also known as the predictant (accumulated total reservoir inflow during time t + 1 to t + $T_{0\text{-}EE}$), and $Rank_{x_i}$ and $Rank_{y_i}$ are the sort values of x_i and y_i in their individual time-series of the variable, respectively. The most correlated candidate predictors for forecasting accumulated total inflow will be selected as optimal inputs, and the selected inputs must subject to hydrological relationships and the r_{rank} must larger than the assigned threshold values.

2.2.3. Assessment Index of Forecast Models

The performance of the forecast models was evaluated using the mean absolute error (MAE) and correlation coefficient (CC) criterion in the present study. The other feasible alternatives are root mean square error (RMSE), R^2, and coefficient of efficiency (CE). However, RMSE and R^2 are respectively similar to MAE and CC, and CE cannot assess the time delay effect of the forecast. Hence, the other alternatives are not adopted. The computational equations of MAE and CC are expressed as follows:

$$MAE = \frac{\sum_{t=1}^{n} \left| \hat{Y}(t) - Y(t) \right|}{n} \tag{13}$$

$$CC = \frac{N\sum \hat{Y}(t)Y(t) - \sum \hat{Y}(t)\sum Y(t)}{\sqrt{\sum \hat{Y}^2(t) - \frac{(\sum \hat{Y}(t))^2}{n}} \sqrt{\sum Y^2(t) - \frac{(\sum Y(t))^2}{n}}} \tag{14}$$

where $\hat{Y}(t)$ is the forecasted value at time t; $Y(t)$ is the actual value at time t; and n is the number of data. Smaller values of MAE imply a higher accuracy of the forecast model, and larger CC values indicate a closer coupling between the forecasted and measured series.

2.3. Heuristic Construction of RTRLNN

RTRLNN is a dynamic neural network with a stable routing mechanism and algorithm. The dynamic characteristics of a RTRLNN could be illustrated by the outputs of time-series based on an instantaneous impulse to the

RTRLNN. The network structure is different from the traditional static and feed-forward neural networks in that it allows recurrence between neurons and offers the function of local and temporal memory in the network, so the RTRLNN can simulate complex and time-varying systems that previous static neural networks could not handle effectively [25,29]. RTRLNN generally contains one or several recurrent loops. The RTRLNN network structure that we adopt in this study is shown in Figure 5. It is a multilayer perceptron and is composed of a concatenated input-output layer, a processing layer, and an output layer. The recurrent loops are recurrent from the output vector of the processing layer to the concatenated input-output layer. Hence, the concatenated input-output layer not only includes the input factor of the outer environment, but also stores the processed information from the processing layer before the current time. This allows the network to establish a temporal mutual connection and a dependent relationship between input variables because of the inner recurrent connection relationship, so the structure and mechanism can effectively learn the connection of the time-series (Elman, 1990 [30]).

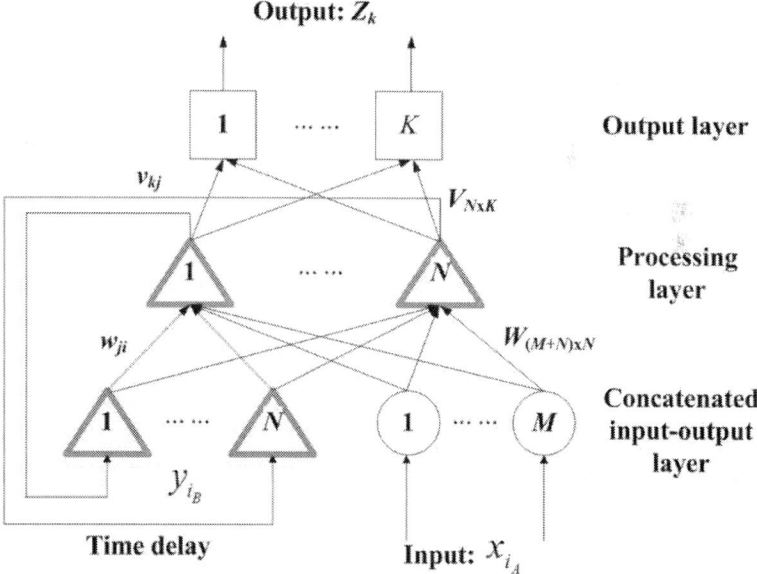

Figure 5. Structure of a RTRLNN.

The input vector of the concatenated input-output layer contains actual input variables x_{iA} and recurrent input variables y_{iB} (iA and iB are the number of actual and recurrent inputs, respectively):

$$u_i = [x_{i_A}, y_{i_B}] \quad for \quad i_A = 1,...,M \quad i_B = 1,...,N \tag{15}$$

where M and N are the total numbers of actual and recurrent inputs, respectively. The feed-forward propagation of the network first multiplies the input vector (ui) with the corresponding weights (wji) to obtain $netj$, then transfers $netj$ by a transfer function ($f(\cdot)$) to obtain the output of the processing layer (yj):

$$net_j = \sum_{i \in i_A \cup i_B} w_{ji} u_i$$

(16)

$$y_j = f(net_j)$$

(17)

where i, j, and k are the neuron numbers of the concatenated input-output layer, the processing layer, and the output layer, respectively. Multiplying yj with the corresponding weights (vkj) and summing them gives $netk$, and transfer $netk$ by a transfer function ($f(\cdot)$) gives the output of the output layer (zk):

$$net_k = \sum v_{kj} y_j$$

(18)

$$z_k = f(net_k)$$

(19)

In this study, the feasible transfer functions of the processing layer include tan-sigmoid (expressed in Equation (20)), linear, log-sigmoid, radial basis function, and symmetric saturating linear function, while the output layer is linear. The best suitable transfer function for the forecast model is extracted fully by trail results.

$$y_j = \frac{e^{net_j} - e^{-net_j}}{e^{net_j} + e^{-net_j}}$$

(20)

During RTRLNN training, the network not only continuously executes the message handling, but also revises each connected weighted vector in real-time according to the simulated error that belongs to the learning algorithm. Set $dk(t)$ as the target value of neuron k at time t. Then we define a time-varying K × 1 error vector $ek(t)$, whose kth element is:

$$e_k(t) = d_k(t) - z_k(t)$$

(21)

Then we define the instantaneous overall network error ($E(t)$) at time t as

$$E(t) = \frac{1}{2} \sum_{k=1}^{K} e_k^2(t)$$

(22)

The total cost function (*Etotal*) is obtained by summing $E(t)$ over all time T

$$E(t) = \frac{1}{2} \sum_{k=1}^{K} e_k^2(t)$$

23)

To minimize the cost function, this study applies the recursive steepest gradient descent method and the conjugate gradient algorithm to adjust the weights (**V** and **W**) along the negative of ∇E_{total}. The other feasible alternative is the Quasi-Newton method which is more time-consuming than the others, so the method is not adopted. Because the total error is the sum of the errors at the individual time-steps, we compute this gradient by accumulating the value of ∇E for each time-step along the trajectory. The weight change for any particular weight ($\Delta vkj(t)$) can thus be written as

$$E_{total} = \sum_{t=1}^{T} E(t)$$

(24)

whereη_1 is the learning-rate parameter. In Equation (24), $\dfrac{\partial E(t)}{\partial v_{kj}(t)}$ can be written as

$$\frac{\partial E(t)}{\partial v_{kj}(t)} = -e_k(t) f'(net_k(t)) y_j(t)$$

(25)

The same method can also be implemented for the specific weight *wmn*, that is

$$\Delta w_{mn}(t-1) = -\eta_2 \frac{\partial E(t)}{\partial w_{mn}(t-1)}$$

(26)

where$\eta2$ is the learning-rate parameter. The partial derivative $\partial E(t)\partial wmn(t-1)$ can be obtained by the chain rule for differentiation as follows:

$$\frac{\partial E(t)}{\partial w_{mn}(t-1)} = \left[\sum_{k=1}^{K} -e_k(t) f'(net_k(t)) v_{kj}(t) \right] \frac{\partial y_j(t)}{\partial w_{mn}(t-1)} \tag{27}$$

$$\Rightarrow \frac{\partial y_j(t)}{\partial w_{mn}(t-1)} = f'(net_j(t)) \frac{\partial net_j(t)}{\partial w_{mn}(t-1)} \tag{28}$$

$$\Rightarrow \frac{\partial net_j(t)}{\partial w_{mn}(t-1)} = \sum_{k \in (i_A \cup i_B)} \frac{\partial(w_{ji}(t-1) u_i(t-1))}{\partial w_{mn}(t-1)} \tag{29}$$

$$\Rightarrow \frac{\partial net_j(t)}{\partial w_{mn}(t-1)} = \sum_{i \in (i_A \cup i_B)} \left[w_{ji}(t-1) \frac{\partial u_i(t-1)}{\partial w_{mn}(t-1)} + \frac{\partial w_{ji}(t-1)}{\partial w_{mn}(t-1)} u_i(t-1) \right] \tag{30}$$

subject to

$$\frac{\partial w_{ji}(t-1)}{\partial w_{mn}(t-1)} = \begin{cases} 1, & when \quad (j=m) \cap (i=n) \\ 0, & else \end{cases} \tag{31}$$

Equation (30) can be rewritten as

∂

$$\frac{\partial net_j(t)}{\partial w_{mn}(t-1)} = \sum_{i \in (i_A \cup i_B)} w_{ji}(t-1) \frac{\partial u_i(t-1)}{\partial w_{mn}(t-1)} + \delta_{mj} u_n(t-1) \tag{32}$$

subject to

$$\delta_{mj} = \begin{cases} 1, & if \quad j = m \\ 0, & else \end{cases} \tag{33}$$

where δmj is the Kronecker delta. From the definition of $ui(t)$, we also note that

$$\frac{\partial u_i(t-1)}{\partial w_{mn}(t-1)} = \begin{cases} 0, & when \quad i \in i_A \\ \dfrac{\partial y_i(t-1)}{\partial w_{mn}(t-1)}, & when \quad i \in i_B \end{cases} \tag{34}$$

According to the propagation mechanism of RTRLNN, the initial state of the network at time t = 0 has no functional dependence on the synaptic weights, that is

$$\frac{\partial y_j(0)}{\partial w_{mn}(0)} = 0 \tag{35}$$

$$\frac{\partial y_j(t)}{\partial w_{mn}(t-1)} = f'(net_j(t)) \left[\sum_{i \in t_B} w_{ji}(t-1) \frac{\partial y_i(t-1)}{\partial w_{mn}(t-1)} + \delta_{mj} u_n(t-1) \right] \tag{36}$$

$$\frac{\partial y_j(t)}{\partial w_{mn}(t)} = \left\{ \pi_{mn}^j(t) \middle| (\forall j \in i_B) \cap (\forall m \in i_B) \cap [\forall n \in (i_A \cup i_B)] \right\} \approx \frac{\partial y_j(t)}{\partial w_{mn}(t-1)} \tag{37}$$

where $\pi_{mn}^j(t)$ are the triple indexed sets of variables which describe a dynamic system. For each time step t and all appropriate m, n, and j, the dynamics of the system are governed by

$$\pi_{mn}^j(t) = f'(net_j) \left[\sum_{i \in i_B} w_{ji}(t-1) \pi_{mn}^i(t-1) + \delta_{mj} u_n(t-1) \right]$$

$$I.C.: \quad \pi_{mn}^j(0) = 0 \tag{38}$$

Then the weight changes can be computed as

$$\Delta w_{mn}(t-1) = \eta_2 \left[\sum_k e_k(t) f'(net_k(t)) v_{kj}(t) \right] \pi_{mn}^j(t) \tag{39}$$

$$\Delta v_{kj}(t) = \eta_1 e_k(t) f'(net_k(t)) y_j(t) \tag{40}$$

2.4. Heuristic Construction of ANFIS

ANFIS was proposed by Jang (1993) [31], and is based on a fuzzy inference system constructed by combining the self-organization characteristics of a neural network. Hence, ANFIS integrates two algorithms to improve its accuracy, and solves for the best parameters by employing capabilities of learning and self-adaption. ANFIS is composed of an input layer, a rule layer, a normalization layer, a consequent layer, and an output layer, as shown in Figure 6. The modeling tool can transform the fuzzy-complex process and phenomenon into an artificial logic language that is therefore a potential approach for typhoon precipitation forecasting. The computation and transmission of each layer is described as follows.

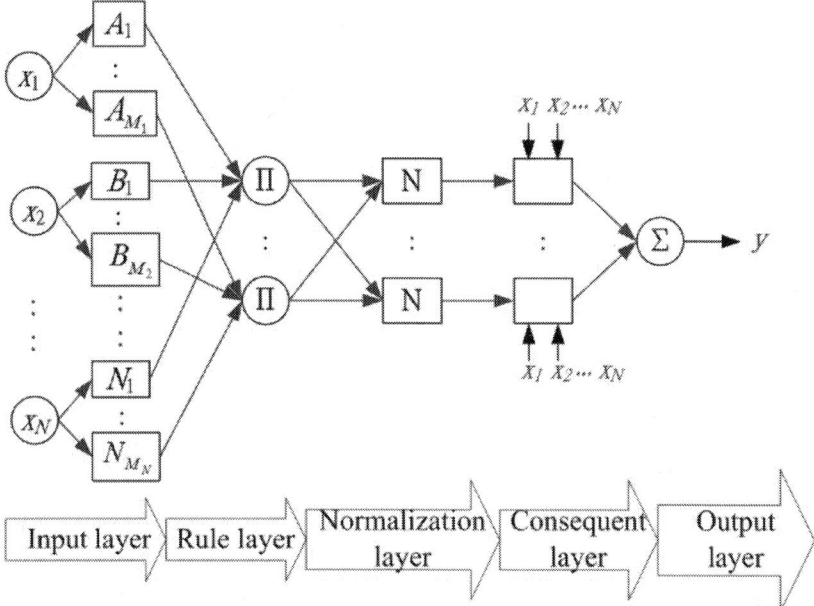

Figure 6. Structure of an ANFIS.

(1) Input layer

This layer projects input to a group of fuzzy sets and estimates the values of a group of membership functions. The most common types of membership functions are triangular, trapezoidal, Gaussian, generalized bell-shaped, and sigmoid functions. To retrieve the parameters of the input layer efficiently, this study adopts a group of Gaussian functions as the membership functions with subtractive clustering (SC), which can be expressed as follows:

$$O_{1,ji} = u_{ji}(x_i) = \exp\left(-\frac{\left\|x_i - c_{ji}\right\|^2}{2\sigma_{ji}^2}\right) \quad i=1,2,...,N \quad j=1,2,...,M_i$$

(41)

where $u_{ji}(x_i)$ is the membership function; c_{ji} and σ_{ji} are the antecedent parameters; N is the number of inputs; and M_i is the number of the fuzzy membership functions of input i.

(2) Rule layer

This layer precedes the antecedent match of the fuzzy logic rule between variables, and then applies a T-norm product operation to obtain the weighted value of each rule, that is,

$$O_{2,p} = w_p = \prod_{i=1}^{N} u_{pi}(x_i) \quad p=1,...,P$$

(42)

where w_p is the weighted value; and P is the number of rules.

(3) Normalization layer

The node of this layer computes the output ratio between the node and all other nodes, that is,

$$O_{3,p} = \overline{w}_p = \frac{w_p}{\sum\limits_{p=1}^{P} w_p}$$

(43)

(4) Consequent layer

The output of the consequent layer node is the product of the outputs of the normalization layer and the Sugeno fuzzy model (Takagi and Sugeno, 1983 [32]), that is,

$$O_{4,p} = \overline{w}_p f_p = \overline{w}_p \left(\sum_{i=0}^{N} r_{pi} x_i\right)$$

(44)

where r_{pi} represents the consequent parameters; and x_0 is equal to 1.

(5) Output layer

This layer sums the outputs of the previous layer to compute the model output, that is,

$$O_{5,p} = \sum_{p=1}^{P} \overline{w}_p f_p = \frac{\sum\limits_{p=1}^{P} w_p f_p}{\sum\limits_{p=1}^{P} w_p}$$

(45)

ANFIS is a feed-forward neural network and is constructed by supervised learning. The network parameters can be divided into antecedent parameters (nonlinear parameters: c_{ji}, σ_{ji}) and consequent parameters (linear parameters: r_{pi}), and the model structure is determined by setting the number of membership functions in the input layer and the number of nodes in the rule layer. The parameters can be solved by the steepest gradient descent method and Newton's method, for example. However, the methods would be slow and would produce a worse convergence and drop-in local optimum if the searching problem was more complex. To decrease the time for model construction in obtaining the best network structures and parameters, this study constructs ANFIS using hybrid algorithms including subtractive clustering (SC) and a least square estimator (LSE). The input and output vectors were first classified by subtractive clustering before training the model. The number of clusters obtained from the classification was set as the number of membership functions for node fuzzification at the various input layers and the number of nodes at the rule layers. After determining the network structures, the center point and standard deviation of each cluster were taken as the initial parameters of the input layer membership functions (Gaussian function). The training data were then fed into the network with the consequent linear parameter set and the antecedent nonlinear parameter set solved by the least squares estimator and the gradient steepest descent method, respectively. The corresponding algorithm flowchart of the model construction is shown in Figure 7. The network structure significantly reduces the time required to retrieve the optimal number of fuzzy membership functions, number of rules, and network parameters; the optimal network structure and parameters can be obtained after simply setting the adjacent radius in subtractive clustering between 0 and 1 (Jang, 1993 [31]).

Subtractive clustering was employed in the present study to construct fuzzy if-then rules in order to reduce the number of parameters of the fuzzy membership function in the ANFIS model. This was performed to establish a suitable rule base in the fuzzy inference system. Subtractive clustering was proposed by Chiu (1994) [33], in which every data point is treated as the candidate of the cluster center. Subtractive clustering is a fast and independent clustering method: the computational complexity is proportional to the number of data and is independent of the system dimension. For example, $xi(i=1,2,...,n)$ are n sets of data in an M-dimensional space and the corresponding density measures D are defined as

$$D_i = \sum_{j=1}^{n} \exp\left(-\frac{\left\| x_i - x_j \right\|^2}{\left(r_a / 2 \right)^2} \right)$$

(46)

where the adjacent radius ra is a positive number representing the distance near the center, and the data points outside the radius have minimum impact on the density measure. The density measure is calculated for each data point (x_i), and the one with the highest density (D_{c1}) is selected as the first cluster center (x_{c1}). The definition of the density measure is then modified to select the next cluster center. Setting that x_{ck} is the cluster center selected at the kth round, and the corresponding density measure is D_{ck}, the modified formula is as follows:

$$D_i = D_i - D_{ck} \exp\left(-\frac{\left\|x_i - x_{ck}\right\|^2}{(r_b/2)^2}\right)$$

(47)

where radius rb has the same definition as ra and is usually set as $1.5ra$ so that the selected center will not be too close to that of the previous one. The above procedure of cluster center selection is repeated until a termination condition is reached, or there is a sufficient number of cluster centers.

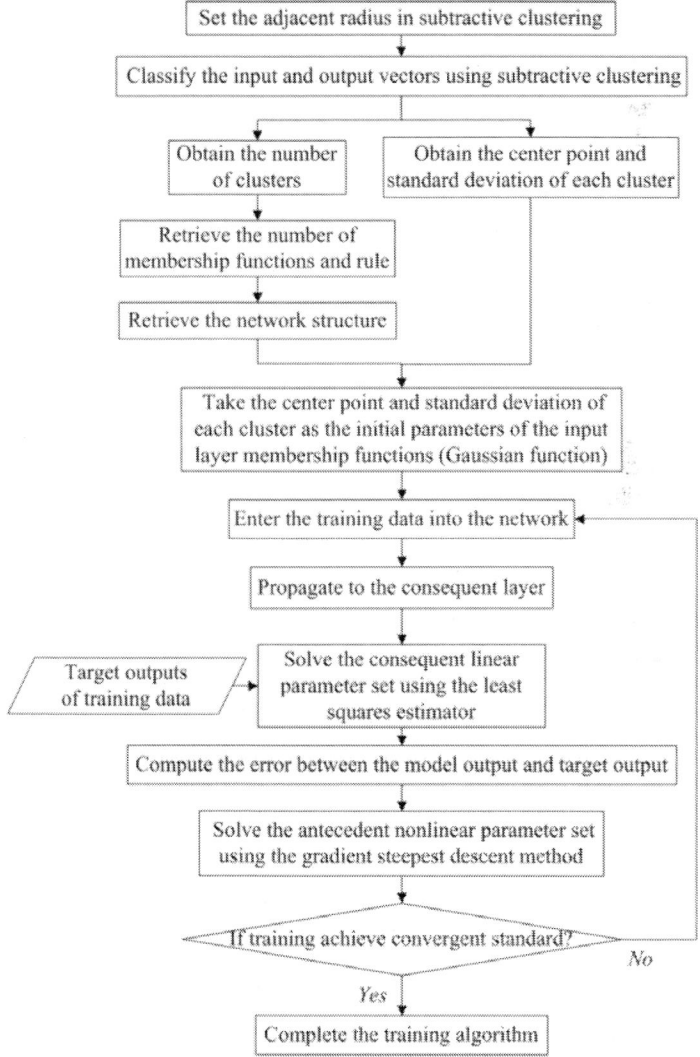

Figure 7. Flowchart of training the parameter and structure of ANFIS.

2.5. Analysis of Temporal and Spatial Forecasted Error Feature of the Developed Long Lead-Time Models

In this research, RTRLNN and ANFIS were used to study four types of coupled heuristic and non-heuristic forecast models (RTRLNN-CHI, RTRLNN-NHI, ANFIS-CHI and ANFIS-NHI) for long lead-time forecast of the total reservoir inflow. To evaluate the forecast accuracy and applicability of the four models on typhoon invasion, analyses were conducted on the characteristics of the temporal and spatial forecast errors for the most optimal forecast case of the four models. Assessments were made as to which model had the best forecast performance. For the analysis of the temporal forecast error, calculations were made for each forecast model for the absolute error between the forecasted time and the forecasted total reservoir inflow during the verification phase of the typhoon event at each field. The errors were then used to assess the capability and limits of the model for the long lead-time forecasting of the total reservoir inflow, which could be calculated as follows:

$$AEP(\Delta t) = \frac{\displaystyle\sum_{p=1}^{P} \frac{\left|\hat{Y}_p(\Delta t) - Y_p(\Delta t)\right|}{Y_p(\Delta t)}}{P} \times 100\% \tag{48}$$

where $AEP(\Delta t)$ is average error percentage for forecasted lead-time Δt, $Y_p(x,y)$ and $Y(\Delta t)p$ are the forecasted and actual accumulated total reservoir inflow on typhoon event number p for forecasted lead-time Δt, respectively; and P is the total number of typhoon events.

For each forecast model, the analysis of the spatial forecast error included calculation of the absolute error on the forecasted total reservoir inflow at the spatial position of the typhoon center during the verification phase of the typhoon event at each field. These errors were used to discuss the capability and limits of the long lead-time forecasting of the total reservoir inflow for each model when the typhoon center moved to each of the spatial grids, which could be expressed as

$$AEP(x,y) = \frac{\displaystyle\sum_{p=1}^{P} \frac{\left|\hat{Y}_p(x,y) - Y_p(x,y)\right|}{Y_p(x,y)}}{P} \times 100\% \tag{49}$$

where $AEP(x,y)$ is the average error percentage while the typhoon center is located at longitude x and latitude y, $Y_p(x,y)$ and $Y(x,y)p$ are the forecasted and actual accumulated total reservoir inflow on typhoon event number pwhile the typhoon center is located at longitude x and latitude y, respectively, and P is the total number of typhoon events.

2.6. Output Sensitivity Analysis of Single or Combined Heuristic Inputs Due to Future Forecasted Uncertainty

The coupled heuristic model in this research can forecast the rainfall-runoff hydrology under a specific movement path for the future typhoon, which increases the long lead-time forecast accuracy of the accumulated total reservoir inflow. However, if this model was applied to real-time forecasting, the uncertainty of the meteorology and hydrology for the long lead-time typhoon in the future would be unacceptably high. There would be cases with unavoidable forecast errors on quantities such as the long lead-time total rainfall in the future, the delay from the current time to the maximum rainfall (T_{0-MP}), the delay to the end of the direct runoff (T_{0-DRE}), and the delay to the end of the water retreat in the typhoon event (T_{0-EE}). When such heuristic information is coupled with the input of the heuristic model, it is possible that unexpected errors will be generated on the forecast output of the model. Thus, in order to evaluate the feasibility, applicability, and accuracy of the heuristic model for real-time forecasting, sensitivity analysis was conducted on the effects on the output when forecast errors exist in the heuristic input of the most optimal heuristic model. The above analysis was used to judge whether the forecast accuracy of the heuristic model was better than that of the non-heuristic model for real-time forecasting when errors exist in the input. The expression for the analysis is as shown below:

$$\hat{Y}_i^H(t) = f_{\left[\substack{RTRLNN-CHI \\ ANFIS-CHI}\right]}\left(x_i^H(t) \pm EP \cdot x_i^H(t), X^{NH}(t)\right) \tag{50}$$

$$AEP(i, \pm EP) = \frac{\sum_{t=1}^{n} \frac{\left|\hat{Y}_i^H(t) - Y(t)\right|}{Y(t)}}{n} \times 100\% \tag{51}$$

where $\hat{Y}_i^H(t)$ is the forecasted value at time t under entering error into heuristic input number i; $f_{\left[\substack{RTRLNN-CHI \\ ANFIS-CHI}\right]}$ is the developed coupled heuristic forecast model; $x_i^H(t)$ is the input value of heuristic input number i at time t; $\pm EP$ is the average error percentage based on previous studies; $X^{NH}(t)$ is the value of non-heuristic input at time t; $Y(t)$ is the actual value at time t; and n is the number of data.

3. APPLICATION

3.1. Study Area

The methodology proposed in the present study was applied to the Shihmen Reservoir catchment area, which measures approximately 763.4 km^2. The main stream within this area is the Dahan Creek, which is the upper stream of the Tamsui River. The effective capacity is approximately 2.098×10^8 cubic meters. The annual average rainfall in the catchment area is approximately 2350 mm, with 80% of the annual rainfall concentrated in the period between May and October. Most of the rainfall is from typhoon precipitation. The annual inflow of the Shihmen Reservoir is approximately 1.510 billion tons. The study area is shown in Figure 8.

3.2. Data Used in Model Construction

This study used instantaneous observed non-heuristic and coupled heuristic information to forecast accumulated total inflow during current time to the event recessional ending. The output variable was taken as the future accumulated total inflow forecast in the Shihmen Reservoir catchment area. In this study, the end of the typhoon flood event was defined as the moment that simultaneously satisfied the following conditions: (1) the Meteorology Bureau lifted the alarm for typhoon on land and over the sea; (2) rainfall completely stopped in the catchment areas; (3) the reservoir inflow decreased below 300 cms. The model construction included two stages, namely the training and validation stages. The adopted typhoon events for training are Aere, Matsa, Talim, Long-Wang, Wipha, Fung-Wong, Sinlaku, Morakot, Megi, and Meari, and for validation they are Haitang, Sepat, Krosa, Jangmi, and Parma. The total data number for training is 459, and for validation it is 211. The adopted typhoon events for model construction among the training and validation stages are shown in Table 1, and the moving paths of the typhoons used in model construction are shown in Figure 9.

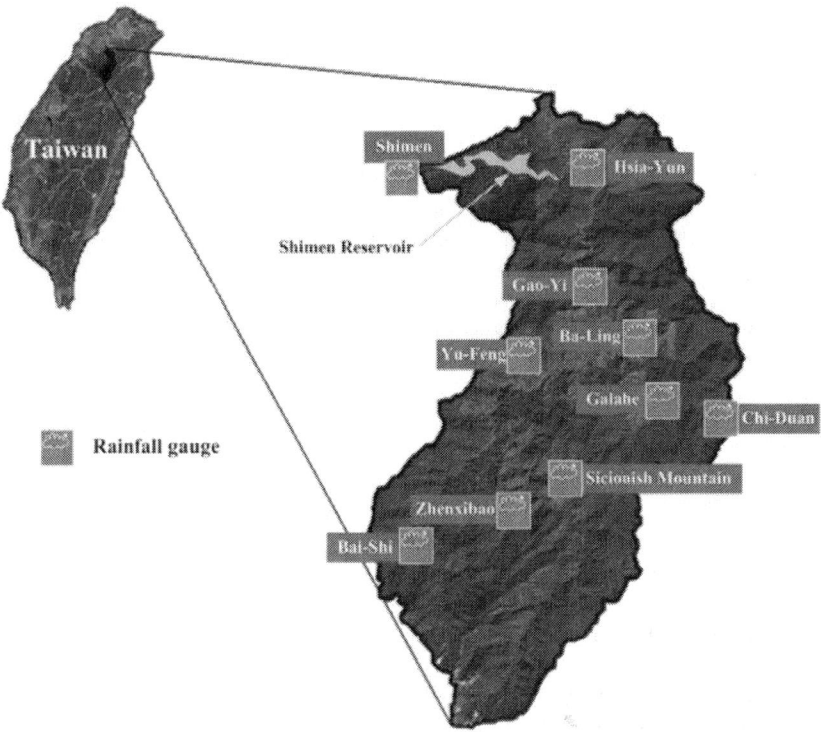

Figure 8. Study area.

Table 1. The adopted typhoon events for model construction among the training and validation stages.

Construction Stage	Typhoon Name	Time Period	Total Reservoir Inflow (m^3)	Data Number	Total Data Number
Training	Aere	23–26 August 2004	748, 936, 728	58	459
	Matsa	4–6 August 2005	541, 872, 324	61	
	Talim	31 August 2005–2 September 2005	201, 308, 580	33	
	Long-Wang	2–3 October 2005	68, 596, 704	24	
	Wipha	18–20 September 2007	186, 601, 752	48	
	Fung-Wong	28–29 July 2008	103, 422, 564	33	
	Sinlaku	13–16 September 2008	554, 322, 600	75	
	Morakot	7–10 August 2009	205, 435, 980	71	
	Megi	21–22 October 2010	54, 991, 728	37	
	Meari	25 June 2011	44, 826, 012	19	
Validation	Haitang	17–20 July 2005	237, 416, 256	53	211
	Sepat	18 August 2007	128, 935, 224	20	
	Krosa	6–8 October 2007	409, 855, 824	53	
	Jangmi	28–30 September 2008	220, 301, 136	53	
	Parma	6 October 2009	40, 997, 340	18	
	Fanapi	19 September 2010	33, 694, 956	14	

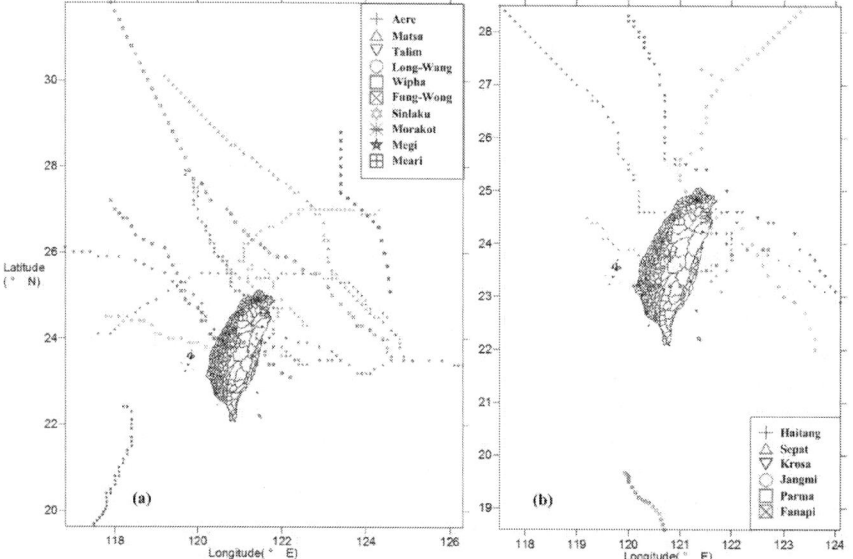

Figure 9. Moving paths of typhoons used in model construction: (**a**) training; (**b**) validation.

3.3. Results and Discussion

3.3.1. Model Inputs Selection

Correlation analysis was applied in the present study to assess the correlation coefficient between each input factor and the future accumulated total inflow for the Shihmen Reservoir. The selected heuristic model inputs and corresponding correlation coefficients are the future accumulated total basin precipitation (r_{rank}: 0.926), duration from current time to the end of the flood event (r_{rank}: 0.960), duration from current time to the time of DRE (r_{rank}: 0.751), duration from current time to the time of maximum precipitation (r_{rank}: 0.548), and observed-predicted inflow increase/decrease rate (r_{rank}: 0.401). These selected variables are the most correlated inputs among all heuristic candidate predictors and the r_{rank} value of all the selected inputs must be larger than 0.4. Furthermore, the selected non-heuristic model inputs and corresponding correlation coefficients are the observed hourly basin precipitation at the current time (r_{rank}: 0.672), hourly reservoir inflow (r_{rank}: 0.509), typhoon central longitude (r_{rank}: 0.610), central wind speed (r_{rank}: 0.650), and central pressure (r_{rank}: 0.639); these selected variables are the most correlated inputs among all non-heuristic candidate predictors and the r_{rank} values must all be larger than 0.5. Research conducted by Lin and Chen (2005) [34] revealed that excessive model inputs could introduce additional noise into the model, therefore 10 input factors were selected as a maximum based on the correlation coefficients. Based on the above analytical results, the developed heuristic forecast model includes 10 inputs including both heuristic and non-heuristic inputs, and the non-heuristic model only includes five non-heuristic inputs. The heuristic parameters are considered to be essential for use in eliminating the forecasting uncertainty, and for characterizing future long lead-time accumulated total inflow.

3.3.2. Results of Model Construction

In this study, RTRLNN and ANFIS were used to construct coupled heuristic and non-heuristic forecast models for long lead-time forecasting of the future accumulated total inflow for the Shihmen Reservoir. The choice of the particular set of parameters of the optimal model is retrieved by applying an intelligent heuristic searching strategy on the setting structure parameters (i.e., neuron numbers of the processing layer for RTRLNN and the adjacent radius for ANFIS), and the values of the connected parameters corresponding to the setting structure are calibrated by using the heuristic algorithm described in Section 2.3 and Section 2.4. The searching strategy first constructs models by setting a series of neuron numbers (1–15) and adjacent radius (0–1) equally from the feasible domain with the reliable representative amount (10 per neuron number and 100 for adjacent radius), and then the structure parameter of the best model among the equally distributed sampling process was strengthened by construction with more experimental frequency to retrieve the optimal model efficiently. The forecasted outcomes of the most optimal model of the four types of forecast architecture (RTRLNN-CHI, RTRLNN-NHI, ANFIS-CHI, and ANFIS-NHI) are shown in Table 2. The best training and verification results for the RTRLNN-CHI model, ANFIS-CHI model, and RTRLNN-NHI model are shown in Figure 10, Figure 11 and Figure 12, respectively. The MAE values for the verification stage of the RTRLNN-CHI model, RTRLNN-NHI model, ANFIS-CHI model, and ANFIS-NHI model were respectively 11,721,556 m^3, 30,475,270 m^3, 14,429,374 m^3, and 53,236,429 m^3, while the CC values for the verification were 0.979, 0.876, 0.975, and 0.658, respectively. The results indicate that the respective forecast accuracy and stability of the RTRLNN-CHI and ANFIS-CHI models are significantly higher than those of the RTRLNN-NHI and ANFIS-NHI models. This shows that the proposed heuristic forecast model may be highly accurate in its forecast of the total reservoir inflow under the following conditions: (1) when the input includes key inputs such as the future accumulated total precipitation and the delays from the current moment to the key hydrology points of the hydrograph (maximum precipitation, direct runoff ending, and event recessional ending); (2) with assistance of the comprehensive simulation of the real-time observed atmospheric factors and rainfall-runoff factors of the typhoon.

Table 2. Best assessment indexes values of the four kinds of constructed models.

Structure Parameters/Assessment Indexes	RTRLNN-CHI Model	RTRLNN-NHI Model	Structure Parameters/Assessment Indexes	ANFIS-CH I Model	ANFIS-NH I Model
Best node number of hidden layer	3	9	Best adjacent radius/rule number	0.922/2	0.836/3
MAE of training (m³)	4587459	22430139	MAE of training (m³)	7249160	59066261
MAE of validation (m³)	11721556	30475271	MAE of validation (m³)	14429375	53236429
CC of training	0.999	0.980	CC of training	0.998	0.867
CC of validation	0.980	0.876	CC of validation	0.976	0.659

The input information for the RTRLNN-NHI and ANFIS-NHI models only included the current real-time observed conditions of the typhoon atmosphere and rainfall-runoff status. The average forecasted accuracy of the RTRLNN-NHI and ANFIS-NHI models is respectively worse than that of the RTRLNN-CHI/ANFIS-CHI models by 1.6/1.11 times and 3.54/2.68 times, and average forecasted stability is worse by 10.5%/10.2% and 32.8%/32.5%. Hence, these two non-heuristic models could not be used to accurately and physically simulate the accumulated total reservoir inflow after long lead-time changes in the meteorology and hydrology. The reason is that the model inputs only have initial conditions (observed rainfall-runoff variables) but do not have boundary conditions for future periods to simulate the shape and duration of future inflow hydrographs. This obviously caused the forecasting accuracy and stability of the accumulated total inflow at the moment of maximum rainfall in the early stage of the event to be inferior to that at the later stages after the flood peaked. Moreover, the activation function in the hidden layer of the most optimal RTRLNN-based forecast model for the total reservoir inflow was a tan-sigmoid transfer function, while the activation function in the output layer was a linear activation function. The reason is that the shape of the tan-sigmoid function is similar to the cumulative distribution function (CDF) of the inflow hydrograph, and the shape of the CDF is exactly the inverse of the accumulated total inflow hydrograph. Hence, the tan-sigmoid function can simulate future accumulated total inflow better than the other shapes of functions. Furthermore, the numbers of neurons for the processing layer of the most optimal models in the RTRLNN-CHI and RTRLNN-NHI models were three and nine, respectively, and the rule numbers for the most optimal ANFIS-CHI and ANFIS-NHI models were two and three, respectively. This indicates that there was insufficient input information to represent future boundary conditions of the typhoon rainfall-runoff relationship over a long lead-time in the non-heuristic model. Therefore, a more complicated and time-consuming model is required to simulate the accumulated total reservoir inflow in the future.

In addition, from the evaluation-index point of view, for the two heuristic models, the overall forecast accuracy and stability of the RTRLNN-CHI model for forecasting the total reservoir inflow was slightly better than that of the ANFIS-CHI model by 0.23 times of MAE and 0.41% of CC. The reason is that there is a fixed-ratio real-time feedback calculation mechanism in the structure of the RTRLNN model. When the simulation mechanism and characteristics of the model were applied to forecasting time-varying output targets with a long lead-time and high uncertainty, better forecasting results were obtained than with the ANFIS model, which lacks flexibility in the input and rule layers that were used.

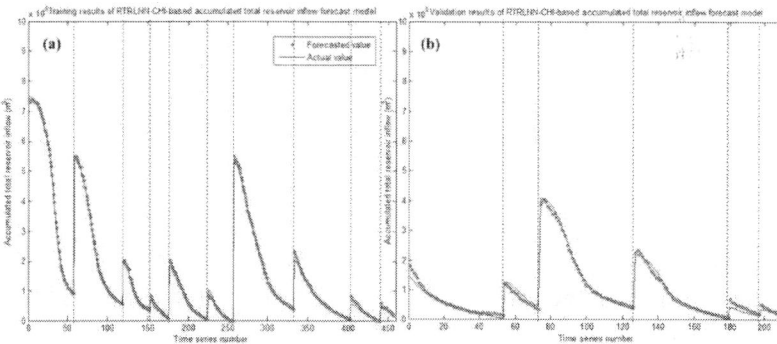

Figure 10. Training and validation results of RTRLNN-CHI-based accumulated total reservoir inflow forecast model: (**a**) training stage; (**b**) validation stage.

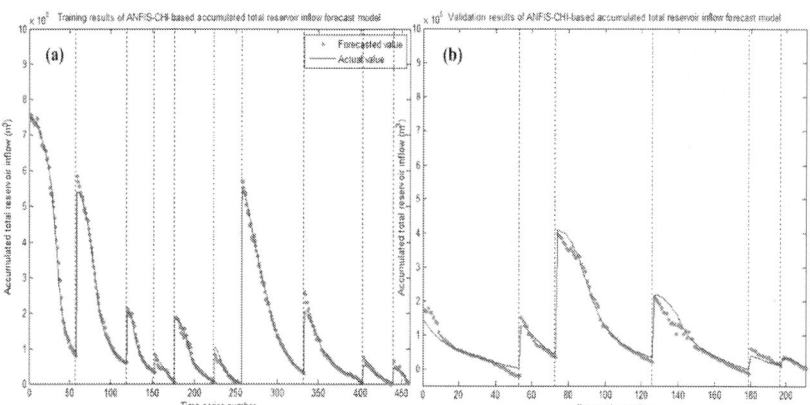

Figure 11. Training and validation results of ANFIS-CHI-based accumulated total reservoir inflow forecast model: (**a**) training stage; (**b**) validation stage.

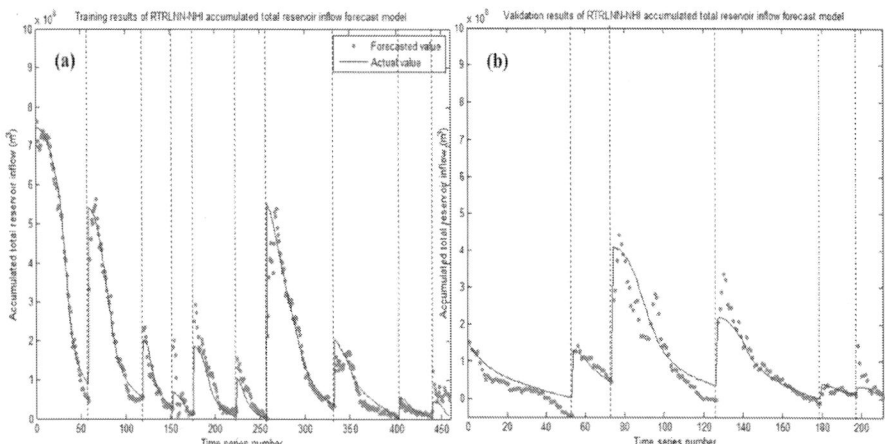

Figure 12. Training and validation results of RTRLNN-NHI-based accumulated total reservoir inflow forecast model: (**a**) training stage; (**b**) validation stage.

3.3.3. Analytical Results of Temporal and Spatial Forecasted Error Feature of the Developed Models

In this research, characteristic analyses were made on the temporal and spatial forecast errors of the four forecast models (RTRLNN-CHI, RTRLNN-NHI, ANFIS-CHI, and ANFIS-NHI) for long lead-time and total reservoir inflow. The analysis targets were six typhoon events for verification, which were used to evaluate the limits and applicability of the four models for long lead-time forecasting. Judgments were also made on the range of the typhoon center where the future total reservoir inflow may be accurately and appropriately forecasted. The average error percentages of the four developed accumulated total inflow forecast models across long lead-times are shown in Table 3, and a comparison of the temporal forecasted error features of the developed models across long lead-times is given in Figure 13. The percentages of the average forecast errors in the forecast time range of 24 to 48 h for the four models (RTRLNN-CHI, RTRLNN-NHI, ANFIS-CHI, and ANFIS-NHI), $\left(\dfrac{\sum\limits_{\Delta t=24}^{48} AEP(\Delta t)}{24} \right)$, were 4.6%, 16.7%, 7.7%, and 27.1%, respectively. The average error percentages (AEP) for the period of 48 to 72 h, $\left(\dfrac{\sum\limits_{\Delta t=48}^{72} AEP(\Delta t)}{24} \right)$, were 9.3%, 16.1%, 12.1%, and 39.3%, respectively. The AEP in the period of 20 to 79 h, $\left(\dfrac{\sum\limits_{\Delta t=20}^{79} AEP(\Delta t)}{60} \right)$, were 6.3%, 15.2%, 9.2%, and 31.8%, respectively. The results indicate that the long lead-time forecast accuracy of the heuristic model is obviously better than that of the non-heuristic model by 9%–23% for a lead-time of 24–48 h and 4%–30%

for a lead-time of 48–72 h. The forecast error of the heuristic model did not significantly increase along with the increasing of the forecasted lead-time. The hydrograph patterns of the reservoir inflow for the meteorology and hydrology of the future typhoon can be appropriately simulated ahead of time with heuristic inputs. Therefore, the accumulated total inflow can be accurately forecasted for the long lead-time future. Furthermore, there was a highly complicated nonlinear relationship between the rainfall from the typhoon meteorology and the runoff in the catchment area from hydrology. There was better flexibility in the calculation mechanism of the RTRLNN model than in the ANFIS model. RTRLNN had detailed linkage between various types of inputs and outputs. It also calculated feedback information in real-time. Moreover, the accuracies of the various forecasted time intervals using the RTRLNN-CHI model were better than those obtained using the ANFIS-CHI model. The forecast error was less than 10% when the forecasted lead-time reached three days. The AEP for a forecasted lead-time of an average 49 h, $\left(\dfrac{\sum\limits_{\Delta t=20}^{79} AEP(\Delta t)}{60} \right)$, was only 6.3% for the RTRLNN-CHI model. Hence, a long lead-time forecast model for the accumulated total reservoir inflow was successfully developed using the heuristic technique in this study.

Table 3. Average error percentage of the four developed models across long lead-times.

Forecasted Lead-Time	RTRLNN-CHI Model	RTRLNN-NHI Model	ANFIS-CHI Model	ANFIS-NHI Model
During 24 to 48 h	4.6%	16.7%	7.7%	27.1%
During 48 to 72 h	9.3%	16.1%	12.1%	39.3%
During 20 to 79 h	6.3%	15.2%	9.2%	31.8%

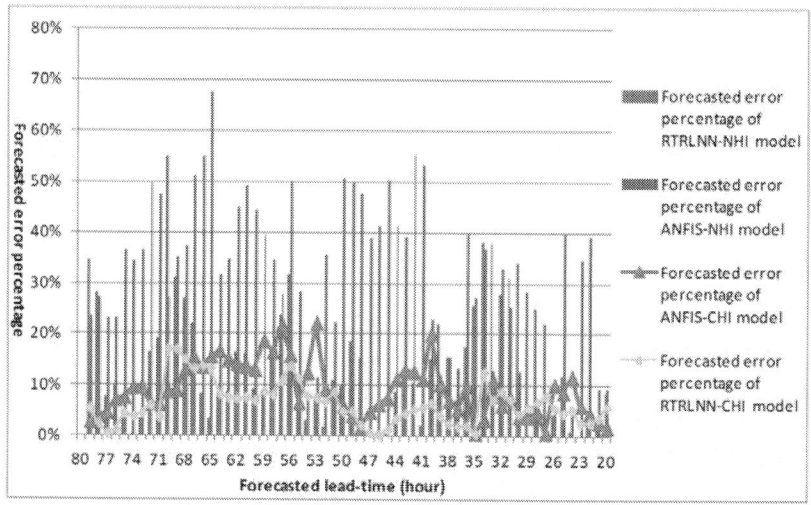

Figure 13. Comparison of temporal forecasted error features of the four developed accumulated total inflow forecast models across long lead-times.

The comparison of the spatial forecasted error feature of the four developed accumulated total inflow forecast models for the Shihmen Reservoir with relation to the central location of the typhoon is shown in Figure 14. This figure represents the absolute error percentage ($AEP(x,y)$) on the forecasted total reservoir inflow in the catchment area of the Shihmen Reservoir when the typhoon center was moving in the vicinity of any region in Taiwan (in the range of longitude 118–124 degrees, latitude 22–28 degrees). It can be seen from the spatial distributions of the forecast errors of the four models that the overall error space and range of the heuristic models (e.g., Figure 14a for the RTRLNN-CHI model, and Figure 14c for the ANFIS-CHI model) are much smaller than those of the non-heuristic models (e.g., Figure 14b for the RTRLNN-NHI model, and Figure 14d for the ANFIS-NHI model). In the error spatial distribution map of the RTRLNN-CHI model, the range where the $AEP(x,y)$ was less than 10% (which was about 61% of the researched range) was much bigger than that of the ANFIS-CHI model (which was about 47% of the researched range). The range where the $AEP(x,y)$ was greater than 20% (which was about 13% of the researched range) was much smaller than that of the ANFIS-CHI model (which was about 32% of the researched range). Figure 14 confirms that the forecast accuracy and stability of the RTRLNN-CHI model were better than those of other models when typhoon invasion was at the basin of the Shihmen Reservoir. When the typhoon center was in the southeast of Taiwan, the structure and meteorology field after the typhoon passed Taiwan was destroyed by the terrain. There was also the co-existing effects of the monsoon. Therefore, there was a significant difference between the future typhoon meteorology and rainfall-runoff conditions and the observed data when the typhoon center was in the southeast of Taiwan. As a result, the forecast error on the total reservoir inflow of the typhoon when it is located in the southeast of Taiwan is greater than that when the typhoon center is located elsewhere. However, after the typhoon center passed Taiwan, the circulation structure was not affected by the terrain. Hence, the forecast error on the total reservoir inflow was relatively small.

3.3.4. Sensitivity Analysis Results of Output with Relation to Heuristic Inputs Due to Future Forecasted Uncertainty

In order to evaluate the feasibility, applicability, and accuracy of the heuristic model when applied to real-time forecasting, a sensitivity analysis was conducted on the forecasted error effects of the outputs with relation to single or combined heuristic inputs for the most optimal heuristic model (RTRLNN-CHI). According to the previously developed short lead-time hydrological forecast models [35,36], an average forecast error ($\pm EP$) of 10% was assumed on each heuristic input, which was inputted into the model to simulate the absolute error percentage ($AEP(i,\pm EP)$) of the forecast. The results from the analysis are shown in Figure 15, where Heuristic input 1 (HI1) is the future accumulated total precipitation, HI2 is the duration from the current time to the flood recessional ending time, HI3 is the duration from the current time to the DRE time, and HI4 is the duration from the current time to the maximum precipitation time. The results displayed show that when $\pm 10\%$ error was inputted with HI2, the maximum output error would appear regardless of whether the case was that of a single input or combinations of multiple inputs. This indicates that the flood duration is the most important factor in forecasting the future accumulated total reservoir inflow, which is also the most sensitive input for the output of the model. Among all the combinations of heuristic input errors, the average forecast error was 9.98% for a 10% overestimation on the input, with a maximum of 13.6% (when there are 10% errors on all of HI1, HI2, HI3, and HI4). For the case of underestimation with a 10% error on the input, the average forecast error was 11.01%, with a maximum of 13.6%. There was an additional 1.03% error on the output for the 10% underestimation error on the heuristic inputs when compared to the case of 10% overestimation on the input. These results indicate that the absolute error percentage (13.6%) of the heuristic model with a 10% error on all the heuristic inputs was still lower than that of the most optimal non-heuristic model (RTRLNN-NHI), $AEP = 15.2\%$. This shows that the real-time forecast accuracy of the RTRLNN-CHI model is still better than that of the non-heuristic models (RTRLNN-NHI and ANFIS-NHI) even when there are errors on the heuristic inputs.

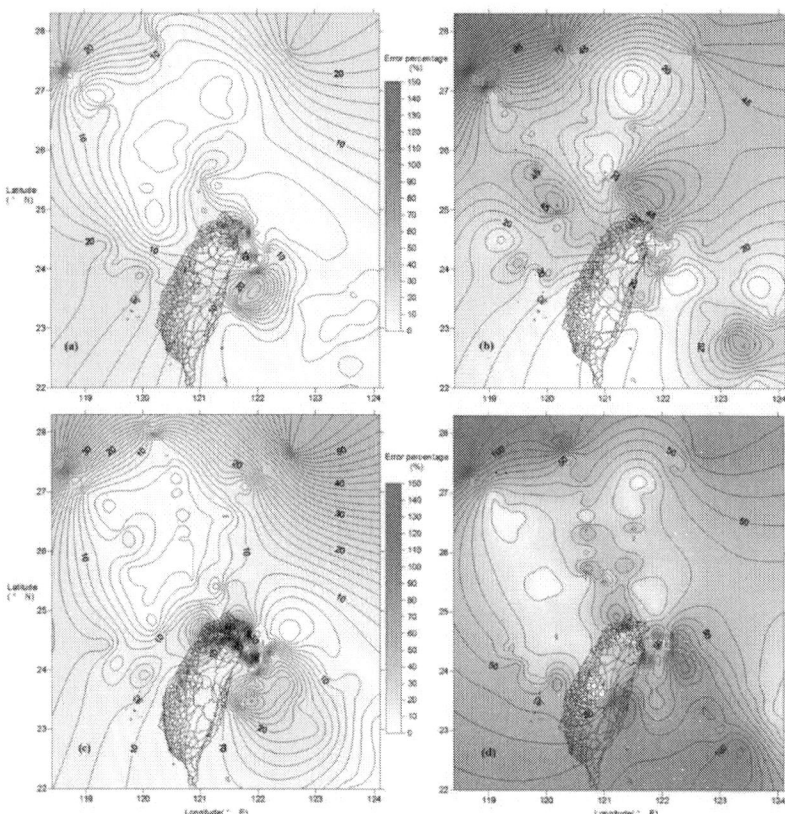

Figure 14. Comparison of spatial forecasted error feature of the four developed accumulated total inflow forecast models for the Shihmen Reservoir with relation to the typhoon central location: (**a**) RTRLNN-CHI model; (**b**) RTRLNN-NHI model; (**c**) ANFIS-CHI model; and (**d**) ANFIS-NHI model.

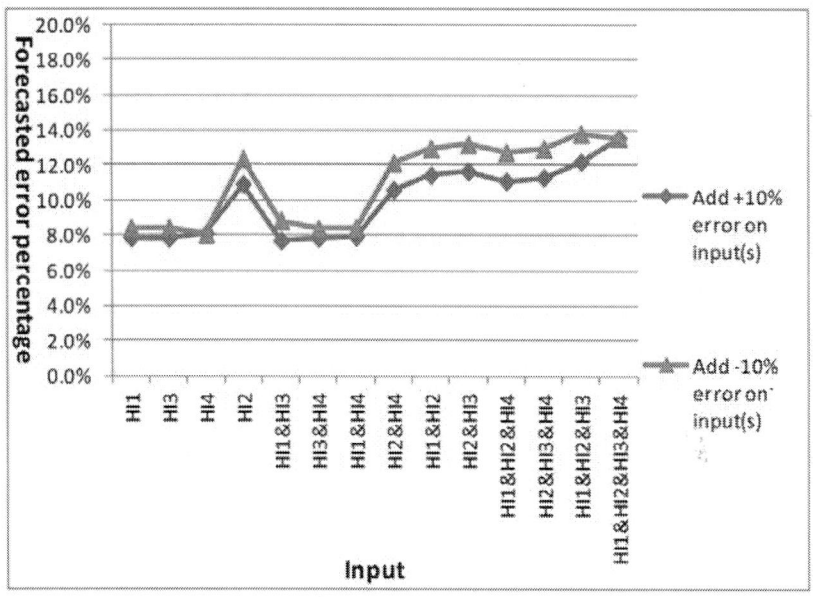

Figure 15. Sensitivity analysis results of model output with relation to single or combined heuristic inputs due to future forecasted uncertainty.

3.3.5. Construction Results of Heuristic Forecast Database for Heuristic Inputs

In this study, heuristic data mining techniques were used to construct a forecast database for the relationship between the position of the typhoon center, the rainfall hyetograph, and the inflow hydrograph in the catchment area from historical samples. The database was constructed to facilitate forecasting of the future accumulated total precipitation. Similarly, heuristic data mining techniques can also be applied on the delay of the future typhoon invasion by constructing a forecast database from relationships on the spatial position distribution of the typhoon center when the water has completely retreated, the moment of maximum rainfall, and the end time of direct runoff in historical samples. The characteristic map of the typhoon center position vs. the rainfall in the Shihmen Reservoir basin is as shown in Figure 16, which is the contour map after Kriging interpolation of the spatial sample information (X axis (typhoon central longitude), Y axis (typhoon central latitude), Z axis (basin precipitation of Shihmen Reservoir)). In this figure, strong rainfalls occurred when the typhoon center was at the southeast of Taiwan because the typhoon was under the influence of the Coriolis Effect. The air exhibits counterclockwise rotation in the Northern Hemisphere. When the typhoon was in the southeast of Taiwan, the typhoon rain belt entrained by the wind field under counterclockwise rotation was not blocked by the terrain of the mostly flat lands before entering the catchment area of the Shihmen Reservoir. After the rain belt entered the catchment area of the Shihmen Reservoir from the northwest to the southeast, it was blocked by the Snow-Capped mountain range. The catchment area of the Shihmen Reservoir belonged to the upwind side and heavy rain would happen then. In contrast, when the typhoon was not in the southeast of Taiwan, the rain belt entrained by the wind field was blocked by the Snow-Capped mountain range, the Central mountain range, and the Yusan mountain range before entering the catchment area of the Shihmen Reservoir. The catchment area of the Shihmen Reservoir was at leeward and there were no heavy rains in the Shihmen Reservoir at this time. With the assistance of this figure, the rainfall hyetograph in the catchment area of the Shihmen Reservoir during the future whole typhoon flood event can be obtained from the combined information of the hourly forecasted positions of the future typhoon center, the real-time estimated amount of rainfall, and the correction from the observed amount. The desired forecast of the total rainfall for the total reservoir inflow was obtained by summing over the rainfall hyetograph.

In this study, derivations on the spatial characteristics of the position distribution of the typhoon center were made for the moment when the maximum rainfall occurred, the direct runoff ended, and the water retreat ended in the typhoon flood events. First, the basin of the Shihmen Reservoir was located as the ellipse E_b in Figure 16, and then the terrain factors that might affect the rainfall in the reservoir basin were identified. Then, an axis (Line M_1–M_2) was marked along the direction of the Snow-Capped mountain range. The second axis (Line P_1–P_2), was defined as being perpendicular to Line M_1–M_2. Using these two axes as the reference, a contour map was created of the spatial distribution of the positions of the typhoon center when the water retreated below 300 cm of the reservoir inflow for the periods when a typhoon alarm is historically issued on land until the alarm is lifted. This resulted in the elliptical distribution line (E_E), from which the starting time of the forecast and the ending time of the water retreat could be determined. Regarding the model construction event, a contour map could be made for the spatial distribution of the positions of the typhoon center between the start of the rainfall and the end of the direct runoff for historical typhoon flood events. This resulted in the elliptical distribution line E_{DRE}. Further, the elliptical distribution line E_{MP} could be obtained from the contour map of the spatial distribution of the typhoon center when the maximum rainfall occurred in the historical typhoon flood events. The time of the maximum rainfall may be determined from this distribution line (E_{MP}) and the contour lines of rainfall.

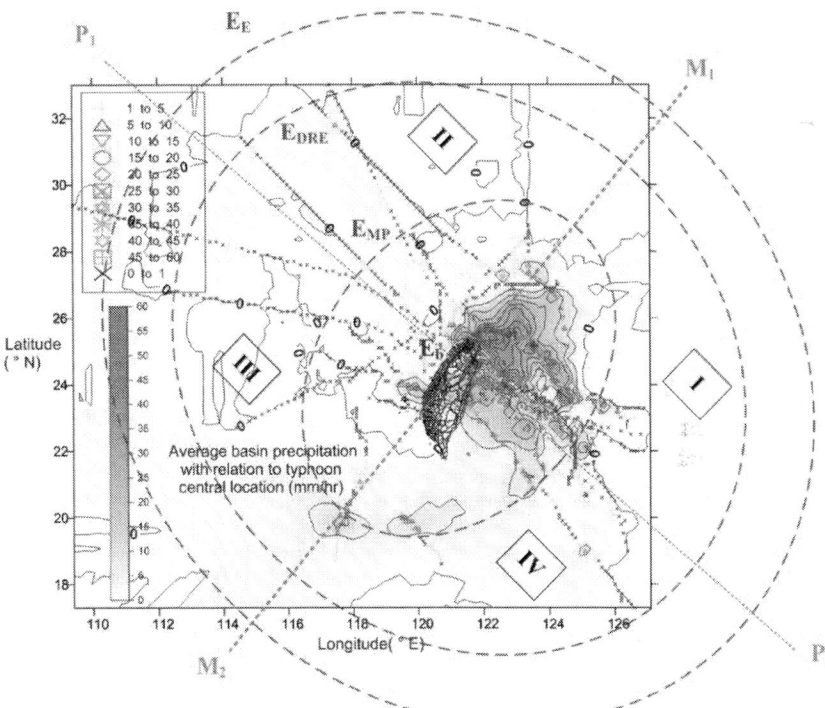

Figure 16. Construction results of heuristic forecast database for the Shihmen basin precipitation and duration characteristics curves with relation to the typhoon central location.

In Figure 16, the distribution lines were mainly elliptical, which was due to the Coriolis effect and the Terrain effect during the movements of the typhoon. When the path and direction of typhoon movement and the spatial distribution of terrain height are non-uniform, the distribution lines are elliptical instead of circular. The main axis of the ellipse was related to the direction of the terrains and mountains, so the elliptical distribution was a rotation of the main axis. Besides, the delay of the hydrograph pattern of rainfall was mainly dominated by the moving path of the typhoon. Because the Coriolis effect acted on the typhoon, it moved toward the direction of 270° to 360°, while the synthetic moving direction coincided with Line P_1–P_2, i.e., the perpendicular line of the Snow-Capped mountain range. Therefore, the long axes of E_E and E_{DRE} are along Line P_1–P_2, while the long axis of E_{MP} was in the direction of the Snow-Capped mountain range (Line M_1–M_2) because whether or not strong rainfall occurred was mainly related to the angle and position between the direction from which the rain belt of typhoon entered the reservoir basin and the direction of the Snow-Capped mountain range. In addition, from a monsoon climatology point of view, Taiwan is mainly affected by southwest monsoons from mid-March to mid-

September, and is predominantly under the effects of the northeast monsoon at other times. When the typhoon center was located in quadrants I and II in Figure 16, the wind field with counterclockwise rotation easily accompanied the northeast monsoon in the basin direction of the Shihmen Reservoir; when it was located in quadrants III and IV, the wind field was easily accompanied by the southwest monsoon. For typhoon invasion in quadrants I and II after mid-September and in quadrants III and IV during mid-March and mid-September, the typhoon was easily accompanied by co-existing effects of the monsoon to increase rainfall duration and precipitation. The contour lines of E_E, E_{DRE}, and E_{MP} can be expressed with the following equations:

$$\frac{\left[(x\cos 49^\circ + y\sin 49^\circ)\Delta d_{lon.} - 13481.3_{(E.km)}\right]^2}{\left[1113.92_{(km)}\right]^2} + \frac{\left[(-x\sin 49^\circ + y\cos 49^\circ)\Delta d_{lat.} - 2758.7_{(N.km)}\right]^2}{\left[1179.209_{(km)}\right]^2}$$
$$= 1$$

$$(52)$$

$$\frac{\left[(x\cos 49^\circ + y\sin 49^\circ)\Delta d_{lon.} - 13481.3_{(E.km)}\right]^2}{\left[885.2_{(km)}\right]^2} + \frac{\left[(-x\sin 49^\circ + y\cos 49^\circ)\Delta d_{lat.} - 2758.7_{(N.km)}\right]^2}{\left[942.7_{(km)}\right]^2}$$
$$= 1$$

$$(53)$$

$$\frac{\left[(x\cos 49^\circ + y\sin 49^\circ)\Delta d_{lon.} - 13481.3_{(E.km)}\right]^2}{\left[587.4_{(km)}\right]^2} + \frac{\left[(-x\sin 49^\circ + y\cos 49^\circ)\Delta d_{lat.} - 2758.7_{(N.km)}\right]^2}{\left[425.6_{(km)}\right]^2}$$
$$= 1$$

$$(54)$$

where $\Delta d_{lon.}$ and $\Delta d_{lat.}$ are the representative distance of each longitude and latitude, respectively.

4. CONCLUSIONS

Typhoon long lead-time rainfall-runoff is characterized as a chaotic, fuzzy, highly uncertain, and nonlinear system. The routing mechanism and characteristics of the real-time recurrent learning neural network (RTRLNN) and the adaptive network-based fuzzy inference system (ANFIS) have the potential ability to reason and learn using deterministic real-time recurrent routing and fuzzy logic. Therefore, the present study applied RTRLNN and ANFIS combined with multiple artificial intelligence-based heuristic techniques to develop coupled heuristic long lead-time accumulated total reservoir inflow forecast models (RTRLNN-CHI and

ANFIS-CHI), in order to improve the accuracy and stability of long-term accumulated total inflow forecasting. The proposed system was evaluated by a comparison with the RTRLNN- and ANFIS-based non-heuristic models (RTRLNN-NHI and ANFIS-NHI). The inputs of the heuristic models are composed of coupled observed non-heuristic inputs (typhoon characteristics factors, hourly basin precipitation, hourly reservoir inflow) and forecasted heuristic inputs (future accumulated total precipitation, duration from the current time to the time of maximum precipitation, direct runoff ending and event recessional ending, and observed-predicted inflow increase/decrease rate). The present study first employed non-parametric correlation analysis to assess the most appropriate input variables for long lead-time non-heuristic and heuristic models. This study also analyzed temporal and spatial forecasted error features to assess the goodness and applicability of the developed four long lead-time models, and we also analyzed the output sensitivity of single or combined heuristic inputs to determine whether the developed heuristic model can suffer the impact of future forecasted uncertainty and error on inputs.

The proposed method was applied to Taiwan's Shihmen Reservoir catchment area with a study period from 2004 to 2012. The results showed lead us to the following conclusions. (1) The accuracy and stability of the RTRLNN-based long lead-time accumulated total reservoir inflow prediction model are better than that of the ANFIS-based model. This is because RTRLNN incorporates a real-time recurrent deterministic routing mechanism with a more elastic and fine connection than ANFIS. (2) Under the synthesized simulation using key heuristic inputs of future total precipitation, flooding duration, and OPIID rate with other real-time observed hydrometeorological factors, the coupled heuristic RTRLNN-based model (RTRLNN-CHI, average error percentage (AEP): 6.3%, average forecasted lead-time: 49 h) and ANFIS-based model (ANFIS-CHI, AEP: 9.2%) could achieve a better prediction than the non-heuristic model (RTRLNN-NHI, AEP: 15.2%; ANFIS-NHI, AEP: 31.8%) because of the full consideration of different runoff/infiltration scenarios and initial/boundary conditions in each time step. (3) The hydrograph pattern of the reservoir inflow for the future typhoon meteorology and hydrology could be appropriately simulated ahead of time by using heuristic inputs. The accuracy of the long lead-time (24–72 h) total inflow forecast at the typhoon center during the invasion period in Taiwan (longitude 118–124 degrees, latitude 22–28 degrees) of the heuristic model was obviously better than that of the non-heuristic model. (4) When there were 10% errors on all the heuristic inputs, the AEP (13.6%) of the heuristic model was still lower than that of the most optimal non-heuristic model (RTRLNN-NHI, 15.2%). This indicates that the real-time forecast accuracy of the RTRLNN-CHI model even with errors on the heuristic inputs is still higher than that of the non-heuristic models (RTRLNN-NHI and ANFIS-NHI).

The key factors to effectively forecast long lead-time accumulated total reservoir inflow under a complex typhoon effect in real-time rely on the predicted accuracy of the meteorological-hydrological heuristic inputs and the associated data-preprocessing process. Future study can focus on improving the predicted accuracy of the heuristic inputs by coupling with novel numerical weather forecast models as a basis to provide future rainfall-runoff boundary conditions for a soft-computing model.

REFERENCES

1. Lin, G.F.; Chen, G.R.; Huang, P.Y.; Chou, Y.C. Support vector machine-based models for hourly reservoir inflow forecasting during typhoon-warning periods. J. Hydrol.**2009**, 372, 17–29. Lin, G.F.; Wu, M.C.; Chen, G.R.; Tsai, F.Y. An RBF-based model with an information processor for forecasting hourly reservoir inflow during typhoons. Hydrol. Process.**2009**, 23, 3598–3609.

2. Hsu, N.S.; Huang, C.L.; Wei, C.C. Real-time forecast of reservoir inflow hydrographs incorporating terrain and monsoon effects during typhoon invasion by novel intelligent numerical-statistic impulse techniques. J. Hydrol. Eng.**2015**, 20.

3. Webster, P.J.; Holland, G.J.; Curry, J.A.; Chang, H.R. Changes in tropical cyclone number, duration, and intensity in a warming environment. Science**2005**, 309, 1844–1846.

4. Wu, L.; Wang, B.; Geng, S. Growing typhoon influence on East Asia. Geophys. Res. Lett.**2005**, 32.

5. Knutson, T.R.; McBride, J.L.; Chan, J.; Emanuel, K.; Holland, G.; Landsea, C.; Held, I.; Kossin, J.P.; Srivastava, A.K.; Sugi, M. Tropical cyclones and climate change. Nat. Geosci.**2010**, 3, 157–163.

6. Bertoni, J.C.; Tucci, C.E.; Clarke, R.T. Rainfall-based real-time flood forecasting. J. Hydrol.**1992**, 131, 313–339.

7. Lardet, P.; Obled, C. Real-time flood forecasting using a stochastic rainfall generator. J. Hydrol.**1994**, 162, 391–408.

8. Toth, E.; Brath, A.; Montanari, A. Comparison of short-term rainfall prediction models for real-time flood forecasting. J. Hydrol.**2000**, 239, 132–147.

9. Anderson, M.; Chen, Z.; Kavvas, M.; Feldman, A. Coupling HEC-HMS with atmospheric models for prediction of watershed runoff. J. Hydrol. Engine**2002**, 7, 312–318.

10. Collischonn, W.; Haas, R.; Andreolli, I.; Tucci, C.E.M. Forecasting river uruguay flow using rainfall forecasts from a regional weather-prediction model. J. Hydrol.**2005**, 305, 87–98.

11. Dahlke, H.E.; Easton, Z.M.; Fuka, D.R.; Walter, M.T.; Steenhuis, T.S. Real-time forecast of hydrologically sensitive areas in the salmon creek watershed, New York state, using an online prediction tool. Water**2013**, 5, 917–944.

12. Brath, A.; Montanari, A.; Toth, E. Neural networks and non-parametric methods for improving real-time flood forecasting through conceptual hydrological models. Hydrol. Earth Syst. Sci.**2002**, 6, 627–639.

13. Hsu, N.S.; Wei, C.C. A multipurpose reservoir real-time operation model for flood control during typhoon invasion. J. Hydrol.**2007**, 336, 282–293.

14. Kitanidis, P.K.; Bras, R.L. Real-time forecasting with a conceptual hydrologic model: 1. Analysis of uncertainty. Water Resour. Res.**1980**, 16, 1025–1033.

15. Georgakakos, K.P.; Bras, R.L. Real-time, statistically linearized, adaptive flood routing. Water Resour. Res.**1982**, 18, 513–524.

16. Ho, J.-Y.; Lee, K.T. Grey forecast rainfall with flow updating algorithm for real-time flood forecasting. Water**2015**, 7, 1840–1865.

17. Thirumalaiah, K.; Deo, M. Hydrological forecasting using neural networks. J. Hydrol. Engine**2000**, 5, 180–189.

18. Xu, Z.X.; Li, J.Y. Short-term inflow forecasting using an artificial neural network model. Hydrol. Process.**2002**, 16, 2423–2439.

19. Wu, J.; Han, J.; Annambhotla, S.; Bryant, S. Artificial neural networks for forecasting watershed runoff and stream flows. J. Hydrol. Engine**2005**, 10, 216–222.

20. Aqil, M.; Kita, I.; Yano, A.; Nishiyama, S. Neural networks for real time catchment flow modeling and prediction. Water Resour. Manag.**2007**, 21, 1781–1796.

21. Wu, C.L.; Chau, K.W.; Li, Y.S. Methods to improve neural network performance in daily flows prediction. J. Hydrol.**2009**, 372, 80–93.

22. Pan, T.Y.; Wang, R.Y. State space neural networks for short term rainfall-runoff forecasting. J. Hydrol.**2004**, 297, 34–50.

23. Chau, K.; Wu, C.; Li, Y. Comparison of several flood forecasting models in Yangtze river. J. Hydrol. Eng.**2005**, 10, 485–491.

24. Chang, F.J.; Chang, L.C.; Huang, H.L. Real-time recurrent learning neural network for stream-flow forecasting. Hydrol. Process.**2002**, 16, 2577–2588.

25. Liu, Y.B.; Gebremeskel, S.; de Smedt, F.; Hoffman, L.; Pfister, L. A diffusive approach for flow routing in GIS based flood modeling. J. Hydrol.**2003**, 283, 91–106.

26. Molnar, P.; Ramirez, J.A. Energy dissipation theories and optimal channel characteristics of river networks. Water Resour. Res.**1998**, 34, 1809–1818.

27. Spearman, C. The proof and measurement of association between two things. Amer. J. Psychol.**1904**, 15, 72–101.

28. Haykin, S. Neural Networks: A Comprehensive Foundation, 2nd ed.; Prentice Hall: Upper Saddle River, NJ, USA, 1999.

29. Elman, J.L. Finding structure in time. Cogn. Sci.**1990**, 14, 179–211.

30. Jang, J.S.R. ANFIS: Adaptive Network-based Fuzzy Inference System. IEEE Trans. Syst. Man Cybern.**1993**, 23, 665–685.

31. Takagi, T.; Sugeno, M. Derivation of Fuzzy Control Rules from Human Operator's Control Actions. In Proceedings of the IFAC Conference on Fuzzy Information, Marseille, France, 19–21 July 1983; pp. 55–60.

32. Chiu, S.L. Fuzzy model identification based on cluster estimation. J. Intell. Fuzzy Syst.**1994**, 2, 267–278.

33. Lin, G.F.; Chen, L.H. Application of an artificial neural network to typhoon rainfall forecasting. Hydrol. Process.**2005**, 19, 1825–1837.

34. Chau, K.W.; Wu, C.L. A hybrid model coupled with singular spectrum analysis for daily rainfall prediction. J. Hydroinform.**2010**, 12, 458–473.

35. Wang, W.C.; Chau, K.W.; Xu, D.M.; Chen, X.Y. Improving forecasting accuracy of annual runoff time-series using ARIMA based on EEMD decomposition. Water Resour. Manag.**2015**, 29, 2655–2675.

Index